Advances in Lithium-Ion Battery Safety and Fire

Advances in Lithium-Ion Battery Safety and Fire

Guest Editors

Zhi Wang

Tong Liu

Mingzhi Jiao

MDPI

Basel • Beijing • Wuhan • Barcelona • Belgrade • Novi Sad • Cluj • Manchester

Guest Editors

Zhi Wang
School of Safety Engineering
China University of Mining and Technology
Xuzhou
China

Tong Liu
School of Safety Engineering
China University of Mining and Technology
Xuzhou
China

Mingzhi Jiao
CUMT-IoT Perception Mine Research Center
China University of Mining and Technology
Xuzhou
China

Editorial Office
MDPI AG
Grosspeteranlage 5
4052 Basel, Switzerland

This is a reprint of the Special Issue, published open access by the journal *Batteries* (ISSN 2313-0105), freely accessible at: https://www.mdpi.com/journal/batteries/special_issues/Q3WUDF336X.

For citation purposes, cite each article independently as indicated on the article page online and as indicated below:

Lastname, A.A.; Lastname, B.B. Article Title. *Journal Name* **Year**, *Volume Number*, Page Range.

ISBN 978-3-7258-6396-9 (Hbk)
ISBN 978-3-7258-6397-6 (PDF)
https://doi.org/10.3390/books978-3-7258-6397-6

Contents

Preface

In recent years, the safety issues of lithium-ion batteries have received widespread attention. The risk of fire and explosion in lithium-ion batteries is inevitable. With this Special Issue, we wanted to gather scientific articles of international standing that would give an overview of this topic that is as complete as possible. The result was the publication of relevant manuscripts (articles and reviews) based on four main topics: battery material safety, battery performance degradation, battery thermal runaway, and battery thermal runaway suppression.

Zhi Wang, Tong Liu, and Mingzhi Jiao
Guest Editors

batteries

MDPI

Editorial

Advances in Lithium-Ion Battery Safety and Fire Prevention

Zhi Wang

School of Safety Engineering, China University of Mining and Technology, Xuzhou 221116, China; zhiwang@cumt.edu.cn

With the rapid development of new energy technologies, lithium-ion batteries (LIBs) have become one of the core technologies in modern energy storage and electric mobility. With their advantages of high energy density, long life cycle, and light weight, LIBs are increasingly used in portable electronic devices, electric vehicles, and energy storage systems. However, the safety of LIBs, particularly related to their aging and the fires and explosions caused by thermal runaway (TR), is still one of the major challenges in their widespread application.

Firstly, changes in environmental factors, such as temperature, humidity, vibration, etc., have an impact on the internal ion migration and chemical reactions of LIBs, which can lead to the degradation of battery performance, as evidenced by the loss of capacity and voltage drop [1]. Secondly, due to the inherent instability of LIBs, the temperature inside the battery may rise sharply under abnormal conditions such as impact, overcharging, external heat sources, etc., triggering the TR phenomenon. TR not only accelerates the chemical reactions inside the battery and generates excessive temperature, but also may release large amounts of toxic gases, such as HF, CO, CO_2, etc. These gases have high toxicity and flammability, which may further trigger a fire or an explosion, resulting in serious casualties and property losses [2]. Therefore, in-depth studies on the decay laws of battery performance, TR mechanisms, and effective measures for preventing and controlling the risk of fires and explosions caused by TR in batteries have become a necessity in battery safety research.

To meet this challenge, in recent years, academia and industry have carried out a large number of studies on the safety of LIBs. These studies mainly focus on the following aspects: first, the performance degradation laws and aging mechanisms of LIBs in complex environments (e.g., temperature, humidity, vibration, etc.) [3,4]; second, the mechanisms of TR in LIBs, exploring the laws of temperature, pressure, and changes in the composition of the gases [5]; third, the design and optimization of battery materials and structure, such as through the use of flame-retardant electrolytes, the design of self-closing diaphragms, and other measures to enhance the intrinsic safety of batteries [6], while at the same time also developing efficient fire extinguishing technologies and gas elimination technologies to reduce heat propagation and the generation of harmful gases during thermal runaway [7,8]; and fourthly, the development of new types of battery management systems or early warning models to reduce the occurrence of TR through intelligent monitoring and control.

Despite some progress being made in this field, there are still a number of research gaps that need to be addressed urgently. Therefore, this Special Issue of the journal *Batteries*, "Advances in Lithium-Ion Battery Safety and Fire Prevention", brings together 11 research papers on LIB safety, covering a wide range of areas from basic research to applied technologies. These studies provide valuable references for improving battery safety. Specifically, in terms of battery performance, Sabeel et al. (contribution 1) reviewed the effects of vibration environments on battery performance, pointing out that vibration causes damage to the

 https://doi.org/10.3390/batteries11120455

battery structure and promotes dendrite formation, self-discharge, and lithium plating. In terms of TR of batteries, Jin et al. (contribution 2) investigated the evolution of TR in LIBs under different environmental pressures, and Sun et al. (contribution 3) investigated the effect of environmental pressure on the packaging forms of batteries (e.g., cylindrical and pouch commercial batteries), which provide guidance for the safe application of batteries under different environmental pressures. Hoelle et al. (contribution 4) designed a novel experimental setup to help the researchers to gain an in-depth understanding of the particle ejection process during TR. Mulder et al. (contribution 5) explored the electrical properties of TR gases, especially the flash-arc phenomenon that it may lead to. Mao et al. (contribution 6) summarized the flame behavior during TR in batteries, which provides a theoretical basis for further optimization of TR prevention and control techniques. In terms of enhancing the intrinsic safety of batteries, Santiago-Alonso et al. (contribution 7) and Al-Hamdani et al. (contribution 8) developed a new type of electrolyte to enhance the safety and efficiency of batteries, and Mao et al. (contribution 9) investigated the effect of different barrier materials on the propagation of TR in batteries and analyzed the barrier effects of different materials and their influencing factors, which provided fire prevention and the extinguishing of lithium batteries with theoretical support. In terms of monitoring and early warning technology, Pu et al. (contribution 10) studied an early warning method for TR in LIBs based on an electronic nose and a machine learning algorithm. Xie et al. (contribution 11) proposed a multi-parameter fusion early warning method based on a cloud model and Dempster–Shafer evidence theory, which effectively improves the accuracy of the TR risk assessment.

In summary, this Special Issue shows the latest research results in the field of LIB safety, covering research on battery performance, TR, intrinsic safety, fire prevention and extinguishing, monitoring and early warning systems, and other dimensions, providing valuable references to enhance the safety of LIBs. With the continuous improvement of battery energy density and the increasing diversification of application scenarios, LIB safety research is facing greater challenges. Future research will further explore the impact of complex environments on battery performance and TR, analyze the intrinsic evolution mechanisms, and develop new battery materials or fire extinguishing materials to ensure battery stability and safety. At the same time, combined with mechanical learning technology, the accuracy of TR warning systems for LIBs will be improved to cope with more complex safety issues.

Funding: This work is supported by the National Natural Science Foundation of China (No. 52204253), the Natural Science Foundation of Jiangsu Province (BK20242088), the China Postdoctoral Science Foundation (2023M733766), the Key-Area Research and Development Program of Guangdong Province (No. 2024B1111080003), and the Key Science and Technology Program of the Ministry of Emergency Management (2025EMST110401).

Acknowledgments: As the editor of this Special Issue, I wish to extend my heartfelt gratitude to all the authors for their invaluable contributions and to the reviewers for their diligent efforts in assessing the relevance and quality of the papers. Furthermore, I am deeply appreciative of the unwavering assistance and support provided by the editorial team of Batteries throughout the review process.

Conflicts of Interest: The author declares no conflicts of interest.

List of Contributions:

1. Sabeel, K.; Al-Greer, M.; Bashir, I. Advancements in Vibration Testing: Effects on Thermal Performance and Degradation of Modern Batteries. *Batteries* **2025**, *11*, 82.

2. Jin, Y.; Meng, D.; Zhao, C.-X.; Yu, J.-L.; Wang, X.-H.; Wang, J. Experimental Study on Thermal Runaway Characteristics of High-Nickel Ternary Lithium-Ion Batteries under Normal and Low Pressures. *Batteries* **2024**, *10*, 287.
3. Sun, Q.; Liu, H.; Wang, Z.; Meng, Y.; Xu, C.; Wen, Y.; Wu, Q. Investigation on Thermal Runaway Hazards of Cylindrical and Pouch Lithium-Ion Batteries under Low Pressure of Cruise Altitude for Civil Aircraft. *Batteries* **2024**, *10*, 298.
4. Hoelle, S.; Kim, H.; Zimmermann, S.; Hinrichsen, O. Lithium-Ion Battery Thermal Runaway: Experimental Analysis of Particle Deposition in Battery Module Environment. *Batteries* **2024**, *10*, 173.
5. Mulder, B.; Birke, K.P.; Obry, B.; Wigger, S.; Kozakov, R.; Smirnov, P.; Schein, J. Thermal Propagation Test Bench for the Study of the Paschen Curve and Lightning Arcs of Venting Gas. *Batteries* **2024**, *10*, 397.
6. Mao, Y.; Chen, Y.; Chen, M. Review of Flame Behavior and Its Suppression during Thermal Runaway in Lithium-Ion Batteries. *Batteries* **2024**, *10*, 307.
7. Santiago-Alonso, A.; Sánchez-Pico, J.M.; Emeterio, R.S.; Villanueva, M.; Salgado, J.; Parajó, J.J. Pyrrolidinium-Based Ionic Liquids as Advanced Non-Aqueous Electrolytes for Safer Next Generation Lithium Batteries. *Batteries* **2024**, *10*, 319.
8. Al-Hamdani, N.; Saravia, P.V.; Di Salvo, J.L.; Paz, S.A.; De Luca, G. Unravelling Lithium Interactions in Non-Flammable Gel Polymer Electrolytes: A Density Functional Theory and Molecular Dynamics Study. *Batteries* **2025**, *11*, 27.
9. Mao, Y.; Chen, Y.; Ye, Y.; Chen, Y.; Chen, M. Comparative Study of Thermal Runaway Propagation and Material Barrier Effect of Lithium-Ion Batteries. Batteries **2025**, *11*, 214.
10. Pu, Z.; Yang, M.; Jiao, M.; Zhao, D.; Huo, Y.; Wang, Z. Thermal Runaway Warning of Lithium Battery Based on Electronic Nose and Machine Learning Algorithms. *Batteries* **2024**, *10*, 390.
11. Xie, Z.; Zhang, Y.; Wang, H.; Li, P.; Shi, J.; Zhang, X.; Li, S. The Multi-Parameter Fusion Early Warning Method for Lithium Battery Thermal Runaway Based on Cloud Model and Dempster–Shafer Evidence Theory. *Batteries* **2024**, *10*, 325.

References

1. Wang, Z.; Zhao, Q.; Wang, S.; Song, Y.; Shi, B.; He, J. Aging and post-aging thermal safety of lithium-ion batteries under complex operating conditions: A comprehensive review. *J. Power Sources* **2024**, *623*, 235453. [CrossRef]
2. Fan, H.; Zhang, J.; Zhang, G.; Jiang, L.; Jiang, W.; He, Z.; Wang, X.; Wen, Y.; Xu, N. Thermal runaway studies of power lithium-ion batteries under various abusive conditions: A review. *Appl. Therm. Eng.* **2026**, *283*, 128945. [CrossRef]
3. Huang, R.; Wei, G.; Wang, X.; Jiang, B.; Zhu, J.; Chen, J.; Wei, X.; Dai, H. Revealing the low-temperature aging mechanisms of the whole life cycle for lithium-ion batteries (nickel-cobalt-aluminum vs. graphite). *J. Energy Chem.* **2025**, *106*, 31–43. [CrossRef]
4. Zhang, T. Performance degradation and sealing failure analysis of pouch lithium-ion batteries under multi-storage conditions. *J. Power Sources* **2024**, *613*, 234891. [CrossRef]
5. Liu, J.; Jin, X.; Zhang, H.; Zhao, K.; Wang, Z. Thermal runaway and combustion of LiFeO4 lithium-ion battery induced by electric-thermal coupling abuse. *J. Energy Storage* **2026**, *141*, 119271. [CrossRef]
6. Lyu, Y.; Xiang, Z.; Chen, L.; Li, Z.S.; Wang, J.; Chen, K.; Ying, D.; Chen, B.; Wu, C. Solvent-free fabrication of TPU-reinforced PE/carbon composites for high-performance positive temperature coefficient materials in lithium-ion battery safety. *RSC Adv.* **2025**, *15*, 32071–32079. [CrossRef] [PubMed]
7. Lei, S.; Huaiyu, L.; Bojun, T.; Chunfeng, Z.; Xiaojun, Z.; Liyang, W. Experiments on liquid-immersed thermal runaway management for large energy-stored lithium-ion battery modules. *Therm. Sci. Eng. Prog.* **2025**, *68*, 104347. [CrossRef]
8. Wang, H.; Chen, Z.; Li, X.; Liu, Y.; Xiao, Y.; Yang, L.; Wang, P. The study on thermal runaway gas adsorption and health state estimation of lithium-ion battery. *Comput. Theor. Chem.* **2025**, *1253*, 115390. [CrossRef]

batteries

Article

Lithium-Ion Battery Thermal Runaway: Experimental Analysis of Particle Deposition in Battery Module Environment

Sebastian Hoelle [1,2,*], Hyojeong Kim [2,3], Sascha Zimmermann [2] and Olaf Hinrichsen [1,4]

1 Department of Chemistry, TUM School of Natural Sciences, Technical University of Munich, Lichtenbergstraße 4, 85748 Garching near Munich, Germany
2 BMW Group, 80937 Munich, Germany
3 Institute for Applied Materials (IAM-AWP), Karlsruhe Institute of Technology, 76344 Eggenstein-Leopoldshafen, Germany
4 Catalysis Research Center, Technical University of Munich, Ernst-Otto-Fischer-Straße 1, 85748 Garching near Munich, Germany
* Correspondence: sebastian.hoelle@tum.de

Abstract: In this paper, a novel experimental setup to quantify the particle deposition during a lithium-ion battery thermal runaway (TR) is proposed. The setup integrates a single prismatic battery cell into an environment representing similar conditions as found for battery modules in battery packs of electric vehicles. In total, 86 weighing plates, positioned within the flow path of the vented gas and particles, can be individually removed from the setup in order to determine the spatial mass distribution of the deposited particles. Two proof-of-concept experiments with different distances between cell vent and module cover are performed. The particle deposition on the weighing plates as well as the particle size distribution of the deposited particles are found to be dependent on the distance between cell vent and cover. In addition, the specific heat capacity of the deposited particles as well as the jelly roll remains are analyzed. Its temperature dependency is found to be comparable for both ejected particles and jelly roll remains. The results of this study help researches and engineers to gain further insights into the particle ejection process during TR. By implementing certain suggested improvements, the proposed experimental setup may be used in the future to provide necessary data for simulation model validation. Therefore, this study contributes to the improvement of battery pack design and safety.

Keywords: Li-ion battery; thermal runaway; cell venting; particle emission; particle size distribution; specific heat capacity

1. Introduction

The thermal runaway (TR) of lithium-ion batteries attracts increasingly more attention as one of the most critical safety issues related to electric vehicles (EVs). On the one hand, this is caused by the occasionally emerging fire accidents of EVs that attract public attention [1–3]. On the other hand, the phenomenon is extensively studied within the scientific community, as shown in the increasing number of annual publications [4].

The main characteristics of a TR can be summarized as significant heat release caused by self-sustaining exothermic reactions, accompanied by gas and particle ejection. A TR can be triggered by either mechanical, electrical, or thermal abuse and may lead to fire or explosion of the battery cell [3,5]. Within an EV battery pack, the occurrence of a single cell TR results in the risk of so-called TR propagation, which is the process of TRs propagating from one cell to neighboring ones, resulting in catastrophic consequences [6–8]. Feng et al. proposed two different pathways for this failure development: an "expected" and an "unexpected" failure sequence. The expected sequence is primarily driven by heat conduction and consequently the TR first propagates from the triggering cell to its neighbors within the module. Secondly, the TR propagates to adjacent modules. The unexpected sequence is primarily caused by the gas–particle flow vented out of the trigger cell. The ignition of the venting gases may lead to fire and the particle accumulations can result in short circuits between cells or modules. Such fires and short circuits cause uncertainties and therefore the propagation sequence is difficult to predict for the unexpected pathway [8].

The gas–particle flow during TR as the origin of the unexpected pathway has been investigated by several researchers in recent years [9]. Experimental and numerical studies have been conducted to investigate the behavior of the ejected gas and particles in order to explore potential strategies to prevent or mitigate the unexpected failure sequence. On the experimental side, previous studies focused on different topics such as the characteristics of the jet stream out of the vent [10–13], the elemental composition of the venting gas [14–16], the size distribution of the ejected particles [12,16–20], and the elemental composition of the vented particles [14–17,20–22]. For example, Zou et al. investigated the jet manifestation, temperature and velocity of a prismatic 38 Ah cell triggered by external heating into TR [10]. Garcia et al. focused on optical jet observations by Schlieren visualization and Natural Luminosity of a 2.6 Ah cylindrical cell [11]. Wang et al. used the planar laser scattering technique to observe the jet flow of a prismatic 27 Ah cell and estimate characteristics such as gas and particle velocities [12]. Ding et al. conducted experiments with a cylindrical 5 Ah cell and investigated the influence of a low-pressure/high-altitude environment on the mass loss and flame height during TR [13]. The size distribution of the particles ejected during TR was investigated among others by Zhang et al. in Refs. [16,17]. They triggered a prismatic 50 Ah cell by external heating into TR within a sealed chamber. Wang et al. used the same cell within the same setup and additionally investigated the thermal oxidation characteristics of the ejected particles depending on their size [18]. Wang et al. analyzed the particle size distribution as well as the particle deposition of a prismatic 27 Ah cell within a combustion chamber [12]. Premnath et al. investigated battery modules containing either cylindrical 2.3 Ah cells or prismatic 60 Ah cells. However, their focus was on particles with a diameter smaller than 2.5 μm [19]. Liao et al. investigated the elemental composition of the ejected particles during TR for a cylindrical 2.4 Ah cell. They used an autoclave and focused on the influence of a varying state of charge on the venting behavior [14]. Chen et al. also evaluated the elemental composition of the ejected particles during TR for a cylindrical 2.4 Ah cell. They varied the state of charge, heating temperature, as well as heating power and additionally analyzed the thermal properties of the collected powder by thermogravimetric analysis—differential scanning calorimetry (TGA-DSC) [20]. Essl et al. analyzed the elemental composition of the ejected particles during TR of a 41 Ah pouch cell [15]. Zhang et al. and Wang et al. also determined the elemental composition of the ejected particles within their studies [16,17,22]. Held et al. focused on the contamination after a battery fire. Therefore, they triggered a module containing 12 prismatic cells with 90 Ah each into TR and analyzed the elemental composition of the fire soot depositing on the surroundings [21].

On the numerical side, there is only a small number of publications dealing with the venting behavior of lithium-ion batteries. Coman et al. were one of the first to introduce isentropic flow equations in their TR models to account for the heat dissipation due to gas and particle ejection [23]. They also extended their model from the single-cell level to the battery-pack level in the following years [24]. However, their model did not simulate the gas flow itself but the influence of the venting process on the thermal behavior of the solid body [23,24]. Ostanek et al. performed a similar study investigating the influence of the electrolyte vaporization and gas generation due to decomposition reactions on the cell body modeled as a solid [25]. Li et al. studied the flow through different vent geometries for 18,650 cells by performing CFD simulations. The results provide insights into the jet structure as well as the turbulence levels and can be used to estimate the influence on combustion and heat transfer to the surrounding cells [26]. Kim et al. focused on the internal pressure increase, the resulting venting process and the subsequent gas-phase dynamics of 18,650 cells [27]. However, both models consider a single cell only and consequently are not capable of predicting the thermal impact of the vented gas and particles on the neighboring cells. Citarella et al. were one of the first to consider the gas flow within their model to investigate the TR propagation behavior of a battery pack [28]. Another study of Mishra et al. focused on the spread of the gas flow within a battery pack by studying parameters such as the cell-to-cell gap, the distance between cell vent and housing, as well as the location of the vent [29].

The models mainly focused on the gas phase of the venting process, although the ejected particles are assumed to have a significant influence on the TR propagation process. To address this issue, Wang et al. built a multiphase model including both fluid flow and particles. They investigated not only the jet behavior and particle ejection, but also the particle deposition after TR [12]. However, there is a lack of experimental data that are needed to validate such simulation models.

This publication focuses on the experimental determination of the particle deposition during a lithium-ion battery TR. The objective is to develop an experimental setup that allows for quantification of the mass distribution of ejected particles in a battery module-like environment. Therefore, a novel experimental setup for prismatic batteries is proposed and two proof-of-concept experiments are performed with different distances between battery cell and cover. In addition, the particle size distribution of depositing particles is analyzed and specific heat capacity measurements are conducted. To the authors' knowledge, an experimental determination of the mass distribution of depositing particles during TR has not been the subject of any scientific publication. In addition, specific heat capacity measurements of ejected particles and jelly roll remains after a TR are not yet available in the literature. The results of this study help engineers or researchers to gain further insights into the particle ejection process during TR. By implementing certain suggested improvements, the proposed experimental setup may be used in the future to provide necessary data for simulation model validation. Therefore, this study contributes to the improvement of battery pack design and safety.

2. Experimental Setup

In this study, a novel experimental setup to quantify the particle deposition during a lithium-ion battery TR is proposed. The setup integrates a single prismatic battery cell into a battery module-like environment as found in EV battery packs. Within the flow path of the vented gas and particles, there are several weighing plates that can be individually removed from the setup in order to determine the spatial mass distribution of the deposited particles. Afterwards, the collected particles as well as the remains of the jelly roll are further analyzed to determine mass loss of the battery cell during TR, size distribution of the vented particles, and specific heat capacity measurements of both vented particles and jelly roll remains.

2.1. Geometry

The experimental setup used in this study is shown in Figure 1. The main component is the investigated battery cell (dimensions 180 mm × 32 mm × 72.5 mm) that is compressed within a steel frame. In order to reduce the heat transfer from the cell to its surroundings, a thermal insulation material is used on both sides of the cell (thickness of 10 mm). In addition, there is a built-in load cell to adjust the compression force to specified values. The steel frame can be attached to a base plate, resulting in the substructure shown in Figure 1a. As shown in Figure 1b, an intermediate sheet metal surrounding the battery cell can be attached to the base plate. Grooves are milled into this sheet metal, so that in total 86 weighing plates (T-shape with an upper surface of 32 mm × 45 mm) can be inserted. Afterwards, two fixation plates are screwed to each side of the intermediate sheet metal as shown in Figure 1c. The result is a flat surface that is on the level of the top of the battery cell. Finally, a cover is attached to the base plate via telescopic legs, allowing for setting different distances between the battery cell's vent and the cover. Figure 1d shows the resulting assembly that can be integrated into a steel box as shown in Figure 1e. Figure 1f schematically shows the nailing process for TR initiation.

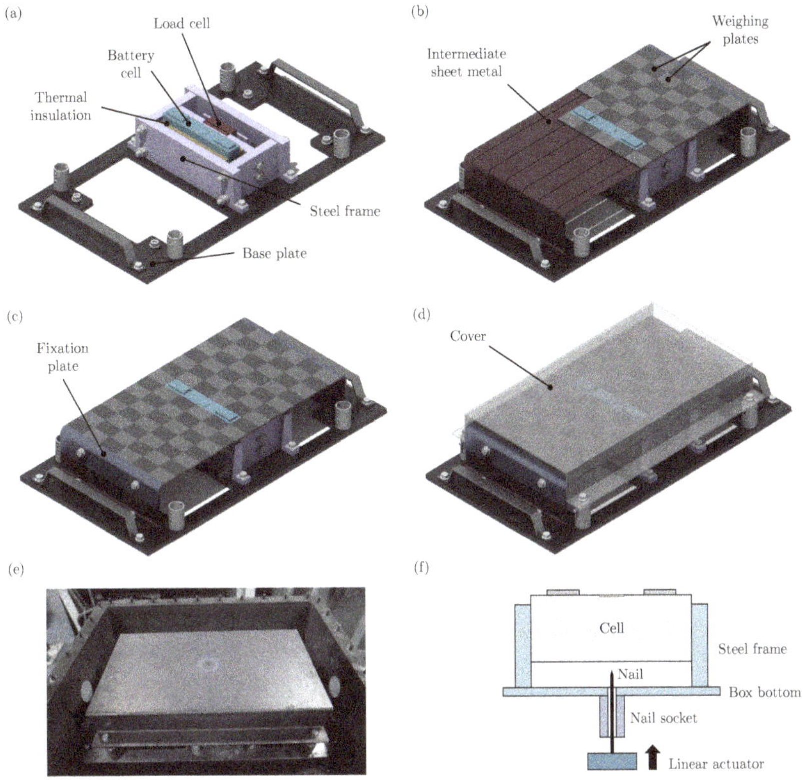

Figure 1. (**a**) Substructure of the experimental setup including a steel frame to fix the position of the battery cell. (**b**) Intermediate sheet metal that is used to position the weighing plates. (**c**) Fully assembled and fixed weighing plates. (**d**) Full experimental setup with cover. (**e**) Full view of the setup within a steel box. (**f**) Schematic representation of linear actuator for nailing process.

2.2. Test Procedure and Evaluation Methodology

Two proof-of-concept experiments with different distances between cell vent and cover are performed. Therefore, the cell is installed into the steel frame and compressed with a force of 3 kN in order to ensure a fixed position during nail penetration and simulate conditions as found in a battery module for EV battery packs. Afterwards, the intermediate sheet metal is screwed to the base plate and the weighing plates are inserted into the grooves. Subsequently, the position of the weighing plates is fixed by mounting the fixation plates to both ends of the intermediate sheet metal. The cell is then charged from a state of charge of SoC = 30% (used during logistics) to SoC = 100% with the constant current constant voltage (CCCV) charging method. In the final step, the cover is attached at two specified heights h:

- Test 1: $h = 24$ mm distance between cell vent and cover, which aims to represent EV battery pack conditions.
- Test 2: $h = 40$ mm distance between cell vent and cover, which aims to show the influence of the distance on particle deposition.

The full setup is attached to the steel box, which comes with a nail actuator that is used to penetrate the bottom side of the battery cell with a steel nail (3.2 mm diameter) as shown in Figure 1f. The penetration depth is 15 mm with a penetration speed of 7 mm s^{-1}. The nail is not retracted from the battery after TR initiation.

The nail penetration triggers a TR and consequently results in gas and particles being ejected out of the cell vent. Since the cover is flushed with the long sides of the intermediate sheet metal, the resulting flow path is directed towards both fixations plates. Vented particles that are carried by the gas flow eventually deposit either on one of the weighing plates or within/outside of the steel box.

In preparation for the disassembly of the test setup and the evaluation of the results, 86 sealable glasses are labeled and weighed in order to collect the particles deposited on each individual weighing plate. As soon as the setup is cooled down, the cover and the fixation plates are dismounted. Next, each weighing plate is individually taken out of the intermediate sheet metal and deposited particles are collected in the prepared glasses. To do so, loose particles on each weighing plate are first poured into the corresponding glass with the aid of a funnel as exemplary shown in Figure 2a. Then, the weighing plates are wiped off with the aid of a brush to collect non-adherent particles. Particles that are sticking to the weighing plates and therefore cannot be collected by pouring or wiping are not collected in this step, but scraped off in the second step. This helps to differentiate the mass of particles that are sticking to the weighing plates from particles that are loose as shown in Figure 2b. After collecting all particles, the intermediate sheet metal is dismounted from the base plate, and the tested cell can be removed from the steel frame for further analysis.

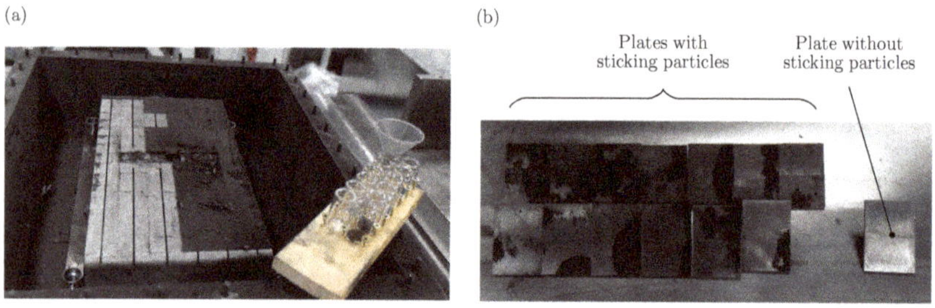

Figure 2. (**a**) Process of particle collecting with sealable glasses and funnel. (**b**) Examples of plates with sticking particles and a plate without sticking particles.

2.2.1. Particle Deposition and Particle Size Distribution

After each test, the mass loss during TR of the battery cell Δm_{cell} is calculated by

$$\Delta m_{\text{TR}} = m_{\text{cell}} - m_{\text{cell,postTR}}, \tag{1}$$

with m_{cell} being the mass of the cell before the test and $m_{\text{cell,postTR}}$ being the mass of the cell after the test. The mass of deposited particles $m_{\text{particles},i}$ is determined for each weighing plate i by

$$m_{\text{particles},i} = m_{\text{glass},i,\text{postTR}} - m_{\text{glass},i}, \tag{2}$$

where $m_{\text{glass},i}$ is the mass of each glass in its empty state and $m_{\text{glass},i,\text{postTR}}$ is the mass of each glass with the collected particles inside. The weighing of the glasses is performed twice: first, the glasses with collected loose particles are weighed. Afterwards, the sticking particles are scraped off the weighing plates and added to the corresponding glass. The second weighing results in the mass of sticking particles $m_{\text{particles},i,\text{stick}}$ for each weighing plate i:

$$m_{\text{particles},i,\text{stick}} = m_{\text{glass},i,\text{scraped}} - m_{\text{glass},i,\text{postTR}}, \tag{3}$$

with $m_{\text{glass},i,\text{scraped}}$ being the mass of the glass with added scraped-off particles. The total sum of particles deposited on the weighing plates $m_{\text{particles,total}}$ is the sum of all loose particles $m_{\text{particles,loose}}$ and all sticking particles $m_{\text{particles,stick}}$:

$$\begin{aligned} m_{\text{particles,total}} &= m_{\text{particles,loose}} + m_{\text{particles,stick}} \\ &= \textstyle\sum_i m_{\text{particles},i} + \sum_i m_{\text{particles},i,\text{stick}}. \end{aligned} \tag{4}$$

After the mass distribution on the weighing plates is determined, the particle size distribution is analyzed. Therefore, the particles (diameter d) are separated into four different size ranges by sieving:

1. $d < 80\,\mu m$,
2. $80\,\mu m \leq d < 200\,\mu m$,
3. $200\,\mu m \leq d < 800\,\mu m$,
4. $d \geq 800\,\mu m$.

2.2.2. Specific Heat Capacity Measurements

The specific heat capacity of the four particle size samples is measured via the differential scanning calorimetry (DSC) apparatus 204 F1 Phoenix manufactured by Netzsch®, Selb, Germany. In addition, four samples of the jelly roll remains, i.e., the burnt remains inside the cell after the TR process from four different positions as shown in Figure 3 are analyzed. Therefore, three probes of approximately 10 mg of each sample are loaded in aluminum crucibles under air atmosphere. In order to ensure specimens containing all chemical components, the three particle samples with $d > 80\,\mu m$ as well as all four samples out of the jelly roll remains are ground in a ball mill before loading the crucibles (four times of milling for 2 min with 2500 rpm and a 1 min pause in between). The measurements are performed in a temperature range of 25–500 °C with a temperature rate of $10\,K\,min^{-1}$. The blank measurement is performed with an empty crucible and the calibration measurement with a sapphire standard (mass: 24.38 mg, diameter: 5 mm, thickness: 0.5 mm).

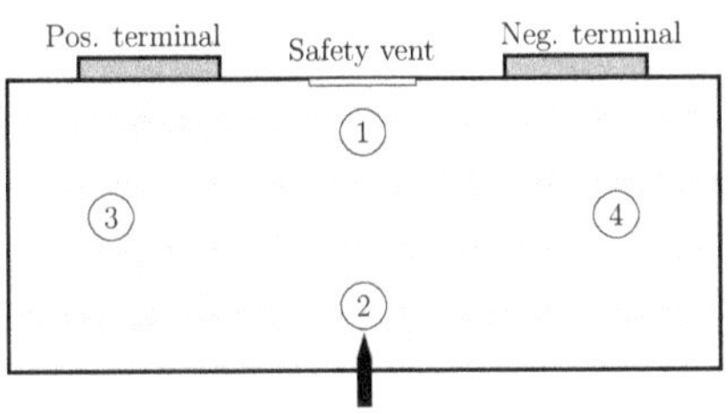

Figure 3. Schematic representation of the battery cell with the four positions of sample extraction out of the jelly roll remains after TR: (1) beneath the safety vent, (2) above the nailing position, (3) middle of the jelly roll beneath the positive terminal, and (4) middle of the jelly roll beneath the negative terminal.

2.3. *Investigated Cell*

The properties of the investigated battery cell are summarized in Table 1. The cells are prismatic prototype lithium-ion batteries with a nominal capacity between 60 Ah and 70 Ah. The cell consists of a $LiNi_{0.8}Mn_{0.1}Co_{0.1}O_2$ (NMC811) cathode and a graphite anode. The electrolyte is based on lithium hexafluorophosphate ($LiPF_6$) conducting salt with ethylene carbonate (EC), ethyl methyl carbonate (EMC), diethyl carbonate (DEC), and dimethyl carbonate (DMC) solvents. The cells are in a fresh/unused aging state.

Table 1. Basic properties of the investigated cell.

Parameter	Value
Capacity	60–70 Ah
Mass	980.2 ± 4.2 g
Dimensions	180 mm × 32 mm × 72.5 mm
Cathode	$LiNi_{0.8}Mn_{0.1}Co_{0.1}O_2$ (NMC811)
Anode	Graphite
Electrolyte	$LiPF_6$ with EC, EMC, DEC, and DMC

3. Results and Discussion

Two proof-of-concept experiments with different distances between the battery cell's vent and the experimental setup's cover were performed to investigate the suitability of the setup for quantifying the particle deposition after a lithium-ion battery TR. In both experiments, the cell was triggered into TR by nail penetration and showed an expected venting behavior with gas release and particle ejection (only) through the vent, i.e., post-test analysis did not show any indications of gas or particle venting through the nailing hole. In the following, the particle deposition on the weighing plates as well as the mass loss of the cells during TR, the size distribution of the collected particles, as well as the specific heat capacity of both collected particles and samples out of the jelly roll remains are analyzed and discussed. The authors would like to point out that all results analyzed and discussed in the following are based on just the two proof-of-concept experiments and consequently the reliability of the statements made is limited.

3.1. Particle Deposition

Figure 4 shows the view on the weighing plates after cover removal for Test 1 with a 24 mm distance between cell vent and cover (Figure 4a) and Test 2 with a 40 mm distance between cell vent and cover (Figure 4b). In order to facilitate the result description and interpretation, the weighing plates are assigned to six columns as indicated with the numbered circles.

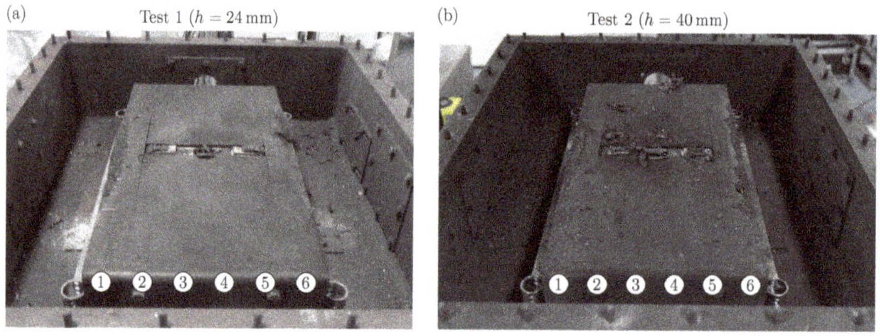

Figure 4. (**a**) View on the weighing plates of Test 1 after cover removal. (**b**) View on the weighing plates of Test 2 after cover removal. The numbered circles indicate the weighing plates' column number.

For Test 1 (see Figure 4a), there is no blank surface visible, which means that all weighing plates are fully covered with particles. On the outer Columns 1 and 6, accumulations of particles can be seen. These seem to concentrate towards the center of the intermediate sheet metal (at the position of the battery cell) and decrease towards the fixation plates. On the weighing plates within the middle Columns 2–5, there are no larger clumps of particles visible. In addition, a significant amount of particles deposits within the steel box. This is also the case for Test 2 (compare with Figure 4b). However, the particle deposition on the weighing plates shows significant differences compared to Test 1. The particle accumulations on the outer Columns 1 and 6 are also visible, but not as clearly as for Test 1. Also, there are bigger particles depositing towards the fixation plates over all the weighing plate columns as well as around the battery cell. In general, the particle deposition on the weighing plates appears random for Test 2.

A possible cause of the observed differences in particle deposition is the varying distance between cell vent and cover. As shown in Figure 5, the lower distance of h = 24 mm for Test 1 compared to h = 40 mm for Test 2 results in the flow path area A being smaller in Test 1. More precisely, the flow path area is $A_1 = 270\,\text{mm} \times 24\,\text{mm} = 6480\,\text{mm}^2$ in Test 1 and $A_2 = 270\,\text{mm} \times 24\,\text{mm} = 10{,}800\,\text{mm}^2$ in Test 2. Because A_1 is 40% smaller than A_2, the gas flow velocities are higher in Test 1 assuming that the volume flow rate of venting gas is the same in both tests. As a consequence of higher flow velocities, the forces acting on particles within the flow, such as drag force or Saffman lift, are higher. Therefore, the flow is capable of carrying larger particles and particles of the same size for a longer distance. Consequently, large particles are carried by the gas flow into the steel box for Test 1, whereas the lower gas velocities in Test 2 lead to the deposition of such large particles on the weighing plates. The accumulations of particles in Columns 1 and 6 are observed for both tests, which is in accordance with the expected flow field. Assuming ideal conditions, three stagnation points form: one above the cell vent at the cover, and two more at each side wall of the intermediate sheet metal (Columns 1 and 6) in the row of the battery cell. Independently of flow path area A, the gas velocity approaches zero towards a stagnation point. Therefore, particles are likely to accumulate next to these positions.

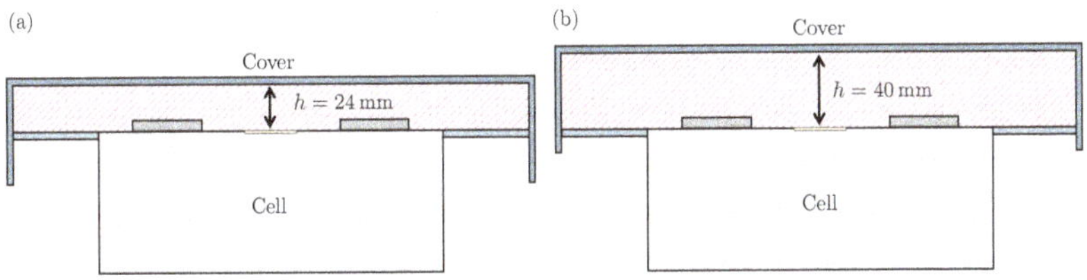

Figure 5. Schematic representation of flow area A during (**a**) Test 1 with $h = 24$ mm and (**b**) Test 2 with $h = 40$ mm.

The results of visual observation are in accordance with the measured mass distributions of deposited particles as shown in Figure 6. For each weighing plate, the mass of collected particles according to Equation (4) is shown for Test 1 in Figure 6a and for Test 2 in Figure 6b.

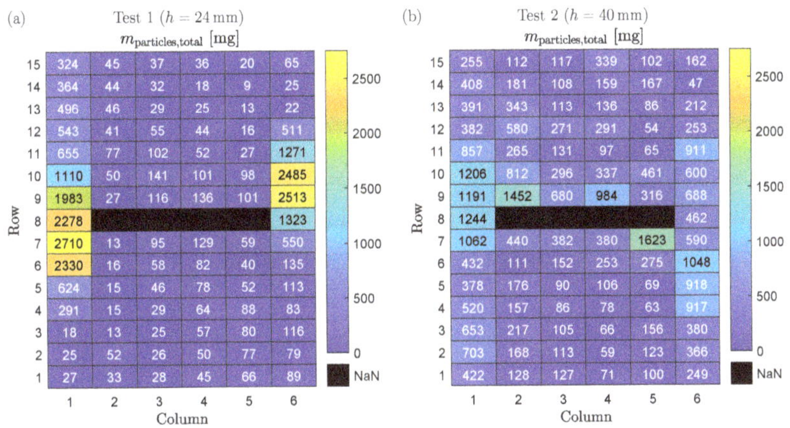

Figure 6. (**a**) Mass of deposited particles on each weighing plate for Test 1. (**b**) Mass of deposited particles on each weighing plate for Test 2.

For Test 1 (see Figure 6a), the maximum mass of particles collected from a single weighing plate is 2710 mg (Row 7, Column 1), whereas the minimum mass is 9 mg (Row 14, Column 5). The accumulation of particles in the outer Columns 1 and 6 as seen in the visual observation is confirmed by the measured mass values. Interestingly, the mass of collected particles in Column 6 decreases to values ≤ 135 mg already in Row 6, while for Column 1 this is the case just from Row 3 to 1. For Rows 9 to 15, a similar observation is made: in Column 1, the minimum collected particle mass is 324 mg (Row 15), whereas in Column 6 the minimum values are ≤ 65 mg for Rows 13 to 15. A possible cause is that particles on weighing plates of the outer columns fall down into the steel box when the cover is removed after the test. There is a conflict of objectives in the design of the experimental setup in this respect. On the one hand, the cover has to be flush with the weighing plates so that no gas or particles escape at the sides of the outer columns. On the other hand, the cover can be under tension due to the high temperatures during the test, which can lead to particles falling from the weighing plates into the steel box when the cover is removed. For the inner Columns 2 to 5, more particle mass seems to deposit next to the battery cell (e.g., Row 9, Columns 3 to 5). However, there exist also outliers such as Rows 7 and 9 in Column 2.

For Test 2 (see Figure 6b), the maximum mass of particles collected from a single weighing plate is 1623 mg (Row 7, Column 5), whereas the minimum mass is 47 mg (Row 14, Column 6). As for Test 1, the particle accumulation in Columns 1 and 6 that has been observed visually is confirmed by the measured mass distribution on the weighing plates. However, a significant higher mass of deposited particles is found in the inner Columns 2 to 5 compared to Test 1. In addition, the two weighing plates with the maximum mass of particles collected are in Columns 2 and 5 and therefore not in the outer columns. As explained above, it is hypothesized that this observed behavior is the result of lower gas flow velocities in Test 2 compared to Test 1 due to the higher distance between cell vent and cover.

The separate evaluation of the measured mass distribution of loose and sticking deposited particles is shown in Figure 7. For each weighing plate, the total mass of loose and sticking particles according to Equations (2) and (3), respectively, is shown for Test 1 in Figure 7a,c, as well as for Test 2 in Figure 7b,d.

For Test 1, the mass distribution of loose particles (Figure 7a) is similar to the mass distribution of all particles (Figure 6a). One exception is the area around the battery cell: the mass of loose particles is increasing from Row 7 or 9 (next to the battery cell) towards Row 1 or 15 (next to the fixation plates). This can be explained by the mass distribution of sticking particles as shown in Figure 7c that accumulate towards the battery cell. The sticking of particles to the weighing plates is caused by high temperature above the melting point of the particle's material composition. Therefore, the high mass of sticking particles is equivalent to the high temperatures of both gas and particles during TR. This is in accordance with the shown results: the further away from the battery cell, the longer the vented gas and particles can cool down, and consequently the less particle mass is sticking to the weighing plates.

As observed in Test 1, the mass distribution of loose particles in Test 2 (Figure 7b) is similar to the mass distribution of all particles in Test 2 (Figure 6b), and sticking particles are only present in the area close to the battery cell (Rows 7 to 11) as shown in Figure 7d. However, the total mass of sticking particles is significantly lower for Test 2 compared to Test 1, which can be related to lower temperatures: as the distance between cell vent and cover is longer for Test 2 compared to Test 1, the pressure loss between the inner battery cell and the environment is lower. Consequently, the total pressure in the area next to the battery cell is also lower, which results in decreased gas temperatures due to a higher expansion of the gas from inside of the cell to outside of the cell.

The sum of collected particle mass on all weighing plates as well as the mass loss during the TR of the cells is summarized in Table 2 for each test. The results are in accordance with the observations made above:

- The battery cell's mass loss during TR Δm_{TR} is lower for the 24 mm distance between cell vent and cover (Test 1) compared to the 40 mm distance between cell vent and cover (Test 2). This behavior can be explained by a higher pressure loss from the inside of the battery cell to the outside of the battery cell for Test 1 due to the smaller flow area, A. However, the difference may also be caused by the statistical variance of nail penetration tests.
- The total mass of particles deposited on the weighing plates, $m_{particles,total}$, is lower for the 24 mm distance between cell vent and cover (Test 1) compared to the 40 mm distance between cell vent and cover (Test 2). This is explained by higher gas flow velocities in Test 1 compared to Test 2 due to the smaller flow area A.

- The total mass of sticking particles deposited on the weighing plates $m_{\text{particles,stick}}$ is higher for the 24 mm distance between cell vent and cover (Test 1) compared to the 40 mm distance between cell vent and cover (Test 2). This is explained by higher gas flow temperatures in Test 1 compared to Test 2.
- Only 7.3% of the total ejected particle mass deposits on the weighing plates for Test 1, assuming that 70.9% of the total mass loss during TR are particles as found in the literature [30]. For Test 2, 8.4% of the total mass of ejected particles deposits on the weighing plates under the same assumption.

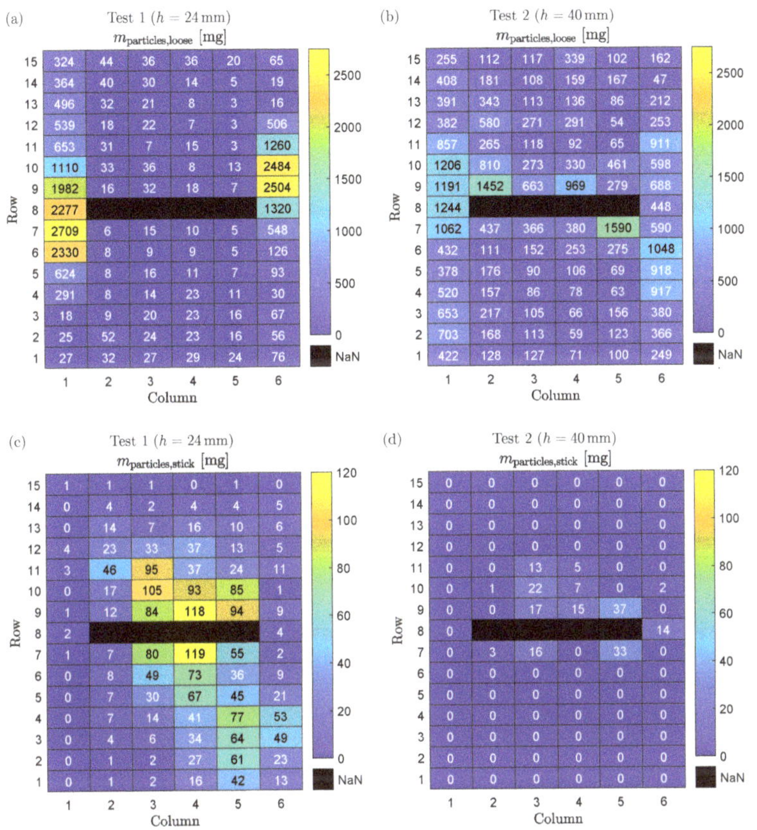

Figure 7. (**a**) Mass of deposited loose particles on each weighing plate for Test 1. (**b**) Mass of deposited loose particles on each weighing plate for Test 2. (**c**) Mass of deposited sticking particles on each weighing plate for Test 1. (**d**) Mass of deposited sticking particles on each weighing plate for Test 2.

Table 2. Experimental results of the two tests with different distance between cell vent and cover.

Parameter Distance Vent to Cover	Test 1 24 mm	Test 2 40 mm
m_{cell}	984.4 g	976.0 g
$m_{\text{cell,postTR}}$	477.7 g	426.0 g
Δm_{TR}	507.1 g/51.5%	550.0 g/56.4%
$m_{\text{particles,total}}$	26.13 g	32.83 g
$m_{\text{particles,loose}}$	23.95 g	32.65 g
$m_{\text{particles,stick}}$	2.18 g	0.18 g

3.2. Particle Size Distribution

Figure 8 shows the mass percentage of the four particle size ranges for both tests of this study (black and red bars) and compares these results to values from the literature [12,16–18]. Zhang et al. (blue bars) triggered a prismatic 50 Ah lithium-ion battery with an NMC622 cathode into the TR within a sealed chamber by external heating and analyzed the ejected solid particles regarding particle size distribution. They found particles within the range of 2.9 µm $< d <$ 15 mm. However, particles with a size of $d <$ 500 µm were found to account for 90% of the total ejected particle mass [16]. Within a second study, Zhang et al. (brown bars) also analyzed the ejected particles for the same cell within the same experimental setup. The maximum particle size was found to be ≈8 mm [17]. Wang et al. (green bars) conducted a third study with the setup of Zhang et al. and found particles within the size range of 1 µm $< d <$ 2000 µm [18]. Wang et al. (violet bars) investigated the vented particles of a prismatic 27 Ah lithium-ion battery with an LFP cathode within a sealed chamber. They used an electric heater to trigger the TR and collected particles with sizes up to $d >$ 2 mm [12]. Note that the mass percentage of Sizes 3 (200 µm $< d <$ 800 µm) and 4 ($d >$ 800 µm) could not be estimated for the results of Wang et al. published in Ref. [12].

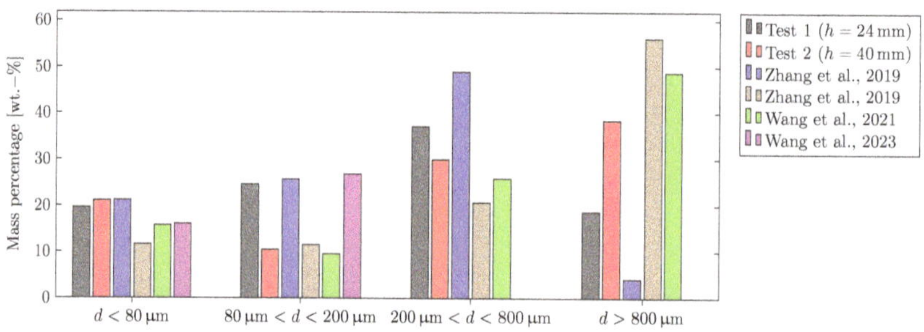

Figure 8. Comparison of the particle size distribution evaluated for both tests in this study with values from the literature [12,16–18]. The mass percentage of Sizes 3 and 4 could not be estimated for the study of Wang et al. (Ref. [12]).

For Test 1 (24 mm distance between cell vent and cover), there is an increase in the mass percentage values from Size range 1 ($d <$ 80 µm: 19.6%) over Size range 2 (80 µm $< d <$ 200 µm: 24.6%) to Size range 3 (200 µm $< d <$ 800 µm: 37.1%). The mass percentage of Size range 4 ($d >$ 800 µm: 18.7%) shows the lowest value. An explanation for these results is that a large fraction of Size range 4 particles is not depositing on the weighing plates but within the steel box due to the high gas flow velocities. It is also possible that particles of Size range 4 shatter into smaller pieces when they impinge into the walls. This can additionally explain the high mass percentage values for particle Size range 3. In addition, the shape of larger particles usually deviates more from the shape of an ideal sphere as shown in previous publications [12,17]. This may lead to higher drag coefficients.

The results of Test 2 (a 40 mm distance between cell vent and cover) show that the mass percentage of Size range 1 particles ($d <$ 80 µm: 21.1%) is close to the results of Test 1. It is therefore hypothesized that the deposition of Size range 1 particles is nearly independent of the gas flow velocities and therefore the distance between cell vent and cover due to their small diameter. In contrast to Test 1, there is an increase in the mass percentage values from Size range 2 (80 µm $< d <$ 200 µm: 10.5%) over Size range 3 (200 µm $< d <$ 800 µm: 30.0%) to Size range 4 ($d >$ 800 µm: 38.4%) for Test 2. Due to the lower gas velocities during Test 2, the forces of the gas flow acting on Size range 4 particles are not high enough to carry them into the steel box.

The results of this study are comparable to the results found in the literature [12,16–18]. For Size range 1, both tests of this study show comparable values to those of the study of Zhang et al. in Ref. [16]. However, comparison of the results of all three studies conducted within the sealed reactor introduced by Zhang et al. reveals that there can be deviations in the particle size distributions for experiments conducted with the same cell and test procedure within the same setup [16–18]. These deviations are also observed for particle Size range 2: while the results of Test 1 are comparable to the results of Refs. [12,16], the results of Test 2 are close to the values published in Refs. [17,18]. For Size range 3, the results of this study's tests are within the range of the literature values, which is also the case for Size range 4 [16–18]. At this point, however, it has to be mentioned that for the results of this study, the size distribution of particles depositing on the weighing plates is analyzed, whereas for the studies in the literature all ejected particles are collected and investigated.

3.3. Specific Heat Capacity Measurements

Figure 9 shows the specific heat capacity over temperature for the four particle size ranges measured via DSC for Test 1 (solid line) and Test 2 (dashed line). The shown curves are the mean values of at least two of the three measurements conducted with probes of each particle size range sample. Some measurements had to be neglected due to significant swelling of the crucible and/or high mass loss (>4%) during the measurement.

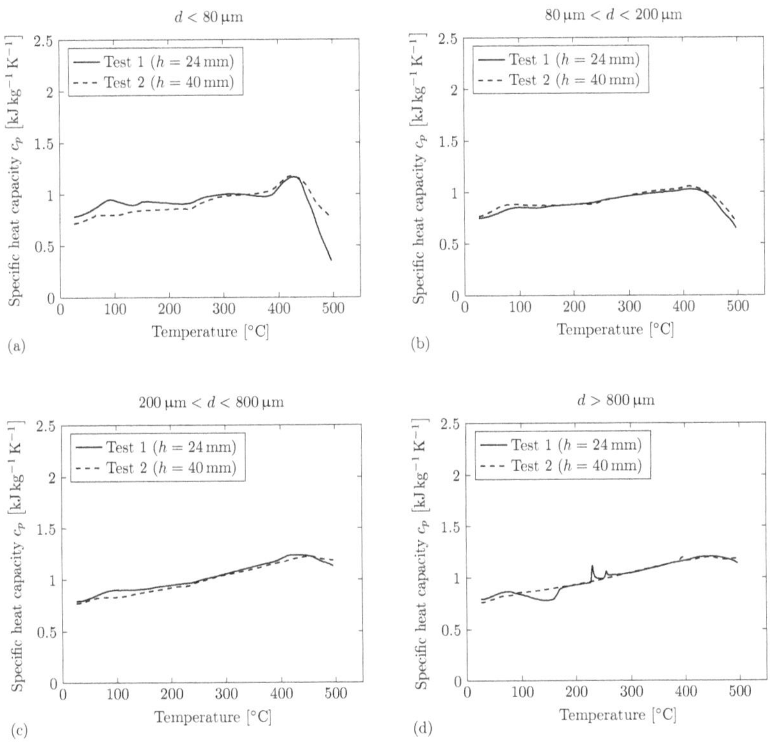

Figure 9. Specific heat capacity of the collected particles depositing on the weighing plates for both tests measured via DSC. The shown curves are the mean values of three probes measured for each size range. (**a**) Particle size range of $d < 80$ µm. (**b**) Particle size range of 80 µm $< d <$ 200 µm. (**c**) Particle size range of 200 µm $< d <$ 800 µm. (**d**) Particle size range of $d > 800$ µm.

For particle Size range 1 ($d < 80$ µm), the specific heat capacity curves over temperature are comparable for both tests. For Test 1, the values of the specific heat capacity are on a slightly elevated level compared to Test 2 for $T < 350$ °C. The curves of both tests show a decreasing trend of the specific heat capacity for $T > 425$ °C, which could be related to an exothermic reaction. For particle Size ranges 2 (80 µm $< d <$ 200 µm) and 3

(200 μm $< d <$ 800 μm), the specific heat capacity curves over temperature are also comparable for both tests. There exist no irregularities within the course of the curves, besides the decreasing trend of the specific heat capacity for $T > 425\,^{\circ}C$ for particle Size range 2. For particle Size range 3, the decreasing trend is not as significant as for particle Size ranges 2 or 1. For particle Size range 4 ($d >$ 800 μm), there exist individual irregularities for the specific heat capacity curves over temperature for each test. For Test 1, an exothermic reaction can be observed for $75\,^{\circ}C < T < 170\,^{\circ}C$. Two endothermic peaks appear at $T \approx 230\,^{\circ}C$ and $T \approx 255\,^{\circ}C$. For Test 2, there exists an endothermic peak at $T \approx 395\,^{\circ}C$.

In conclusion, the specific heat capacity measurements deliver comparable values for all four particle size ranges in both tests. This is an important finding that helps to determine correct parameters for, e.g., simulation models of the gas–particle flow during TR. The found differences between the particle size ranges are potentially caused by the different oxidation characteristics due to different surface areas, which was already investigated in previous studies [18,22]. With increasing particle size, however, the decreasing trend at temperatures of $T > 420\,^{\circ}C$ seems to be less significant. Explicit irregularities are only observed for particle Size range 4. This corresponds with the results of the thermal gravimetric analysis of Wang et al. conducted on ejected particles during TR of a 50 Ah prismatic battery with an NMC622 cathode [18]. They observed a moderate mass loss within the temperature range of $250\,^{\circ}C < T < 600\,^{\circ}C$ for particles of all size ranges. In addition, the results of the smallest particles (10 μm $< d <$ 100 μm) indicated gas that evolved from volatile thermal decomposition or a diffusion of volatiles that were adsorbed in the pores of particles [18].

Figure 10 shows the specific heat capacity over temperature of the four samples out of the jelly roll remains (compare to Figure 3) measured via DSC for Test 1 (solid line) and Test 2 (dashed line). The shown curves are the mean values of at least two of the three measurements conducted with probes of each sample out of the jelly roll remains. Some measurements had to be neglected due to significant swelling of the crucible and/or high mass loss (>4%) during the measurement.

For Position 1 (underneath vent, compare to Figure 3), the specific heat capacity curves over temperature are comparable for both tests. For Test 2, the specific heat is on a slightly elevated level compared to Test 1. In addition, an endothermic reaction can be observed for $60\,^{\circ}C < T < 165\,^{\circ}C$ for Test 2. For Test 1, the curve shows a decreasing trend of the specific heat capacity for $T > 400\,^{\circ}C$. For Position 2 (close to the nail, compare to Figure 3), the specific heat capacity curves over temperature show a comparable behavior for both tests besides individual irregularities. For Test 1, there exists an endothermic reaction for $45\,^{\circ}C < T < 160\,^{\circ}C$. For higher temperatures, the course of the curve is on a lower level compared to Test 2. For Test 2, a small exothermic peak is observed at $T \approx 85\,^{\circ}C$, followed by an endothermic reaction within the temperature range of $90\,^{\circ}C < T < 160\,^{\circ}C$. For Position 3 (positive terminal side, compare to Figure 3), the specific heat capacity curves over temperature show significant irregularities for both tests. An exothermic peak can be observed for $T \approx 90\,^{\circ}C$ for Test 1 and $T \approx 100\,^{\circ}C$ for Test 2. In addition, there exist strong endothermic peaks at $T \approx 395\,^{\circ}C$ for both tests. Further exothermic peaks are observed at $T \approx 425\,^{\circ}C$ and $T \approx 450\,^{\circ}C$. For Position 4 (negative terminal side, compare to Figure 3), the specific heat capacity curves over temperature also show significant irregularities for both tests. For both tests, endothermic peaks can be observed at $T \approx 395\,^{\circ}C$. For Test 2, there exist smaller peaks indicating chemical reactions within the temperature range of $75\,^{\circ}C < T < 155\,^{\circ}C$.

In conclusion, the results show that the specific heat capacity of the jelly roll remains depends on the position within the jelly roll. The specific heat capacity measurements of Position 1 (underneath the vent) and Position 2 samples (close to the nail) delivered comparable results, but there exist significant differences from Position 3 (positive terminal side) and Position 4 (negative terminal side) samples. The latter show a strong endothermic peak at $T \approx 395\,^{\circ}C$. Based on the data available from this study, it is unfortunately not possible to analyze the exact causes of these differences. Further research is therefore recommended in this context.

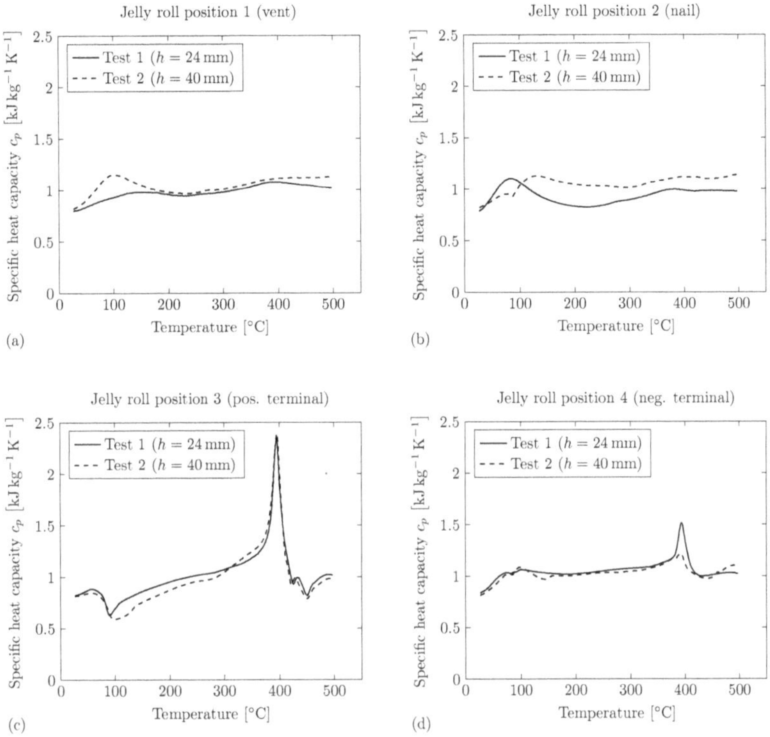

Figure 10. Specific heat capacity of the jelly roll remains for both tests measured via DSC. The shown curves are the mean values of three probes measured for each position (compare to Figure 3). (**a**) Sample extracted from Position 1 (vent), (**b**) Sample extracted from Position 2 (nail), (**c**) Sample extracted from Position 3 (pos. terminal), (**d**) Sample extracted from Position 4 (neg. terminal).

Figure 11 shows the specific heat capacity curves over temperature of the collected particles (solid lines) as well as the jelly roll remains (dashed lines) measured via DSC over temperature for Test 1 (black) and Test 2 (red). The shown curves are the mean values of all four particle sizes and all four jelly roll positions from Figure 9 and Figure 10, respectively.

The specific heat capacity measurements deliver comparable values for the particles ejected in both tests (solid lines). Therefore, it is concluded that there is a negligible dependence between the ejected particle's specific heat capacity and the distance between cell vent and cover. In addition, there is no significant influence of the particle size on the specific heat capacity, as shown in Figure 9. The particle's specific heat capacity can be set independently of the particle size and the battery's installation case within simulation models, which is an important finding. For the jelly roll remains (dashed lines), the specific heat capacity measurements also deliver comparable values in both tests. It is concluded that there is a negligible dependence between the jelly roll remains' specific heat capacity and the distance between cell vent and cover. However, there is an influence of the position within the jelly roll remains on the specific heat capacity, as shown in Figure 10.

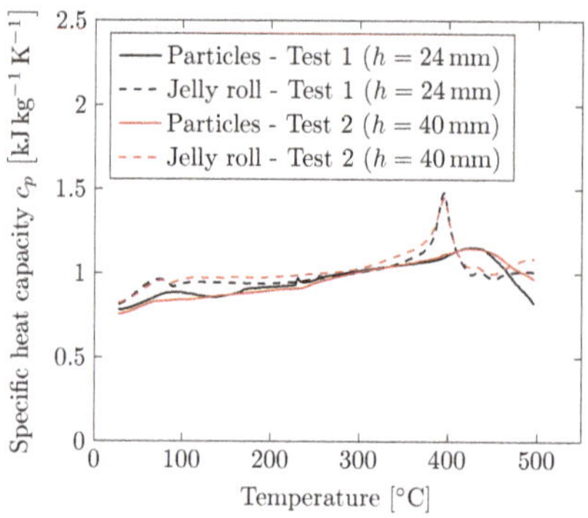

Figure 11. Specific heat capacity of the collected particles depositing on the weighing plates as well as the jelly roll remains for both tests measured via DSC. The shown curves are the mean values of all four particles sizes and all four jelly roll positions from Figure 9 and Figure 10, respectively.

Comparison of the specific heat capacity measurements of the ejected particles with the measurements of the jelly roll remains shows that the jelly roll remains have a slightly higher specific heat capacity. In addition, the occurrence of the endothermic peak at $T \approx 395\,^{\circ}\mathrm{C}$ for the jelly roll remains is a significant difference from the behavior of ejected particles. In this context, further investigations are recommended.

3.4. Suggestions for Further Improvement of the Proposed Setup

Based on the results of the proof-of-concept study presented here, the following suggestions for improvement are recommended for future research:

- As only a small fraction of the total ejected particle mass deposits on the weighing plates, it is recommended to extend the intermediate sheet metal. However, this can also result in higher time consumption for test preparation and evaluation.
- Determining the time-dependent particle deposition on the weighing plates, e.g., by implementing miniature load cells may provide further insights into the particle ejection process. This can also result in a more accurate weight measurement and lesser effort for post-test analysis.
- It is recommended to also collect particles that deposit within the steel box. This can offer further insights into the particle size distribution of all ejected particles.
- The milling process might be improved. There were still Cu particles that were not finely shredded with the method used in this study.
- Further findings could be obtained through the determination of the chemical composition for both ejected particles and jelly roll remains, for example, with inductively coupled plasma optical emission spectroscopy (ICP-OES).
- The presented results serve as proof of concept in the first place. Further tests must be carried out in order to quantify statistical deviations.

4. Conclusions

The presented study proposes a novel experimental setup to quantify the particle deposition during a lithium-ion battery TR. The setup integrates a single prismatic battery cell into an environment representing similar conditions as found for battery modules in EV battery packs. Within the flow path of the vented gas and particles, there are several weighing plates that can be individually removed from the setup in order to determine particle deposition. Two proof-of-concept experiments with a different distance between battery cell vent and cover are conducted. In addition, the size distribution of the vented particles is determined and specific heat capacity measurements are performed via DSC for both the vented particles and the jelly roll remains. The authors would like to point out once again that the results of this study are based on just two proof-of-concept experiments and consequently the reliability of the statements made below is limited.

4.1. Particle Deposition

- The distance between cell vent and cover has a significant influence on the particle deposition on the weighing plates.
- With a 24 mm distance between cell vent and cover, particle accumulations concentrate at the side walls.
- With a 40 mm distance between cell vent and cover, particle accumulations are randomly distributed over the entire area. In addition, larger particles deposit on the weighing plates.

4.2. Particle Size Distribution

- The distance between cell vent and cover has an influence on the size distribution of particles depositing on the weighing plates.
- The mass percentage of small particles ($d < 80$ µm) is found to be nearly independent of the gas flow velocities.

4.3. Specific Heat Capacity Measurements

- The specific heat capacity measurements deliver comparable values for all four particle size ranges in both tests.
- The specific heat capacity measurements for the jelly roll remains indicate influence of the position within the jelly roll and the specific heat capacity.
- The influence of the distance between cell vent and cover on the specific heat capacity is negligible.

In summary, the proposed experimental setup is generally suitable to quantify the particle deposition during a lithium-ion battery TR. However, there is a need for further improvement in order to enhance the quality and reliability of the results. By additionally increasing the number of tests, the results may offer further insights into the particle ejection process during lithium-ion battery TRs. This is of utmost importance, as the ejected particles carry a significant amount of energy and can be responsible for causing short circuits during TR propagation within battery packs. Today, there is a lack of understanding of the role of ejected particles in TR propagation behavior, although they pose a safety risk. This study helps researches and engineers to gain a better understanding of the gas–particle flow occurring during TR. By implementing certain suggested improvements, the proposed experimental setup may in addition be used for simulation model validation. Crucial input parameters of such models are particle size distribution as well as the specific heat capacity of the particles; both are also provided within this study. Therefore, this study contributes to the improvement of battery pack design and safety.

Author Contributions: Conceptualization, S.H. and H.K.; methodology, S.H.; investigation, S.H. and H.K.; data curation, S.H.; writing—original draft preparation, S.H.; writing—review and editing, H.K., S.Z. and O.H.; visualization, S.H.; supervision, S.Z. and O.H. All authors have read and agreed to the published version of the manuscript.

Funding: This research received no external funding.

Data Availability Statement: The original contributions presented in the study are included in the article; further inquiries can be directed to the corresponding author.

Acknowledgments: The authors gratefully acknowledge the support of our BMW Group colleagues from the technology material and process analysis (TWA) in conducting the experiments and the support of our student Yunus Emir who took over a part of the CAD construction. In addition, S. Hoelle acknowledges the support of the TUM Graduate School.

Conflicts of Interest: Authors Sebastian Hoelle, Hyojeong Kim and Sascha Zimmermann were employed by the company BMW Group. The remaining authors declare that the research was conducted in the absence of any commercial or financial relationships that could be construed as a potential conflict of interest.

Abbreviations and Symbols

The following abbreviations are used in this manuscript:

CCCV	constant current constant voltage
DSC	differential scanning calorimetry
DEC	diethyl carbonate
DMC	dimethyl carbonate
EC	ethylene carbonate
EMC	ethyl methyl carbonate
EV	electric vehicle
ICP-OES	inductively coupled plasma optical emission spectroscopy
$LiPF_6$	lithium hexafluorophosphate
NMC	lithium nickel manganese cobalt oxide
SoC	state of charge
TR	thermal runaway

The following symbols are used in this manuscript:

A	flow path area, m^2
c_p	specific heat capacity, $J\,kg^{-1}\,K^{-1}$
d	diameter, m
h	height between cell vent and cover of the test setup, m
m	mass, kg
Δ	difference, -
Indices:	
cell	index representing the full cell
i	index representing the weighing plate i
glas	index representing a glass (used for particle collection and weighing)
loose	index representing loose particles on a weighing plate
particles	index representing the particles vented during TR
postTR	index representing the state after TR
scraped	index representing particles that are scraped off from a weighing plate
stick	index representing sticking particles
total	index representing the total amount of particles on a weighing plate
TR	index representing the TR process

References

1. Sun, P.; Bisschop, R.; Niu, H.; Huang, X. A Review of Battery Fires in Electric Vehicles. *Fire Technol.* **2020**, *56*, 1361–1410. [CrossRef]
2. Huang, W.; Feng, X.; Han, X.; Zhang, W.; Jiang, F. Questions and Answers Relating to Lithium-Ion Battery Safety Issues. *Cell Rep. Phys. Sci.* **2021**, *2*, 100285. [CrossRef]
3. Feng, X.; Ouyang, M.; Liu, X.; Lu, L.; Xia, Y.; He, X. Thermal runaway mechanism of lithium ion battery for electric vehicles: A review. *Energy Storage Mater.* **2018**, *10*, 246–267. [CrossRef]
4. Liu, J.; Li, J.; Wang, J. In-depth analysis on thermal hazards related research trends about lithium-ion batteries: A bibliometric study. *J. Energy Storage* **2021**, *35*, 102253. [CrossRef]
5. Jindal, P.; Bhattacharya, J. Review—Understanding the Thermal Runaway Behavior of Li-Ion Batteries through Experimental Techniques. *J. Electrochem. Soc.* **2019**, *166*, A2165–A2193. [CrossRef]
6. Ouyang, D.; Chen, M.; Huang, Q.; Weng, J.; Wang, Z.; Wang, J. A Review on the Thermal Hazards of the Lithium-Ion Battery and the Corresponding Countermeasures. *Appl. Sci.* **2019**, *9*, 2483. [CrossRef]
7. Wang, Q.; Mao, B.; Stoliarov, S.I.; Sun, J. A review of lithium ion battery failure mechanisms and fire prevention strategies. *Prog. Energy Combust. Sci.* **2019**, *73*, 95–131. [CrossRef]
8. Feng, X.; Ren, D.; He, X.; Ouyang, M. Mitigating Thermal Runaway of Lithium-Ion Batteries. *Joule* **2020**, *4*, 743–770. [CrossRef]
9. Li, W.; Xue, Y.; Feng, X.; Rao, S.; Zhang, T.; Gao, Z.; Guo, Y.; Zhou, H.; Zhao, H.; Song, Z.; et al. Characteristics of particle emissions from lithium-ion batteries during thermal runaway: A review. *J. Energy Storage* **2024**, *78*, 109980. [CrossRef]
10. Zou, K.; Chen, X.; Ding, Z.; Gu, J.; Lu, S. Jet behavior of prismatic lithium-ion batteries during thermal runaway. *Appl. Therm. Eng.* **2020**, *179*, 115745. [CrossRef]

11. Garcia, A.; Monsalve-Serrano, J.; Sari, R.L.; Martinez-Boggio, S. An optical investigation of thermal runaway phenomenon under thermal abuse conditions. *Energy Convers. Manag.* **2021**, *246*, 114663. [CrossRef]
12. Wang, G.; Kong, D.; Ping, P.; Wen, J.; He, X.; Zhao, H.; He, X.; Peng, R.; Zhang, Y.; Dai, X. Revealing particle venting of lithium-ion batteries during thermal runaway: A multi-scale model toward multiphase process. *eTransportation* **2023**, *16*, 100237. [CrossRef]
13. Ding, C.; Zhu, N.; Yu, J.; Li, Y.; Sun, X.; Liu, C.; Huang, Q.; Wang, J. Experimental investigation of environmental pressure effects on thermal runaway properties of 21700 lithium-ion batteries with high energy density. *Case Stud. Therm. Eng.* **2022**, *38*, 102349. [CrossRef]
14. Liao, Z.; Zhang, S.; Li, K.; Zhao, M.; Qiu, Z.; Han, D.; Zhang, G.; Habetler, T.G. Hazard analysis of thermally abused lithium-ion batteries at different state of charges. *J. Energy Storage* **2022**, *27*, 101065. [CrossRef]
15. Essl, C.; Golubkov, A.W.; Gasser, E.; Nachtnebel, M.; Zankel, A.; Ewert, E.; Fuchs, A. Comprehensive Hazard Analysis of Failing Automotive Lithium-Ion Batteries in Overtemperature Experiments. *Batteries* **2020**, *6*, 30. [CrossRef]
16. Zhang, Y.; Wang, H.; Li, W.; Li, C. Quantitative identification of emissions from abused prismatic Ni-rich lithium-ion batteries. *eTransportation* **2019**, *2*, 100031. [CrossRef]
17. Zhang, Y.; Wang, H.; Li, W.; Li, C.; Ouyang, M. Size distribution and elemental composition of vent particles from abused prismatic Ni-rich automotive lithium-ion batteries. *J. Energy Storage* **2019**, *26*, 100991. [CrossRef]
18. Wang, Y.; Wang, H.; Zhang, Y.; Cheng, L.; Wu, Y.; Feng, X.; Lu, L.; Ouyang, M. Thermal oxidation characteristics for smoke particles from an abused prismatic $Li(Ni_{0.6}Co_{0.2}Mn_{0.2})O_2$ battery. *J. Energy Storage* **2021**, *39*, 102639. [CrossRef]
19. Premnath, V.; Wang, Y.; Wright, N.; Khalek, I.; Uribe, S. Detailed characterization of particle emissions from battery fires. *Aerosol Sci. Technol.* **2022**, *56*, 337–354. [CrossRef]
20. Chen, S.; Wang, Z.; Yan, W. Identification and characteristics analysis of powder ejected from a lithium ion battery during thermal runaway at elevated temperatures. *J. Hazard. Mater.* **2020**, *400*, 123169. [CrossRef]
21. Held, M.; Tuchschmid, M.; Zennegg, M.; Figi, R.; Schreiner, C.; Mellert, L.D.; Welte, U.; Kompatscher, M.; Hermann, M.; Nachef, L. Thermal runaway and fire of electric vehicle lithium-ion battery and contamination of infrastructure facility. *Renew. Sustain. Energy Rev.* **2022**, *165*, 112474. [CrossRef]
22. Wang, H.; Wang, Q.; Jin, C.; Xu, C.; Zhao, Y.; Li, Y.; Zhong, C.; Feng, X. Detailed characterization of particle emissions due to thermal failure of batteries with different cathodes. *J. Hazard. Mater.* **2023**, *458*, 131646. [CrossRef] [PubMed]
23. Coman, P.T.; Rayman, S.; White, R.E. A lumped model of venting during thermal runaway in a cylindrical Lithium Cobalt Oxide lithium-ion cell. *J. Power Sources* **2016**, *307*, 56–62. [CrossRef]
24. Coman, P.T.; Darcy, E.C.; White, R.E. Simplified Thermal Runaway Model for Assisting the Design of a Novel Safe Li-Ion Battery Pack. *J. Electrochem. Soc.* **2022**, *169*, 040516. [CrossRef]
25. Ostanek, J.K.; Li, W.; Mukherjee, P.P.; Crompton, K.R.; Hacker, C. Simulating onset and evolution of thermal runaway in Li-ion cells using a coupled thermal and venting model. *Appl. Energy* **2020**, *268*, 114972. [CrossRef]
26. Li, W.; Quiroga, V.L.; Crompton, K.R.; Ostanek, J.K. High Resolution 3-D Simulations of Venting in 18650 Lithium-Ion Cells. *Front. Energy Res.* **2021**, *9*, 788239. [CrossRef]
27. Kim, J.; Mallarapu, A.; Finegan, D.P.; Santhanagopalan, S. Modeling cell venting and gas-phase reactions in 18650 lithium ion batteries during thermal runaway. *J. Power Sources* **2021**, *489*, 229496. [CrossRef]
28. Citarella, M.; Suzzi, D.; Brunnsteiner, B.; Schiffbänker, P.; Maier, G.; Schneider, J. Computational Modelling of Thermal Runaway Propagation in Lithium-Ion Battery Systems. In Proceedings of the 2019 IEEE Transportation Electrification Conference (ITEC-India), Bengaluru, India, 17–19 December 2019. Available online: https://ieeexplore.ieee.org/abstract/document/9080817 (accessed on 22 April 2024).
29. Mishra, D.; Shah, K.; Jain, A. Investigation of the Impact of Flow of Vented Gas on Propagation of Thermal Runaway in a Li-Ion Battery Pack. *J. Electrochem. Soc.* **2021**, *168*, 060555. [CrossRef]
30. Hoelle, S.; Scharner, S.; Asanin, S.; Hinrichsen, O. Analysis on Thermal Runaway Behavior of Prismatic Lithium-Ion Batteries with Autoclave Calorimetry. *J. Electrochem. Soc.* **2021**, *168*, 120515. [CrossRef]

batteries

MDPI

Article

Experimental Study on Thermal Runaway Characteristics of High-Nickel Ternary Lithium-Ion Batteries under Normal and Low Pressures

Ye Jin, Di Meng, Chen-Xi Zhao, Jia-Ling Yu, Xue-Hui Wang * and Jian Wang *

State Key Laboratory of Fire Science, University of Science and Technology of China, Hefei 230026, China; jy330521@mail.ustc.edu.cn (Y.J.); mengl@mail.ustc.edu.cn (D.M.); zhaochenxi@mail.ustc.edu.cn (C.-X.Z.)
* Correspondence: wxuehui@ustc.edu.cn (X.-H.W.); wangj@ustc.edu.cn (J.W.); Tel.: +86-551-6360-1651 (X.-H.W.); +86-551-6360-6463 (J.W.)

Abstract: High-nickel (Ni) ternary lithium-ion batteries (LIBs) are widely used in low-pressure environments such as in the aviation industry, but their attribute of high energy density poses significant fire hazards, especially under low pressure where thermal runaway behavior is complex, thus requiring relevant experiments. This study investigates the thermal runaway characteristics of $LiNi_{0.8}Mn_{0.1}Co_{0.1}O_2$ (NCM811) 18650 LIBs at different states of charge (SOCs) (75%, 100%) under various ambient pressures (101 kPa, 80 kPa, 60 kPa, 40 kPa). The results show that, as the pressure is decreased from 101 kPa to 40 kPa, the onset time of thermal runaway is extended by 28.2 s for 75% SOC and by 40.8 s for 100% SOC; accordingly, the onset temperature of thermal runaway increases by 19.3 °C for 75% SOC and by 33.5 °C for 100% SOC; the maximum surface temperature decreases by 70.8 °C for 75% SOC and by 68.2 °C for 100% SOC. The cell mass loss and loss rate slightly decrease with reduced pressure. However, ambient pressure has little impact on the time and temperature of venting as well as the voltage drop time. SEM/EDS analysis verifies that electrolyte evaporates faster under low pressure. Furthermore, the oxygen concentration is lower under low pressure, which consequently leads to a delay in thermal runaway. This study contributes to understanding thermal runaway characteristics of high-Ni ternary LIBs and provides guidance for their safe application in low-pressure aviation environments.

Keywords: high-Ni ternary lithium-ion battery; thermal runaway; low pressure; fire hazards; aviation applications

1. Introduction

With steady growth in global energy consumption, the intermittent nature of most renewable energies, and the increasing demand for electricity from consumers and industries, the latest developments and opportunities in energy storage have emerged. Compared to other secondary batteries, LIBs offer higher energy density, higher open-circuit voltage, greater output power, lower self-discharge rate, no memory effect, a wide operating temperature range, and a fast charge and discharge speed. Thus, LIBs are widely used in consumer electronics products, military products, and aviation products [1]. In particular, LIBs with high-Ni cathodes, compared to other cathode materials, achieve higher specific capacity and lower cost, making them highly promising. A large number of LIBs and devices containing LIBs are increasingly transported or used in low-pressure environments such as aviation. High-Ni LIBs are being integrated into aviation electronics, including communication systems, flight control systems, and emergency power supplies. They are also transported by air for use in power tools, portable electronic devices, and other applications, either as standalone units or as components of systems. However, Ni does increase the energy density of the cathode and participate in charge–discharge reactions; in the meantime, it introduces safety issues due to (i) thermal phase transitions in the cathode releasing oxygen and (ii) exothermic thermal reactions between the released oxygen species and electrolyte producing significant heat; this accelerates the phase transitions and promotes thermal runaway [2,3]. This has led to numerous fire incidents, causing significant casualties and property damage. According to the U.S. Federal Aviation Administration, from 3 March 2006 to 27 November 2023, there have been as many as 498 fire incidents caused by lithium batteries brought onto aircraft as cargo or luggage [4]. Therefore, it is crucial to address the thermal safety of high-Ni ternary LIBs under low pressure.

Previous research mainly focused on thermal runaway behavior characteristics and propagation characteristics under low ambient pressure. The study of thermal runaway behavior under low pressure primarily revolves around experimental techniques. Chen et al. [5,6] conducted several thermal runaway experiments on LIBs under low pressure: $LiCoO_2$ 18650 LIBs were tested in Hefei and Lhasa, the difference in pressure resulted in the venting temperature, the average burning rate, and the effective combustion heat in Hefei being larger than those in Lhasa. The influence of the pressure on heat release was greater than on mass loss; 18650 LIBs with different cathode materials ($LiFePO_4$, $LiCoO_2$) and SOC were tested in Hefei and Lhasa, finding that total mass loss, total heat release, and the growth rate of unit combustion heat between the two pressures increased with SOC, and total heat release was higher at higher pressure. Xie et al. [7–10] conducted a series of experimental studies on LIBs under low pressure: NCM523 18650 LIBs with different aging cycles were tested under different pressures, the results showed that the onset time and temperature of thermal runaway, as well as the heat release rate, decreased with an increase in cycle number or a decrease in pressure. Thermal runaway was more likely at low pressure because the rupture disk was more easily to open; NCM523 18650 LIBs at different charge–discharge rates were tested under 20 kPa and 95 kPa, and the results showed that higher charge–discharge rates and lower pressures both caused earlier and less intense thermal runaway. The difference in thermal runaway times for cells at different charge–discharge rates at 20 kPa was greater than that at 95 kPa; NCM523 18650 LIBs at different SOC were heated by an electric coil under low pressure, it was found that the decrease in pressure resulted in a delay in voltage drop time, reduced thermal runaway intensity and temperature, and a U-shaped change in thermal runaway time; NCM523 pouch LIBs with different overcharge cycles were tested at 30 kPa and 95 kPa, the results showed that, due to the differential pressure for the release of combustibles, the thermal runaway jet intensity at 30 kPa was stronger. The thermal runaway intensity and heat release at 30 kPa were weaker because of the lower oxygen content. Liu et al. [11] conducted tests on normal ternary 18650 LIBs at 96 kPa and 61 kPa, showing that the thermal runaway onset time and temperature were higher at 61 kPa, while the explosion pressure and TNT equivalent were lower at 61 kPa. Zhang et al. [12] performed experiments on NCM111 LIBs from 101 kPa to 30 kPa, it was found that, as the pressure decreased, the thermal runaway onset time was earlier, CO_2 production was reduced, unsaturated hydrocarbons increased, and the range of the explosion limits of thermal runaway gases increased. Chen et al. [13] conducted thermal runaway tests from 95 kPa to 20 kPa, revealing that the heat release rate, total heat release, surface temperature, CO_2 peak concentration, and mass loss all decreased with lower pressure or heating power, while hydrocarbon (CHx) and CO peaks increased with decreasing pressure. Previous researchers mainly paid attention to the thermal runaway characteristics of ternary LIBs with cathode materials of normal Ni content under low pressure, lacking experimental studies on high-Ni ternary LIBs, and almost none measured the voltage parameter, which respond faster than the surface temperature.

This paper aims to narrow the research gap between high-Ni ternary LIBs and other LIBs under low pressure by conducting thermal runaway experiments on NCM811 18650 LIBs under different ambient pressures. By measuring and discussing key parameters such as the characteristic temperature, voltage, and mass loss, this paper provides insights into thermal runaway characteristics of high-Ni ternary LIBs under low pressure and guides their safe application in such environments, and encourages further research in this area.

2. Materials and Methods

2.1. Battery Samples

The commercial cylindrical graphite/NCM811 LIBs from a battery manufacturer were tested in this study. The basic information of the batteries is shown in Table 1.

Table 1. Basic information of LIBs in experiments.

Item	Specification
Height	65.05 ± 0.15 mm
Diameter	18.35 ± 0.15 mm
Weight	46 ± 2 g
Anode material	Graphite
Cathode material	$LiNi_{0.8}Co_{0.1}Mn_{0.1}O_2$
Nominal capacity	2600 mAh
Nominal voltage	3.6 V
Charge cut-off voltage	4.2 V
Discharge cut-off voltage	2.75 V
Charging cut-off current	0.02 C (52 mA)

Before the experiments, the batteries underwent the following operations: (i) discharged at a constant current of 1 C-rate (C-rate is defined as the ratio of the charge or discharge current to the battery's nominal capacity.) to 2.75 V; (ii) charged at a constant current and constant voltage mode of 0.5 C-rate until the current dropped to 52 mA, then charged to 4.2 V, and then discharged at 1 C-rate to 2.75 V, with these steps cycled three times; (iii) charged at 0.5 C-rate to the specified SOC. After these procedures, the batteries were left to rest for 24 h to return to electrochemical and thermal equilibrium. All batteries used in this study were from the same batch to minimize differences between them.

2.2. Experimental Setup

Figure 1 shows a schematic of the experimental setup. The experiments were conducted in a variable-pressure experimental cabin measuring 3 m in length, 2 m in width, and 2 m in height, providing an ambient pressure range of 20–101 kPa. Once the chamber reaches the set pressure environment by dynamically adjusting the inlet and outlet airflow, the vacuum pump is turned off. After the pressure stabilizes, the experiment is conducted. Due to good sealing, the pressure fluctuation inside the chamber does not exceed 1 kPa by the end of the experiment, which is within the acceptable system error. Inside the chamber, the experimental samples were fixed in a bakelite clamp embedded with insulating sheets. A 150 W heating rod with an appearance similar to the battery sample was used to simulate an overheated cell and trigger thermal runaway in the adjacent cell (the gap between the heating rod and the cell is 0 mm). The cells were placed on thermal insulation wool supported by an iron mesh rack, which was set on a stable table. Two K-type exposed thermocouple beads, each with a 1 mm diameter, were individually affixed to the side surfaces of the heating rod and the cell by high-temperature-resistant Teflon tape to measure the surface temperature, and the data were collected by a temperature-acquisition module (NI 9213). The Ni strip with a thickness of 0.12 mm and a width of 6 mm was spot-welded to the positive electrode of the cell, and the Ni strip was connected to a voltage-acquisition module (CT-4008-5V6A-S1, NEWARE Battery Testing System) to measure the open-circuit voltage. The cell was weighed before and after the experiment by an electronic balance. Thermal runaway processes and fire phenomena were recorded by a video camera (SONY FDR-AX100E). The heating rod was turned off immediately after the cell entered thermal runaway. The conditions of the experiments are shown in Table 2. Each condition was repeated at least three times to reduce random errors.

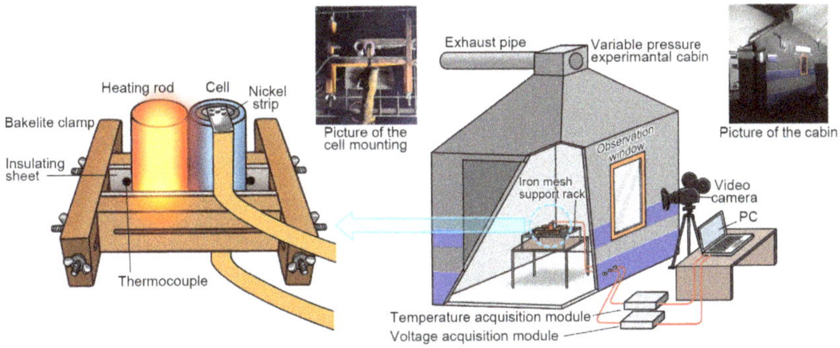

Figure 1. Schematic of the experimental setup.

Table 2. Conditions of experiments.

Testing Group	Testing No.	SOC	Ambient Pressure (kPa)	Heating Power (W)
1	1/2/3/4	75%	101/80/60/40	150
2	5/6/7/8	100%	101/80/60/40	

Following the thermal runaway experiments, all burnt cells were collected and stored in sealed bags. Scanning electron microscopy (SEM)–energy-dispersive X-ray spectroscopy (EDS) analysis was performed on the electrodes of the burnt cells using ZEISS GeminiSEM 450 (Carl Zeiss Microscopy GmbH, Jena, Germany) equipped with AZtecLive EDS software version [4.3] (Oxford Instruments, Abingdon, UK). The information of microscopic morphology and elemental composition was obtained to characterize the internal reactions.

3. Results and Discussion

In this section, the criteria for thermal runaway and venting, which are important events during the experiment, are introduced first. Based on these criteria, the characteristics of thermal runaway are then discussed in detail.

3.1. Criteria for Thermal Runaway and Venting

The criteria for thermal runaway in previous works have often been based on an empirical value of temperature rise rate combined with fire phenomena, given experimental conditions [9,14,15]. In this study, the criteria for thermal runaway and venting were determined based on the second derivative of the cell's surface temperature, independent of specific platforms. That is

$$\Lambda := \left| \frac{\mathrm{d}^2 T_B}{\mathrm{d}t^2} \right| > \Delta \tag{1}$$

where T_B is the temperature of the battery and Δ is the systematic error.

The proof of this criterion is presented below. Firstly, let us consider the heating rod's temperature rise process. The temperature rise rate of the heating rod in the experiment can be approximated as that of an isolated heating rod [16]:

$$\frac{\mathrm{d}T_H}{\mathrm{d}t} = kP - \alpha(T_H - T_0), \tag{2}$$

where $k = 1/(c \cdot m)$, c is the specific heat capacity, m is the mass of the heating rod, P is the power of the heating rod, α is the reduced heat transfer coefficient of the heating rod, and T_0 is the ambient temperature. Solving the differential equation gives:

$$T_H = T_0 + \frac{kP}{\alpha}(1 - e^{\alpha t}) \tag{3}$$

Next, we focus on the cell's temperature rise process, which is influenced primarily by three factors: heat transfer from the heating rod, heat dissipation to the air, and heat released from internal chemical reactions. This analysis leads to the following differential Equation [16]:

$$\frac{dT_B}{dt} = \beta(T_H - T_B) - \alpha'(T_B - T_0) + \gamma, \tag{4}$$

where β is the reduced heat transfer coefficient from the heating rod to the battery, α' is the reduced heat transfer coefficient from the battery to the air, and γ represents the contribution from internal chemical reactions. Note that the contribution of γ before thermal runaway is negligible due to its imperceptibility within the precision of our measurements, its quantification requiring devices, such as a micro-calorimeter [17], a differential scanning calorimeter [2], etc. Therefore, the equation simplifies to:

$$\frac{dT_B}{dt} \approx \beta(T_H - T_B) - \alpha'(T_B - T_0) \tag{5}$$

The second derivative of the temperature of the battery is

$$\frac{d^2T_B}{dt^2} = \beta\frac{dT_H}{dt} - (\alpha' + \beta)\frac{dT_B}{dt} \leq \beta\frac{dT_H}{dt} = \beta kPe^{-\alpha t} \leq \beta kP \tag{6}$$

In this experiment, it can be estimated that

$$\begin{aligned} \beta &\sim 10^{-2} \leq 10^{-1} \\ kP &\sim 10^{-2} \leq 10^{-1} \end{aligned} \tag{7}$$

Therefore, in the absence of venting events and thermal runaway events, we obtain

$$\Lambda = \frac{d^2T_B}{dt^2} \leq \beta kP \leq 10^{-2} \tag{8}$$

Due to the systematic error, the measured second derivative is constrained by

$$\Lambda \leq \Delta = 0.1 \tag{9}$$

When a thermal runaway event occurs, $\Lambda \gg 1$, and during a venting event, $\Lambda > 0.3$ in this experiment. Therefore, $\Lambda > \Delta$ indicates the occurrence of such events. During the temperature rise process before thermal runaway, Λ is smaller than the experimental systematic error, whereas during thermal runaway, Λ significantly exceeds the experimental systematic error. This suggests a criterion for thermal runaway independent of specific platforms, while the criterion for venting events may still vary depending on the specific platform.

3.2. Thermal Runaway Behavior

Figure 2 shows the thermal runaway phenomena of the cells under different pressures. This process can be divided into four stages: (1) The heating stage (not clearly visible in the video): the cell was heated by the heating rod. When a certain temperature was reached, the safety valve of the cell broke, and an unusual sound could be heard at this moment. (2) The jet fire stage: when thermal runaway occurred, the cell first ejected a large amount of high-temperature metal particles upward, immediately triggering combustion or explosion. Once the rapid explosion occurred, it was accompanied by the ejection of the jelly roll. This stage lasted no more than 5 s. (3) The electrolyte combustion stage: after the ejection ended, a sustained flame appeared above the battery due to the burning of the flammable electrolyte. (4) The extinguishing stage: subsequently, as the electrolyte was completely consumed, the flame gradually diminished until it went out.

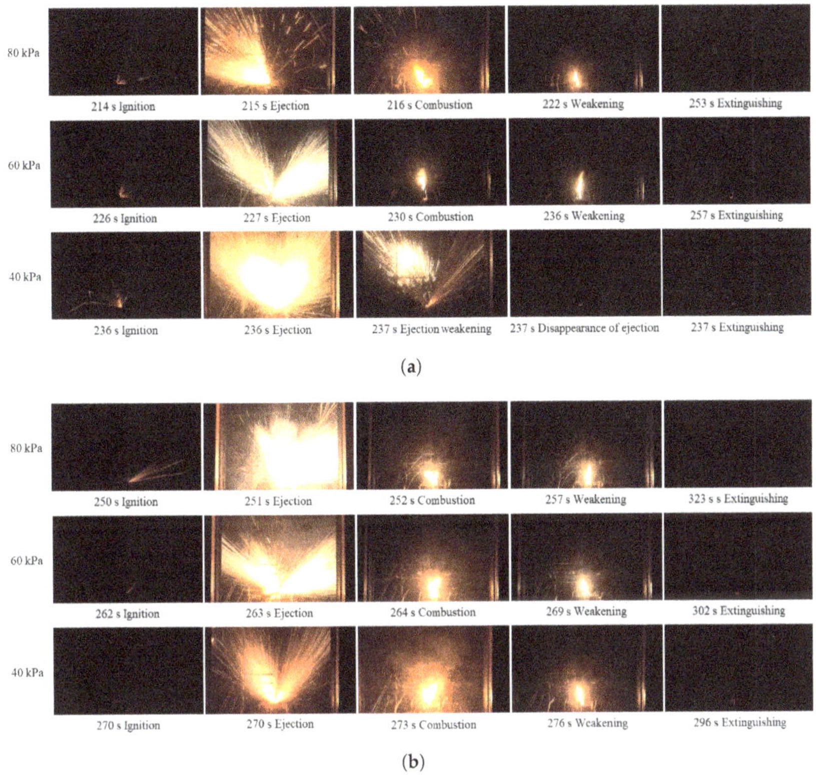

Figure 2. Thermal runaway phenomena of cells at (**a**) 75% SOC and (**b**) 100% SOC under different pressures.

This indicates that, as the ambient pressure gradually decreased, the duration of the electrolyte combustion stage shortened. The duration of electrolyte combustion decreased from 71 s at 80 kPa to 23 s at 40 kPa for 75% SOC. Furthermore, the duration decreased from 37 s at 80 kPa to 0 s at 40 kPa for 100% SOC. This is because the electrolyte evaporates more easily at lower pressures, leaving less residual electrolytes. The duration of electrolyte combustion for the cell at 100% SOC was significantly shorter than that for the cell at 75% SOC. This is because a cell with a higher SOC stores more energy and has a higher amount of lithium intercalation in the anode [18], so it contains more active substances, undergoes more thorough and intense chemical reactions, and generates more heat. Consequently, more electrolyte is consumed during the chemical reactions and more evaporates [19].

To further understand thermal runaway behavior, the temperature evolution curves of cells at 75% SOC under 80 kPa and under 40 kPa during the experiments were displayed by Figure 3. The thermal runaway process of the cell consists of three stages:

- Stage I (Heating stage): The cell is heated, and the temperature increases with an average temperature rise rate of 0.50 °C/s at 80 kPa and of 0.49 °C/s at 40 kPa. Unstable chemical reactions occur inside the cell, gradually raising the surface temperature while the voltage drops sharply.
- Stage II (Venting stage): When internal reactions accelerate and internal temperature rises to a certain level, the safety valve of the cell breaks, entering stage II, which is marked by a decrease in the temperature rise rate. It should be noted that venting does not always lead to a decrease in temperature [15,20–23]. A more accurate description is a decrease in the temperature rise rate, which will be further discussed below. The surface temperature of the cell rises relatively quickly during this stage, with an

average temperature rise rate of 1.05 °C/s at 80 kPa and of 0.87 °C/s at 40 kPa. This stage is very brief, typically lasting within 1 min.

- Stage III (Thermal runaway occurrence stage): The inside of the cell is entirely out of control, with surface temperature rising rapidly to a maximum rate of 235.44 °C/s at 80 kPa and of 72.88 °C/s at 40 kPa, releasing significant energy, finally causing fire or explosion.

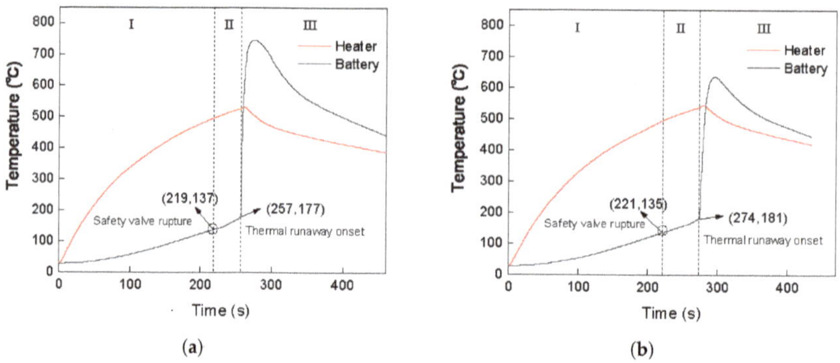

Figure 3. Temperature evolution curves of cells at 75% SOC under (**a**) 80 kPa and under (**b**) 40 kPa. Stage I, II, and III represent heating stage, venting stage, and thermal runaway occurrence stage, respectively.

Here, a discussion on the indicator of venting being a decrease in the temperature rise rate is provided. The temperature T can be expressed as a function of enthalpy H and pressure p [24]:

$$T = T(H, p) \tag{10}$$

So, the contribution of venting to temperature change can be expressed as:

$$dT_{\text{venting}} = \left(\frac{\partial T}{\partial H}\right)_p dH_{\text{venting}} + \left(\frac{\partial T}{\partial p}\right)_H dp_{\text{venting}} = \frac{1}{C_p} dH_{\text{venting}} + \mu_{JT} dp_{\text{venting}} \tag{11}$$

where μ_{JT} is the Joule–Thomson coefficient [24]. Venting is an exothermic process because the internal gas temperature is higher than the external ambient temperature, so $dH_{\text{venting}} < 0$. Previous research show that the vented gases are primarily composed of CO, CO_2, H_2, CH_4, C_2H_4, C_2H_2, C_3H_6, and HF, with H_2 constituting a small proportion [3,25–28]. Except H_2, $\mu_{JT} > 0$ [24], and $dp_{\text{venting}} < 0$; thus, $dT_{\text{venting}} < 0$.

The temperature rise during stage I is contributed by the heating rod and internal chemical reactions:

$$dT_{\text{stage I}} = dT_{\text{heater}} + dT_{\text{reactions}} \tag{12}$$

The temperature rise when venting occurs during stage II is contributed by the heating rod, internal chemical reactions, and venting:

$$dT_{\text{stage II,venting}} = dT_{\text{heater}} + dT_{\text{reactions}} + dT_{\text{venting}} \tag{13}$$

Since $dT_{\text{venting}} < 0$, thus $dT_{\text{stage II,venting}} < dT_{\text{stage I}}$, making the venting marked by a decrease in the temperature rise rate.

To enhance the fundamentals of this study, Figure 4 illustrates the mechanism of thermal runaway process of NCM811 LIBs. The battery consists of cathode material, anode material, separators, electrolyte, and outer shell. Additionally, there is a safety valve near the positive electrode. As the battery heats up, internal thermal expansion occurs, leading to a continuous increase in internal pressure. When the pressure difference between the inside and outside reaches the threshold for safety valve rupture, the safety valve breaks.

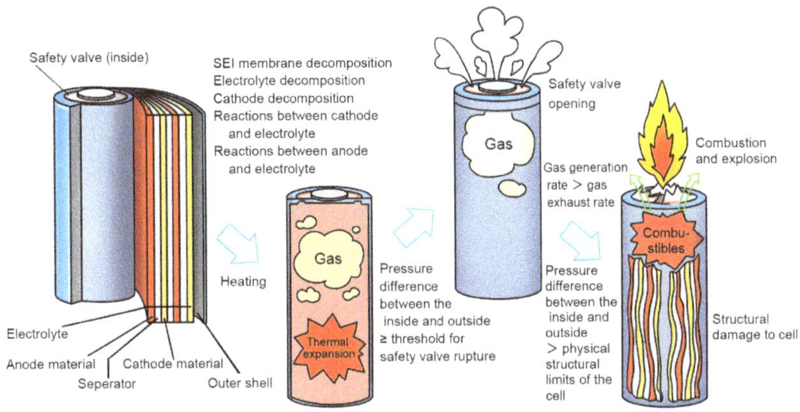

Figure 4. Mechanism diagram of the thermal runaway process of NCM811 LIBs.

Previous studies indicate that NCM811 LIBs undergo various complex chemical reactions during the thermal runaway tests, including solid electrolyte interface (SEI) membrane decomposition, electrolyte decomposition, cathode decomposition, and reactions between cathode and electrolyte, etc. [29,30]. When the temperature is high enough, chemically adsorbed oxygen species at the cathode transform into physically adsorbed oxygen species, leading to the further release of oxygen from the surface, depicted by the following reaction equation:

$$Li_x MO_2 \longrightarrow LiMO_2 + MO_x + O_2 \tag{14}$$

The reaction of oxygen species ($O_2{}^-$, O^-, $O_2{}^{2-}$, etc.) released from the cathode with the electrolyte is considered the triggering reaction for thermal runaway. Highly reactive oxygen species immediately react with the electrolyte, releasing a significant amount of heat:

$$O^n + C_3H_4O_3 \longrightarrow CO_2 + 2\,H_2O \tag{15}$$

This process accelerates the phase transition in the cathode and the release of oxygen, creating a positive feedback loop. In the meantime, there are gas-producing and exothermic reactions of other materials, leading to the accumulation of a large amount of gas and energy, which ultimately cause burning [2]. Moreover, if the internal pressure is high and the vent outlet is blocked, resulting in slow gas venting, and if the pressure difference between the inside and outside exceeds the physical structural limits of the cell, the high-pressure battery may explode [31,32].

3.3. Venting and Thermal Runaway Events

Venting and thermal runaway events are critical and crucial in building a deep understanding of the thermal runaway characteristics of NCM811 LIBs. To facilitate analysis, the temperature evolution curves of cells at 75% SOC and at 100% SOC under different ambient pressures are depicted in Figure 5. For the cells at 75% SOC, under 101 kPa, 80 kPa, 60 kPa, and 40 kPa, thermal runaway triggering occurred at 250 s, 257 s, 273 s, and 274 s, respectively. Correspondingly, triggering temperatures of thermal runaway were 163.9 °C, 176.7 °C, 179.3 °C, and 181.3 °C, respectively. Furthermore, maximum surface temperatures after thermal runaway were 755.2 °C, 745.8 °C, 347.7 °C, and 637.4 °C, respectively. The increased gas generation and potential clogging of the vent outlet of the higher-energy-density cell resulted in an explosion at 60 kPa [2,31,33]. The jelly roll was ejected from the cell, carrying away a significant amount of heat, which caused the maximum surface temperature of the cell to differ from that when combustion alone occurred. For the cells at 100% SOC, under 101 kPa, 80 kPa, 60 kPa, and 40 kPa, thermal runaway triggering occurred at 212 s, 217 s, 220 s, and 239 s, respectively, and triggering temperatures of thermal runaway were 132.3 °C, 139.4 °C, 147.7 °C, and 157.8 °C, respectively, and maximum surface temperatures were 751.9 °C, 728.0 °C, 684.5 °C, and 672.0 °C, respectively. It is observed that onset times and temperatures of thermal runaway for both SOC conditions exhibit a negative correlation with pressures, whereas max surface temperatures show a positive correlation with pressures. This pattern is similar to that reported in the previous articles [15,34].

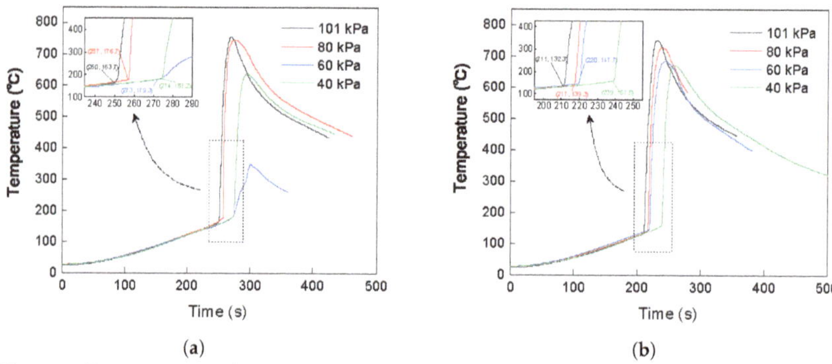

Figure 5. Temperature evolution curves of cells at (**a**) 75% SOC and (**b**) 100% SOC under different pressures.

Venting is a sign for thermal failure of LIBs. Figure 6 shows characteristic times and temperatures of venting under different ambient pressures. The average time and temperature of venting for 75% SOC are 223.8 s and 136.5 °C, respectively, while for 100% SOC, the average time and temperature of venting are 204.4 s and 131.9 °C, respectively. It can be observed that pressure has almost no impact on the time and temperature of venting, but both decrease with an increase in SOC. The pressure difference between inside and outside the cell is given by

$$\Delta p = p_{\text{in}} - p_{\text{out}} \Longleftrightarrow p_{\text{in}} = \Delta p + p_{\text{out}} \quad (16)$$

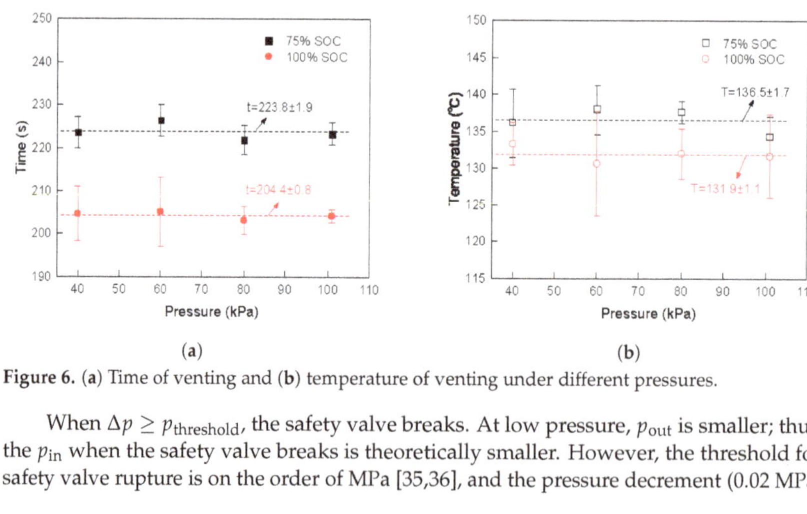

Figure 6. (**a**) Time of venting and (**b**) temperature of venting under different pressures.

When $\Delta p \geq p_{\text{threshold}}$, the safety valve breaks. At low pressure, p_{out} is smaller; thus, the p_{in} when the safety valve breaks is theoretically smaller. However, the threshold for safety valve rupture is on the order of MPa [35,36], and the pressure decrement (0.02 MPa) is too small with respect to the threshold. The cells at 100% SOC have anodes with a higher amount of lithium intercalation [18], resulting in more gas and heat released from chemical reactions. Therefore, venting is more dependent on the increase in p_{in} caused by chemical reactions, electrolyte evaporation, and thermal expansion, making the influence of ambient pressure negligible.

The onset time and temperature of thermal runaway are important parameters for evaluating the thermal stability of LIBs and critical indicators for assessing thermal hazards. It is suggested to increase them to enhance battery safety and to serve as principles for battery design improvements. Figure 7 shows the characteristic onset times and temperatures for thermal runaway under different ambient pressures. It is conveyed that they decrease with increasing pressure or SOC.

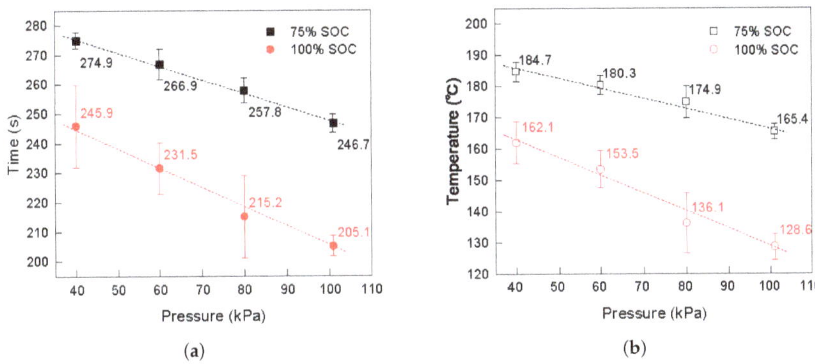

Figure 7. (**a**) Onset time of thermal runaway and (**b**) onset temperature of thermal runaway of cells at 75% SOC and 100% SOC under different pressures.

For cells at 75% SOC, the average onset time of thermal runaway decreases from 274.9 s at 40 kPa to 246.7 s at 101 kPa, a reduction of 28.2 s. Correspondingly, the average onset temperature of thermal runaway decreases from 184.7 °C at 40 kPa to 165.4 °C at 101 kPa, a decrease of 19.3 °C. For cells at 100% SOC, the average onset time decreases from 245.9 s at 40 kPa to 205.1 s at 101 kPa, a reduction of 40.8 s. Similarly, the average onset temperature decreases from 162.1 °C at 40 kPa to 128.6 °C at 101 kPa, a decrease of 33.5 °C.

On the one hand, after venting, the jet flow immediately expands downstream of the nozzle to equilibrate with the ambient pressure, then pressure inside the cell stabilizes at a value close to the ambient pressure [31]. At lower pressures, electrolyte evaporates faster [19], resulting in reduced electrolyte involved in exothermic reactions and slower accumulation of combustible gases and heat. On the other hand, the oxygen concentration decreases as the ambient pressure is decreased, resulting in the reduction of the oxygen participating in the chemical reactions. Therefore, the onset time and temperature of thermal runaway increase at lower pressures.

Cells at 100% SOC have higher active material content than cells of 75% SOC do, allowing them to accumulate combustible gases and heat more quickly, resulting in lower onset time and temperature of thermal runaway. However, the effect of ambient pressure on the onset time and temperature of thermal runaway is more pronounced for 100% SOC. With increasing ambient pressure, the thermal stability of cells at 100% SOC diminishes more significantly than that of cells at 75% SOC. At 101 kPa, the onset temperature of thermal runaway of 100% SOC is even 36.8 °C lower than that of 75% SOC.

While decreasing ambient pressure increases the onset time and temperature of thermal runaway, the lowest ambient pressure explored in this study, down to 40 kPa, is still unable to prevent thermal runaway. Therefore, the thermal safety issues of NCM811 LIBs at low pressures should not be overlooked. Furthermore, compared with previous experimental data on thermal runaway under low pressures [15,34,37], the extent to which the thermal runaway onset time and temperature of NCM811 LIBs increase as the ambient pressure is decreased is smaller in this study. This is because NCM811 LIBs can release more oxygen at lower temperatures [38,39], making them less sensitive to the reduction in oxygen concentration caused by the decrease in pressure compared to LIBs with normal Ni content, such as $LiFePO_4$ LIBs or $LiCoO_2$ LIBs, etc.

The maximum surface temperature after thermal runaway reflects the intensity of the combustion reaction and can assess the hazard caused by thermal runaway to the surrounding environment. Figure 8 summarizes the maximum surface temperatures of cells after thermal runaway under different ambient pressures. The maximum surface temperature increases with increasing pressure; in particular, as the pressure is raised from 40 kPa to 60 kPa, the maximum surface temperature significantly rises from 667.0 °C to 716.1 °C for 75% SOC, and from 687.2 °C to 729.1 °C for 100% SOC, an increase of 49.1 °C for 75% SOC and of 41.9 °C for 100% SOC. This is because oxygen concentration is higher under higher pressures, increasing the reactant concentration in combustion reactions and heat released by them, thereby the maximum surface temperature is higher under higher pressures. This implies that thermal runaway at higher ambient pressures is more hazardous and destructive. However, the maximum surface temperature after thermal runaway remains high at 80 kPa and 60 kPa, and only when the pressure is reduced to 40 kPa is the hazard of thermal runaway somewhat controlled.

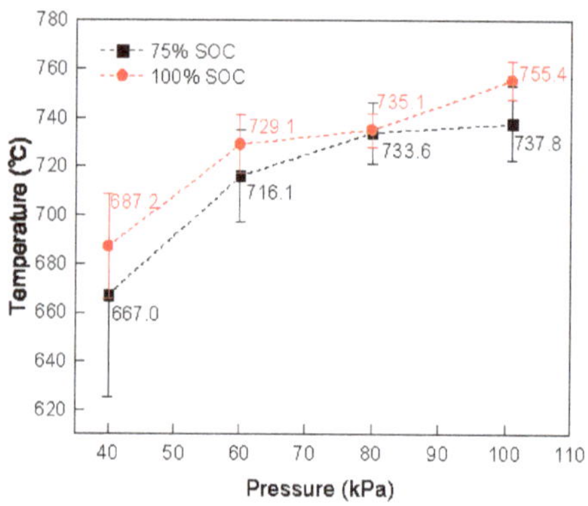

Figure 8. Maximum temperature of cells at 75% SOC and 100% SOC under different pressures.

3.4. Open Circuit Voltage

The open circuit voltage reflects the progress of internal chemical reactions within the cell while the surface temperature is a more delayed indicator. Figure 9 illustrates the voltage evolution during thermal runaway process of the cell. As shown in the figure, the voltage can be divided into four stages:

- Stage I (Voltage stabilization stage): This stage spans from a surface temperature of 28 °C to 55 °C. Before thermal runaway, there is a period during which the voltage remains stable. During this stage, the voltage may exhibit slight fluctuations due to minor heat accumulation, but these changes are not significant. The voltage slightly increases from V_1 (3.9973 V) to V_2 (3.9998 V).
- Stage II (Voltage rise stage): This stage spans from a surface temperature of 55 °C to 113 °C. During the accelerated heating phase, the voltage rises above its initial value, reaching a peak at V_3 (4.0042 V). This may be attributed to the enhanced ionic conductivity of the electrolyte and reduced internal resistance as the temperature increases [40], resulting in a temporary voltage rise.
- Stage III (Voltage decline stage): This stage spans from a surface temperature of 113 °C to 177 °C. After reaching the peak voltage, the voltage begins to sharply decline as severe chemical and structural degradation occurs within the cell. Internal short circuits are caused by the collapse of the separator [41,42]. This sharp voltage drop is

an advanced characteristic indicator that the cell is approaching thermal runaway in contrast to the sharp surface temperature rise.

- Stage IV (Voltage collapse stage): At the onset of thermal runaway, the voltage collapses to an extremely low level, reaching V_4 (−0.0006 V). This indicates that the internal chemical reactions are almost entirely out of control and the internal structure of the battery is completely destroyed.

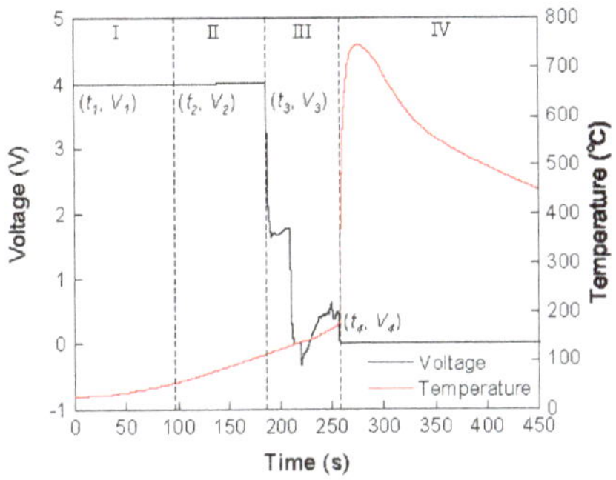

Figure 9. Voltage and temperature evolution curves during the experiment. Stages I, II, III, and IV represent the voltage stabilization stage, voltage rise stage, voltage decline stage, and voltage collapse stage, respectively.

Figure 10 shows the voltage drop times of cells at 75% SOC and 100% SOC under different ambient pressures. The impact of pressure on the voltage drop times of both SOC is slight. The average voltage drop time for 75% SOC is 189.5 s, while for 100% SOC, it is 177.7 s. The voltage drop occurs before the safety valve breaks, during which the cell is a closed system. The collapse of the separator is the direct cause of the sharp voltage drop, and the heat required for the separator to collapse mainly depends on the heat released by internal chemical reactions. Therefore, the voltage drop times are nearly identical under different pressures. Furthermore, the cells at 100% SOC undergo the sharp voltage drop 11.8 s earlier than the cells at 75% SOC do.

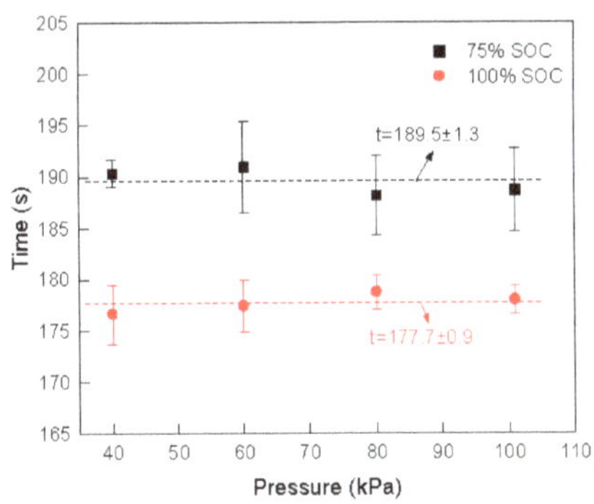

Figure 10. Voltage drop time of cells at 75% SOC and 100% SOC under different pressures.

3.5. Comparison of Characteristic Parameters

To identify the intervals between the events mentioned above and the potential for early warning, Figure 11 summarizes characteristic times of voltage drop, venting, and thermal runaway of cells at 75% SOC and 100% SOC under different ambient pressures. The differences between characteristic times for 75% SOC are significantly larger compared to those for 100% SOC. The average voltage drop time is 34.3 s earlier than the average venting time for 75% SOC, while for 100% SOC, the average voltage drop time is 26.7 s earlier than the average venting time. As the pressure is decreased from 101 kPa to 40 kPa, the difference between the average voltage drop time and the thermal runaway onset time, as well as the difference between the average venting time and the thermal runaway onset time, increases. This indicates that voltage drop and venting can provide earlier warnings of thermal runaway under lower pressures. Therefore, paying attention to voltage drop and venting events under low pressures is of significant importance. It is noteworthy that venting and thermal runaway events are very close for 100% SOC at 101 kPa, differing by just 0.5 s. It shows that cells at higher SOC exhibit more intense internal chemical reactions under higher pressures, and the accumulated combustible gases and heat after venting, plus oxygen, immediately satisfy the conditions required for the battery to ignite. Therefore, only the voltage signal can provide an early warning for the cell at 100% SOC under 101 kPa.

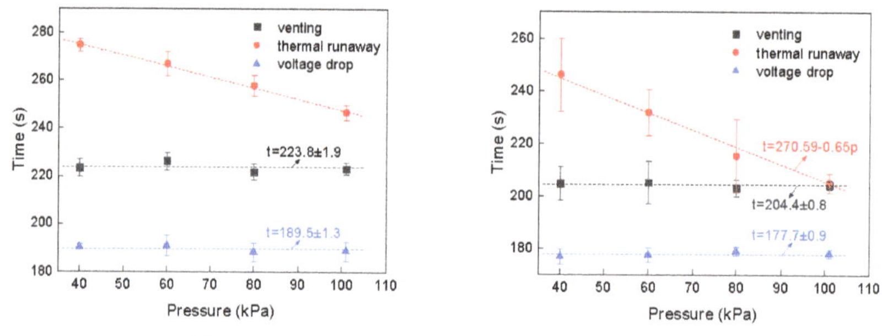

Figure 11. Characteristic times of voltage drop, venting and thermal runaway of cells at (**a**) 75% SOC and (**b**) 100% SOC.

3.6. Mass Loss

To compare changes in cell mass after thermal runaway. Figure 12 shows the mass loss and mass loss fraction of cells. As the ambient pressure is increased from 40 kPa to 101 kPa, the average mass loss for 75% SOC increases from 14.72 g to 17.41 g, with the average loss fraction increasing from 33.6% to 39.7%. For 100% SOC, the average mass loss increases from 18.74 g to 20.87 g, with the average loss fraction increasing from 42.8% to 47.6%. Additionally, the average mass loss and average mass loss fraction for 100% SOC are higher than those for 75% SOC. Mass loss and loss fraction slightly decrease with lower ambient pressures and SOC. It indicates that the intensity of internal chemical reactions is weaker at lower pressures and SOC; this is consistent with the changes in the onset time and temperature of thermal runaway and the maximum surface temperature after thermal runaway discussed above.

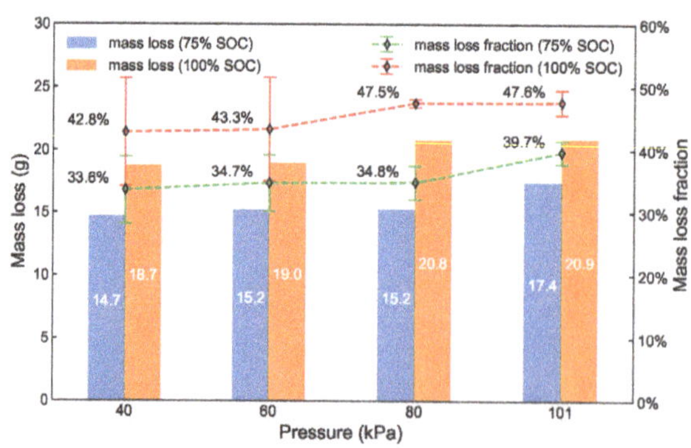

Figure 12. Mass loss and loss rate of cells after thermal runaway under different pressures.

3.7. Post Testing Analysis

Post-testing SEM/EDS analysis was performed to characterize chemical reactions occurring within the electrodes of the cell. SEM/EDS analysis is a mature technical approach used for analyzing battery remnants after thermal runaway [43,44].

As shown in Figure 13, it can be found that particles of the anode and the cathode for the raw cell are both intact, with smooth and clean surfaces. However, after thermal runaway, the integrity of the cell particles is destroyed. The anode surface develops cracks. The deformation of the cathode surface is more severe than that of the anode surface, and there are many pores formed. It suggests that exothermic reactions occurring within the cathode have more pronounced intensity, as the previous research [2] has found that reactions between cathode and electrolyte release tremendous heat.

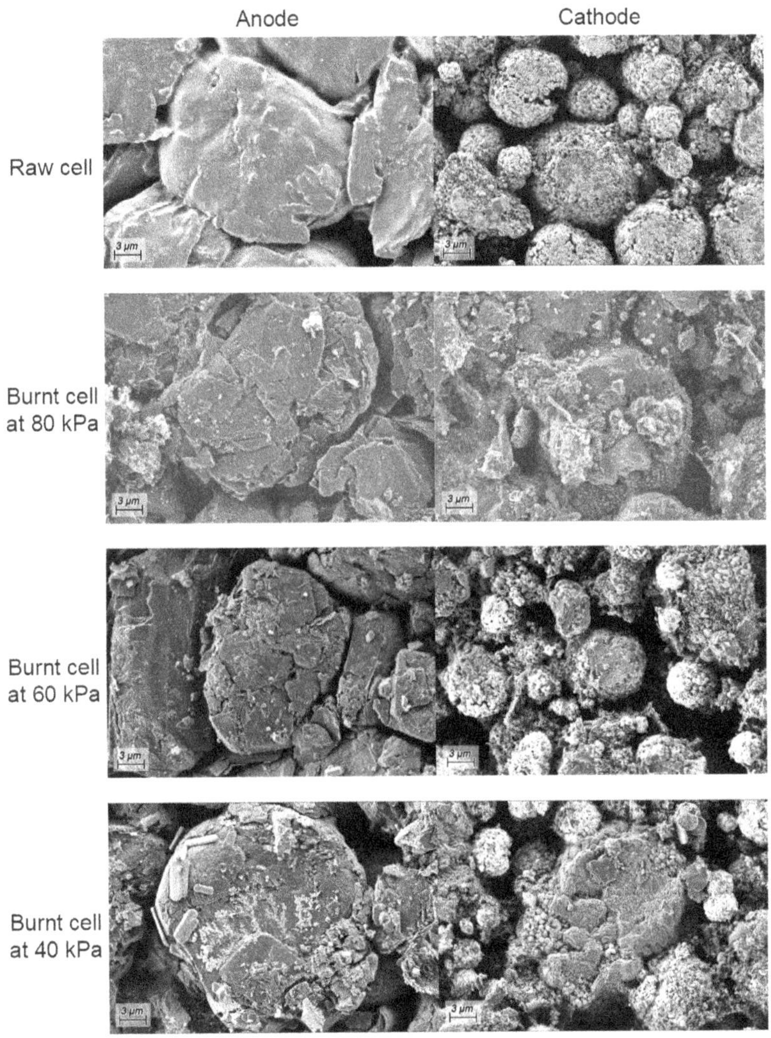

Figure 13. SEM micrographs with a scale bar of 3 micrometers of the raw cell and burnt cells at 100% SOC under different pressures.

As shown in Figure 14, the decrease in the element F at the cathode is significant. At 101 kPa, the F element content is the lowest, indicating the last remaining electrolyte; that is, the chemical reactions within the cathode are the most intense. From 80 kPa to 40 kPa, the F element content gradually decreases, but the internal chemical reactions do not become intense accordingly, suggesting more electrolyte evaporates at lower pressures. As for the anode, there is also a certain decrease in the F element. From 40 kPa to 101 kPa, the F element content gradually decreases, indicating that the chemical reactions within the anode become intense as the ambient pressure increases.

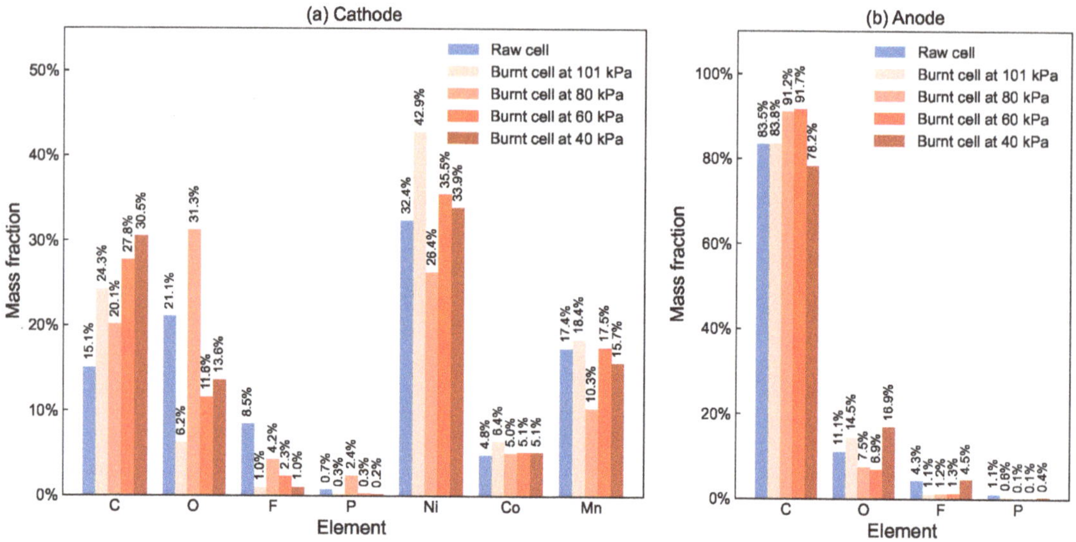

Figure 14. EDS elemental composition diagrams based on SEM micrographs with a scale bar of 10 micrometers of the raw cell and burnt cells at 100% SOC under different pressures.

3.8. Heat Transfer Analysis

To further analyze the impact of ambient pressure on thermal runaway behavior of the cell, a simplified heat transfer model is presented. In this study, the heat sources of the cell include thermal conduction from the heating rod and heat generation from chemical reactions within the cell. The heat sinks of the cell mainly consist of convective and radiative heat dissipation to the surrounding environment. Therefore, the energy balance equation of the cell can be expressed as:

$$cm\frac{\mathrm{d}T_B}{\mathrm{d}t} = \dot{q}_{\mathrm{cond}} + \dot{q}_{\mathrm{reac}} - \dot{q}_{\mathrm{conv}} - \dot{q}_{\mathrm{rad}} \tag{17}$$

where c is the specific heat capacity of the cell, m is the mass of the cell, T_B is the temperature of the cell, $\dot{q}_{\mathrm{cond}}$ is the conduction heat from the heating rod to the cell, $\dot{q}_{\mathrm{reac}}$ is the heat released by reactions, $\dot{q}_{\mathrm{conv}}$ is the convective heat from the cell to the ambient air, and $\dot{q}_{\mathrm{rad}}$ is the radiative heat out to the ambient air.

The conduction heat from the heating rod to the cell can be obtained by

$$\dot{q}_{\mathrm{cond}} = \frac{1}{R}(T_H - T_B) \tag{18}$$

where R is the thermal contact resistance, T_H is the temperature of the heating rod.

The convective heat out to the ambient air is calculated as:

$$\dot{q}_{\text{conv}} = hA(T_B - T_{\text{amb}}) \quad (19)$$

where h is the convective heat transfer coefficient, A is the effective heat dissipation area, T_{amb} is the temperature of the ambient air.

Furthermore, radiative heat out to the ambient air is calculated as:

$$\dot{q}_{\text{rad}} = \epsilon\sigma A(T_B^4 - T_{\text{amb}}^4) \quad (20)$$

where ϵ is the emissivity of the cell surface, σ is the Stefan–Boltzmann constant.

As for the self-heating from chemical reactions, it can be estimated by the following equation:

$$\dot{q}_{\text{reac}} = \sum_i \Delta H_i M_i^{n_i} K_i \exp\left(\frac{-E_{a,i}}{RT_B}\right) \quad (21)$$

where ΔH is the heat release rate of the reaction, M is the initial mass of reactants, n is the reaction order, K is the pre-exponential factor, E_a is the activation energy for the reaction, and R is the universal gas constant. After venting, the internal pressure of the cell quickly reaches equilibrium with the external pressure. Oxygen from the ambient environment enters the cell and is likely to participate in related reactions. According to previous studies [34,45], the lower oxygen concentration under low pressure reduces the rate of internal chemical reactions, as does the faster evaporation of the electrolyte under low pressure discussed above. Therefore, $\dot{q}_{\text{reac}}$ is smaller under lower pressure, which leads to an increase in the onset time and temperature of thermal runaway as the pressure is decreased.

4. Conclusions

This paper systematically studies the thermal runaway behavior of NCM811 LIBs at two highly hazardous SOC (75% and 100%) under different aviation ambient pressures (101 kPa, 80 kPa, 60 kPa, and 40 kPa). Venting and thermal runaway events are two key events in the cell thermal runaway process. Criteria for thermal runaway and venting based on the second derivative of the surface temperature of the cell are proposed. The pressure decrement is negligible relative to the safety valve rupture threshold (on the order of MPa). Therefore, times and surface temperatures of venting are almost the same under different pressures. SEM/EDS analysis and video records of electrolyte burning duration verifies that more electrolyte evaporates under lower pressures. Additionally, the oxygen concentration is lower under lower pressures, these two factors contribute to a reduction in the intensity of chemical reactions; thus, the onset time and temperature of thermal runaway increase under lower pressures. However, the lowest pressure of 40 kPa explored in this paper still cannot prevent the occurrence of thermal runaway. Moreover, compared to other previously studied LIBs, the oxygen release of NCM811 LIBs shifts to lower temperatures and with higher intensity, resulting in a smaller increase in the onset time and temperature of thermal runaway. This indicates that the thermal stability of NCM811 LIBs is relatively poor under low pressure.

Voltage can reflect the progress of internal chemical reactions and is an important parameter that can be observed earlier than the surface temperature to monitor battery status. As the ambient pressure is decreased, the gap between the voltage drop time and the thermal runaway onset time widens, highlighting the significant early warning role of the voltage signal under low pressure.

The thermal runaway behavior of cells at 100% SOC under 101 kPa is the most dangerous and harmful. The time interval between venting and thermal runaway is only 0.5 s, and only the voltage signal can provide an early warning. SEM/EDS analysis verifies that the intensity of internal chemical reactions is the greatest due to the highest amount of active materials and the highest ambient oxygen concentration. Once combustible gases are released, they quickly meet the ignition requirements.

Author Contributions: Conceptualization, Y.J.; methodology, Y.J. and D.M.; software, Y.J.; validation, Y.J., D.M. and X.-H.W.; formal analysis, Y.J.; investigation, Y.J., D.M. and C.-X.Z.; resources, Y.J., D.M., C.-X.Z., J.-L.Y., X.-H.W. and J.W.; data curation, Y.J.; writing—original draft preparation, Y.J.; writing—review and editing, Y.J., D.M., X.-H.W. and J.W.; visualization, Y.J.; supervision, Y.J., D.M., X.-H.W. and J.W.; project administration, Y.J., D.M., C.-X.Z., X.-H.W. and J.W.; funding acquisition, X.-H.W. and J.W. All authors have read and agreed to the published version of the manuscript.

Funding: This work is supported by the Guangdong Basic and Applied Basic Research Foundation (2021B1515130008) and the Science and Technology Program of Fire and Rescue Department Ministry of Emergency Management of China (2022XFZD12).

Data Availability Statement: The data presented in this work will be available on request.

Conflicts of Interest: The authors declare no conflicts of interest.

References

1. Zubi, G.; Dufo-López, R.; Carvalho, M.; Pasaoglu, G. The Lithium-Ion Battery: State of the Art and Future Perspectives. *Renew. Sustain. Energy Rev.* **2018**, *89*, 292–308. [CrossRef]
2. Li, Y.; Liu, X.; Wang, L.; Feng, X.; Ren, D.; Wu, Y.; Xu, G.; Lu, L.; Hou, J.; Zhang, W.; et al. Thermal Runaway Mechanism of Lithium-Ion Battery with LiNi0.8Mn0.1Co0.1O2 Cathode Materials. *Nano Energy* **2021**, *85*, 105878. [CrossRef]
3. Cui, Z.; Manthiram, A. Thermal Stability and Outgassing Behaviors of High-nickel Cathodes in Lithium-ion Batteries. *Angew. Chem. Int. Ed.* **2023**, *62*, e202307243. [CrossRef]
4. Federal Aviation Administration (FAA). Lithium Battery Air Incidents Involving Smoke, Fire, or Extreme Heat. 2024. Available online: https://www.faa.gov/hazmat/resources/lithium_batteries/incidents (accessed on 29 January 2024).
5. Chen, M.; Liu, J.; Ouyang, D.; Wang, J. Experimental Investigation on the Effect of Ambient Pressure on Thermal Runaway and Fire Behaviors of Lithium-ion Batteries. *Int. J. Energy Res.* **2019**, *43*, 4898–4911. [CrossRef]
6. Chen, M.; Ouyang, D.; Weng, J.; Liu, J.; Wang, J. Environmental Pressure Effects on Thermal Runaway and Fire Behaviors of Lithium-Ion Battery with Different Cathodes and State of Charge. *Process Saf. Environ. Prot.* **2019**, *130*, 250–256. [CrossRef]
7. Xie, S.; Ren, L.; Yang, X.; Wang, H.; Sun, Q.; Chen, X.; He, Y. Influence of Cycling Aging and Ambient Pressure on the Thermal Safety Features of Lithium-Ion Battery. *J. Power Sources* **2020**, *448*, 227425. [CrossRef]
8. Xie, S.; Ren, L.; Gong, Y.; Li, M.; Chen, X. Effect of Charging/Discharging Rate on the Thermal Runaway Characteristics of Lithium-Ion Batteries in Low Pressure. *J. Electrochem. Soc.* **2020**, *167*, 140503. [CrossRef]
9. Xie, S.; Sun, J.; Chen, X.; He, Y. Thermal Runaway Behavior of Lithium-ion Batteries in Different Charging States under Low Pressure. *Int. J. Energy Res.* **2021**, *45*, 5795–5805. [CrossRef]
10. Xie, S.; Gong, Y.; Li, G.; Ping, X. Effect of Low-Pressure Environment on Thermal Runaway Behavior of NCM523/Graphite Pouch Cells with Different Overcharge Cycles. *J. Energy Storage* **2022**, *55*, 105444. [CrossRef]
11. Liu, Q.; Zhu, Q.; Zhu, W.; Yi, X.; Han, X. Thermal Runaway Characteristics of 18650 NCM Lithium-ion Batteries under the Different Initial Pressures. *Electrochemistry* **2022**, *90*, 087004. [CrossRef]
12. Zhang, Q.; Bao, F.; Niu, J. Risk analysis method of thermal runaway gas explosion in lithium-ion batteries. *Energy Storage Sci. Technol.* **2023**, *12*, 2263.
13. Chen, X.; Zhang, X.; Wang, H.; Jia, J.; Xie, S.; Zhi, M.; Fu, J.; Sun, Q. Influence of Ambient Pressure and Heating Power on the Thermal Runaway Features of Lithium-Ion Battery. *J. Electrochem. Energy Convers. Storage* **2021**, *18*, 021014. [CrossRef]
14. Fang, J.; Cai, J.; He, X. Experimental Study on the Vertical Thermal Runaway Propagation in Cylindrical Lithium-ion Batteries: Effects of Spacing and State of Charge. *Appl. Therm. Eng.* **2021**, *197*, 117399. [CrossRef]
15. Liu, Y.; Niu, H.; Liu, J.; Huang, X. Layer-to-Layer Thermal Runaway Propagation of Open-Circuit Cylindrical Li-Ion Batteries: Effect of Ambient Pressure. *J. Energy Storage* **2022**, *55*, 105709. [CrossRef]
16. Incropera, F.P.; DeWitt, D.P.; Bergman, T.L.; Lavine, A.S. *Fundamentals of Heat and Mass Transfer*; Wiley: New York, NY, USA, 1996; Volume 6.
17. Ping, P. Lithium Ion Battery Thermal Runaway and Fire Risk Analysis and the Development on the Safer Battery System. Ph.D. Thesis, University of Science and Technology of China, Hefei, China, 2014.
18. Sommer, L.W.; Raghavan, A.; Kiesel, P.; Saha, B.; Schwartz, J.; Lochbaum, A.; Ganguli, A.; Bae, C.J.; Alamgir, M. Monitoring of Intercalation Stages in Lithium-Ion Cells over Charge-Discharge Cycles with Fiber Optic Sensors. *J. Electrochem. Soc.* **2015**, *162*, A2664–A2669. [CrossRef]
19. Kim, H.; Sung, N. The Effect of Ambient Pressure on the Evaporation of a Single Droplet and a Spray. *Combust. Flame* **2003**, *135*, 261–270. [CrossRef]

20. Finegan, D.P.; Scheel, M.; Robinson, J.B.; Tjaden, B.; Hunt, I.; Mason, T.J.; Millichamp, J.; Di Michiel, M.; Offer, G.J.; Hinds, G.; et al. In-Operando High-Speed Tomography of Lithium-Ion Batteries during Thermal Runaway. *Nat. Commun.* **2015**, *6*, 6924. [CrossRef] [PubMed]
21. Jindal, P.; Bhattacharya, J. Review—Understanding the Thermal Runaway Behavior of Li-Ion Batteries through Experimental Techniques. *J. Electrochem. Soc.* **2019**, *166*, A2165–A2193. [CrossRef]
22. Mei, W.; Liu, Z.; Wang, C.; Wu, C.; Liu, Y.; Liu, P.; Xia, X.; Xue, X.; Han, X.; Sun, J.; et al. Operando Monitoring of Thermal Runaway in Commercial Lithium-Ion Cells via Advanced Lab-on-Fiber Technologies. *Nat. Commun.* **2023**, *14*, 5251. [CrossRef]
23. Golubkov, A.W.; Fuchs, D.; Wagner, J.; Wiltsche, H.; Stangl, C.; Fauler, G.; Voitic, G.; Thaler, A.; Hacker, V. Thermal-Runaway Experiments on Consumer Li-ion Batteries with Metal-Oxide and Olivin-Type Cathodes. *RSC Adv.* **2014**, *4*, 3633–3642. [CrossRef]
24. Cao, L.; Zhou, Z. *Thermology Thermodynamics and Statistical Physics*; Science Press: Beijing, China, 2015.
25. Zou, K.; He, K.; Lu, S. Venting Composition and Rate of Large-Format $LiNi_{0.8}Co_{0.1}Mn_{0.1}O_2$ Pouch Power Battery during Thermal Runaway. *Int. J. Heat Mass Transf.* **2022**, *195*, 123133. [CrossRef]
26. Wei, N.; Zhang, F.; Zhang, W.; Li, X. Comparative Study on the Thermal Runaway Characteristics of $Li(Ni_xCo_yMn_z)O_2$ Batteries. *Heliyon* **2024**, *10*, e31203. [CrossRef] [PubMed]
27. Zhang, Q.; Niu, J.; Yang, J.; Liu, T.; Bao, F.; Wang, Q. In-Situ Explosion Limit Analysis and Hazards Research of Vent Gas from Lithium-Ion Battery Thermal Runaway. *J. Energy Storage* **2022**, *56*, 106146. [CrossRef]
28. Wang, H.; Xu, H.; Zhang, Z.; Wang, Q.; Jin, C.; Wu, C.; Xu, C.; Hao, J.; Sun, L.; Du, Z.; et al. Fire and Explosion Characteristics of Vent Gas from Lithium-Ion Batteries after Thermal Runaway: A Comparative Study. *eTransportation* **2022**, *13*, 100190. [CrossRef]
29. Spotnitz, R.; Franklin, J. Abuse Behavior of High-Power, Lithium-Ion Cells. *J. Power Sources* **2003**, *113*, 81–100. [CrossRef]
30. Kim, G.H.; Pesaran, A.; Spotnitz, R. A Three-Dimensional Thermal Abuse Model for Lithium-Ion Cells. *J. Power Sources* **2007**, *170*, 476–489. [CrossRef]
31. Mao, B.; Zhao, C.; Chen, H.; Wang, Q.; Sun, J. Experimental and Modeling Analysis of Jet Flow and Fire Dynamics of 18650-Type Lithium-Ion Battery. *Appl. Energy* **2021**, *281*, 116054. [CrossRef]
32. Wang, Q.; Sun, J.; Chu, G. Lithium Ion Battery Fire Furthermore, Explosion. *Fire Saf. Sci.* **2005**, *8*, 375–382. [CrossRef]
33. Finegan, D.P.; Darcy, E.; Keyser, M.; Tjaden, B.; Heenan, T.M.M.; Jervis, R.; Bailey, J.J.; Vo, N.T.; Magdysyuk, O.V.; Drakopoulos, M.; et al. Identifying the Cause of Rupture of Li-Ion Batteries during Thermal Runaway. *Adv. Sci.* **2018**, *5*, 1700369. [CrossRef]
34. Fu, Y.; Lu, S.; Shi, L.; Cheng, X.; Zhang, H. Ignition and Combustion Characteristics of Lithium Ion Batteries under Low Atmospheric Pressure. *Energy* **2018**, *161*, 38–45. [CrossRef]
35. Qin, P.; Sun, J.; Wang, Q. A New Method to Explore Thermal and Venting Behavior of Lithium-Ion Battery Thermal Runaway. *J. Power Sources* **2021**, *486*, 229357. [CrossRef]
36. Mao, B.; Fear, C.; Chen, H.; Zhou, H.; Zhao, C.; Mukherjee, P.P.; Sun, J.; Wang, Q. Experimental and Modeling Investigation on the Gas Generation Dynamics of Lithium-Ion Batteries during Thermal Runaway. *eTransportation* **2023**, *15*, 100212. [CrossRef]
37. Chen, M.; Liu, J.; He, Y.; Yuen, R.; Wang, J. Study of the Fire Hazards of Lithium-Ion Batteries at Different Pressures. *Appl. Therm. Eng.* **2017**, *125*, 1061–1074. [CrossRef]
38. Bak, S.M.; Nam, K.W.; Chang, W.; Yu, X.; Hu, E.; Hwang, S.; Stach, E.A.; Kim, K.B.; Chung, K.Y.; Yang, X.Q. Correlating Structural Changes and Gas Evolution during the Thermal Decomposition of Charged $Li_xNi_{0.8}Co_{0.15}Al_{0.05}O_2$ Cathode Materials. *Chem. Mater.* **2013**, *25*, 337–351. [CrossRef]
39. Bak, S.M.; Hu, E.; Zhou, Y.; Yu, X.; Senanayake, S.D.; Cho, S.J.; Kim, K.B.; Chung, K.Y.; Yang, X.Q.; Nam, K.W. Structural Changes and Thermal Stability of Charged $LiNi_xMn_yCo_zO_2$ Cathode Materials Studied by Combined *In Situ* Time-Resolved XRD and Mass Spectroscopy. *ACS Appl. Mater. Interfaces* **2014**, *6*, 22594–22601. [CrossRef] [PubMed]
40. Hossain Ahmed, S.; Kang, X.; Bade Shrestha, S.O. Effects of Temperature on Internal Resistances of Lithium-Ion Batteries. *J. Energy Resour. Technol.* **2015**, *137*, 031901. [CrossRef]
41. Venugopal, G. Characterization of Thermal Cut-off Mechanisms in Prismatic Lithium-Ion Batteries. *J. Power Sources* **2001**, *101*, 231–237. [CrossRef]
42. Wang, Q.; Sun, J. Enhancing the Safety of Lithium Ion Batteries by 4-Isopropyl Phenyl Diphenyl Phosphate. *Mater. Lett.* **2007**, *61*, 3338–3340. [CrossRef]
43. Li, W.; Lucht, B.L. Lithium-Ion Batteries: Thermal Reactions of Electrolyte with the Surface of Metal Oxide Cathode Particles. *J. Electrochem. Soc.* **2006**, *153*, A1617. [CrossRef]
44. Saito, Y.; Rahman, M.K. Investigation of Positive Electrodes after Cycle Testing of High-Power Li-ion Battery Cells IV. *J. Power Sources* **2007**, *174*, 877–882. [CrossRef]
45. Weng, J.; Ouyang, D.; Liu, Y.; Chen, M.; Li, Y.; Huang, X.; Wang, J. Alleviation on Battery Thermal Runaway Propagation: Effects of Oxygen Level and Dilution Gas. *J. Power Sources* **2021**, *509*, 230340. [CrossRef]

batteries

MDPI

Article

Investigation on Thermal Runaway Hazards of Cylindrical and Pouch Lithium-Ion Batteries under Low Pressure of Cruise Altitude for Civil Aircraft

Qiang Sun [1], Hangxin Liu [2,*], Zhi Wang [1,3,*], Yawei Meng [1], Chun Xu [2], Yanxing Wen [2] and Qiyao Wu [2]

1 Civil Aircraft Fire Science and Safety Engineering Key Laboratory of Sichuan Province, Civil Aviation Flight University of China, Deyang 618307, China; qiangsun@cafuc.edu.cn (Q.S.); yaweimeng@cafuc.edu.cn (Y.M.)
2 Key Laboratory of Fine Chemical Application Technology of Luzhou, Sichuan Vocational College of Chemical Technology, Luzhou 646099, China; 331205565@qq.com (C.X.); 13292739867@163.com (Y.W.); qiyaowu2024@163.com (Q.W.)
3 School of Safety Engineering, China University of Mining and Technology, Xuzhou 221116, China
* Correspondence: hangxinliu@163.com (H.L.); zhiwang@cumt.edu.cn (Z.W.)

Abstract: Thermal runaway characteristics and hazards of lithium-ion batteries under low ambient pressure in-flight conditions are studied in a dynamic pressure chamber. The influence of ambient pressures (95 kPa and 20 kPa) and packaging forms (cylindrical and pouch commercial batteries) were especially investigated. The results show that the values of heat release, temperature, and CO_2 concentration decrease with the reduction in pressure from 95 kPa to 20 kPa, while the total hydrocarbon and CO increase. Without violent fire, explosion, and huge jet flames, the thermal hazards of TR fire under 20 kPa are lower, but the amount of toxic/flammable gas emissions increases greatly. The amount of CO and hydrocarbons varies inversely with the thermal hazards of fire. Under low-pressure environments of cruise altitude, the thermal hazards of TR fire for pouch cells and the toxic/potentially explosive hazards of gas emissions of cylindrical cells need more attention. The performance of TR hazards for two packaging types of battery is also different. Pouch cells have higher thermal hazards of fire and lower combustible/toxic emitted gases than cylindrical cells. The thermal runaway intensity of individual cells decreases under lower ambient pressure, but the burning intensity increases dramatically when thermal runaway occurs in a battery pack. The open time of a safety valve (rupture of the bag) is shortened, but the trigger time for a thermal runaway varies for different formats of batteries under 20 kPa. Those results may be helpful for the safety warning and hazard protection design of Li batteries under low-pressure conditions.

Keywords: lithium-ion battery; thermal runaway hazards; low ambient pressure; fire thermal hazards; potentially explosive gases

1. Introduction

Lithium-ion batteries (LIBs) have been widely applied in electronic products, vehicles, and aerospace. It is also seen as a promising power source for multi-electric aircraft, all-electric aircraft, and unmanned aerial vehicles, due to the excellent features of lower weight, high energy density, and less frequent maintenance [1,2]. Compared with the cruise altitude flight, when civil aviation aircraft climb to the stratosphere and the landing time is shorter, the aircraft will spend a long period of time at a high altitude of more than 10,000 m; at this time, the atmospheric pressure can be reduced to 20 kPa. Although almost all of the current passenger cabins are equipped with a pressurization system, the majority of the cargo transport aircraft are still exposed to low-pressure environments. However, when the LIBs are misused or encounter the conditions of high temperature [3], overcharging, short-circuit [4], extrusion, or collision [5,6], the thermal runaway (TR) of LIBs occurs occasionally, which is accompanied by heat release, flammable and toxic gas emissions [7], fire, and explosions [8,9]. As functional, auxiliary power or under transportation conditions, the LIB fire incidents of Boeing 787 in 2013–2014, with LIBs as the auxiliary power, as well as the freighter, encompassed crash accidents involving LIBs in the cargo hold, and similar incidents occurred in an Asiana Airlines 747 near South Korea on 28 July 2011, a UPS 747 in Dubai, and UAE on 3 September 2010 [10]. According to the Federal Aviation Administration (FAA), from January 2006 to January 2022, there were more than 375 aviation incidents raised from LIBs. As shown in Figure 1, some of the FAA's incomplete statistics on lithium-ion battery safety incidents in the aviation field were obtained over the past 16 years. The TR hazards of LIB have become the main obstacle for further development in the aviation field, it is particularly necessary to study the characteristics of TR behavior of Li-ion batteries under low-pressure environments.

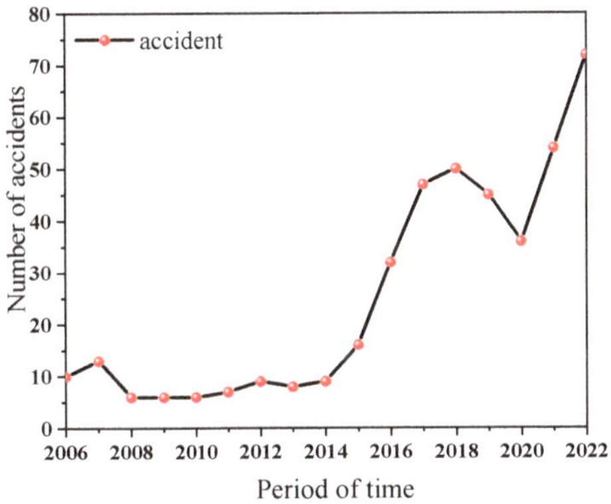

Figure 1. Lithium-ion battery safety incidents in the global aviation sector.

Considerable research has been carried out to illustrate the TR characteristics of LIBs, under thermal and mechanical abuse [5]. Most works focus on revealing the mechanism of chain reactions inside the batteries and the features of TR propagation [11,12]. During the TR, a series of exothermic reactions and massive internal short circuits release a large amount of heat and gases [13]. As the temperature increases, the battery experiences the thermal decomposition and reactions of the SEI film (60–120 °C), separator (>130 °C), electrolyte (>200 °C), binder, and electrode materials [14]. The TR would be triggered by electric energy released by a massive internal short circuit once the separator collapses [15]. The evolution mechanisms of gas emissions under different temperatures during the TR process are established [16–18]. There are more than 100 gaseous products that are identified during the TR, and most of them are hazardous [19]. Seven gas species are confirmed to be the most common within emitted gas products, e.g., CO_2, CO, H_2, CH_4, C_2H_4, C_2H_6, and C_3H_6 [20,21]. As the most critical and toxic gases, fluoride and F-containing batteries are studied, and the quantity of HF is about 15~120 mg/Wh [22]. Besides these toxic gases, the toxic/potentially explosive gases, i.e., CO and hydrocarbon, and the asphyxiant gas of CO_2 also need more attention, which generally accounts for 27.5%, 17.5%, and 24.9% of the total released gas per unit volume, respectively [23]. The combustible and toxic emitted gases are tightly related to the battery materials, capacity, state of charge (SOC), and the TR environments. Higher capacity and a higher SOC lead to an increase in flammable gas composition [24,25]. During the process of TR, visible and audible sparks, jet flame, and deflagration fire can cause huge hazards to human beings [26]. The risks of intense fire and heat associated with toxic/flammable gases and smoke emissions may pose a significant threat to life and property, especially in confined environments, e.g., aircraft, mine shafts, and spacecraft. In a case where the emitted gases are ignited, this can result in an explosion. In addition to the cylindrical LIBs with a hard can, numerous efforts have also been devoted to the TR research of pouch LIBs recently [27]. With a soft aluminum–polymer bag and higher capacity, the TR process of pouch cells is distinct from the cells with hard shells. Yang et al. [28] conducted thermally induced failure tests with 68 Ah pouch LIBs in a 1/2 ISO full test room to investigate the thermal and toxic hazards systematically and observed that the TR process for different stages of charge underwent five burning stages. Akos et al. [29] investigated the initiation of TR for formats of hard-case cells and soft-pouch cells, and revealed the propagation of the initial internal short circuit to the TR. It was determined that only 1% of the battery's electric energy could initiate the TR in different formats of LIBs. The differences between the pouch and hard-case cell formats in solid to gaseous mass loss proportion were discovered and analyzed, and a new method to calculate the volume of released gas was proposed by Sascha et al. [20].

Previous studies have found that ambient pressure can change the TR characteristics of LIBs. According to the literature, environmental pressure has a significant impact on combustion [30]. With the decrease in atmospheric pressure, the combustion efficiency and mass burning rate reduce for common fuel fire; in addition, the characteristics of fire behaviors and temperature are affected by the pressure [31]. The mass loss, HRR, THR, and combustion efficiency decrease, under low pressure, while CO increases, and the ignition time is shortened for cylindrical LIBs, through the TR fire tests in Lhasa (64.3 kPa) and Hefei (100.8 kPa), conducted by Chen et al. [32,33]. Fu et al. [34] concluded that the combustion intensity and the battery surface temperature, as well as flame temperature, decrease under a low-pressure environment of 30 kPa. Xie et al. [35–37] found that as the ambient pressure decreases, the increase in the number of cycles and the expansion of the charge/discharge multiplicity all lead to a decrease in temperature, HRR, THR, and first venting, while the increase in the overcharge cutoff voltage results in a decrease in the thermal stability of Li-ion batteries, by comparing the tests under 20 kPa and 95 kPa. Su et al. [38] established the thermal contribution of decomposition energy and oxidation combustion energy to peak battery temperature in the TR. Other authors found that the decomposition energy released in the TR is decided by the SOC and almost not affected by ambient pressure, and oxidation combustion energy increases in direct proportion to the ambient pressure [39]. Under the ambient pressures of 95 kPa and 20 kPa, the TR characteristics of lithium-ion batteries with different airflow rates are experimented with, and it was found that the lower the ambient pressure at the same airflow rate, the lower the HRR and THR values; while at the same pressure environment, the HRR and THR values increase with the increase in airflow rate. Regarding the effect of low pressures, the Federal Aviation Administration (FAA) has conducted a series of experiments to study the TR fire hazards, vent gas ignition in cargo compartments, and extinguishment and packaging methods to improve the knowledge of LIB safety issues under various application scenarios, especially, the installed Li batteries on transport aircraft [40]. The National Aeronautics and Space Administration (NASA) observed that the vented electrolyte cannot be ignited, while the TR time is shortened under a vacuum environment, through a series of safety tests of 18650 cells used in the spacesuit [41].

The occurrence and development of LIB TR fire in some low-pressure situations, such as the flight phase at high altitudes, are totally different from the ground conditions. The variations in atmospheric pressure and temperature in a flight are quite different from the standard conditions of the land. Especially, in the cruise phase, with an altitude of 9000–13,000 m, the pressure can reach lower than 20 kPa. Most of the time, a flight is at a cruising altitude for the aircraft. Nevertheless, a comprehensive understanding of TR hazards such as fire behaviors, thermal and electric features, and flammable and asphyxiant gas emissions, for different formats of LIBs under low-pressure environments of 20 kPa (the flight cruise phase of civil aviation aircraft), is still not readily available, even though there are some published studies in the literature that have investigated the impact of low ambient pressure (the lowest pressure is 30 kPa) on the TR of LIBs. Typically, the issues of safety and TR hazards for pouch LIBs with high energy density during the cruise phase of aircraft need to be solved. Therefore, it is necessary to gain more insight into the TR process and combustible/toxic emitted gases (especially CO and hydrocarbons from the TR) of hard-case and soft-packing batteries, which is of vital importance for further applications in the aviation field under a low ambient pressure during the cruise phase. This is also significant for the selection of installed Li batteries on aircraft, the preventive measures, and the protection design.

In this work, the TR characteristics and flammable/toxic gas products of the pouch and cylindrical cells were investigated, in a dynamic pressure cabin. Two different pressure environments of 95 kPa (the atmospheric pressure of ground) and 20 kPa (the pressure of cruise altitude) were selected. The parameters of HRR, THR, smoke/fire/battery surface temperatures, and the light transmittance of smoke were obtained to evaluate the fire hazards. The real-time concentrations of CO, CO_2, and THC were also measured. Furthermore, several important parameters were analyzed in detail, such as peak concentration values of emitted gases, values per capacity, and igniting time.

2. Experimental

2.1. Sample

In this study, commercial cylindrical cells, as well as pouch cells from the same batch, were tested. The 18650 cylindrical LIBs have spirally wound layers inside an outer metal cylinder. The anode material is $LiNi_{0.5}Co_{0.2}Mn_{0.3}O_2$ (NCM), and the cathode material is graphite. In the pouch cell, the layers are stacked on top of each other and sealed by the aluminum–polymer bag. The anode and cathode materials are $LiNi_{0.5}Co_{0.2}Mn_{0.3}O_2$ (NCM) and graphite, respectively. The capacity of cylindrical and pouch cells is 2.6 Ah, 10 Ah, and 15 Ah. Figure 2 and Table 1 show the cells and their specifications.

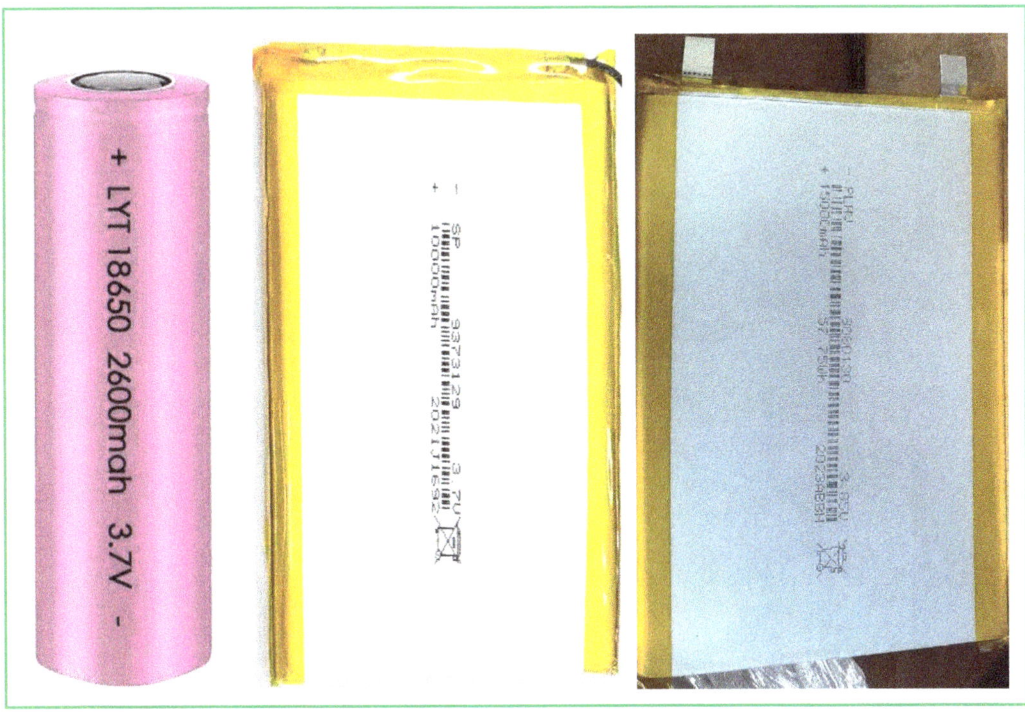

Figure 2. Fresh battery sample diagram.

Table 1. Specifications of the cylindrical and pouch LIBs used in the tests.

Parameter	Cylindrical	Pouch	Pouch
Capacity	2600 mAh	10,000 mAh	15,000 mAh
Anode	$LiNi_{0.5}CO_{0.2}Mn_{0.3}O_2$	$LiNi_{0.5}CO_{0.2}Mn_{0.3}O_2$	$LiNi_{0.5}CO_{0.2}Mn_{0.3}O_2$
Cathode	Graphite	Graphite	Graphite
Electrolyte	EMC + EC + DMC + LiF_6	EMC + EC + DEC + LiF_6	EMC + EC + DMC + LiF_6
Separator	PE	PE	PE
Size	Diameter: 18 mm, height: 65 mm	Length: 100 mm, width: 50 mm, thickness: 6 mm	Length: 150 mm, width: 50 mm, thickness: 6 mm

2.2. Apparatus

The cells were charged and discharged using a LANHE instrument (CT2016D, Wuhan Land Electronic Co. Ltd., Wuhan, China). The charge and discharge rates were kept at 0.5 C. According to the product manual, the LIBs used in the tests were all cycled 10 times and then kept at an SOC of 100%. The charge and discharge of all the cells were conducted at the atmospheric pressure and temperature (25 °C). The TR safety experiments were conducted in a dynamic pressure combustion chamber (size: 2 m × 2 m × 2 m) with variable pressure from 10 kPa to 101 kPa. The pressure accuracy is 0.1 kPa. The experimental equipment mainly consists of a combustion chamber, a pressure control system, an HRR and gas tester (AO2020, Asea Brown Boveri Co., Ltd., Zurich, Switzerland), and an analysis system, as shown in Figure 3a in detail. The chamber pressure was stabilized by controlling the outlet and inlet airflow dynamically. And the airflow rate was set as 120 L/s in all experiments.

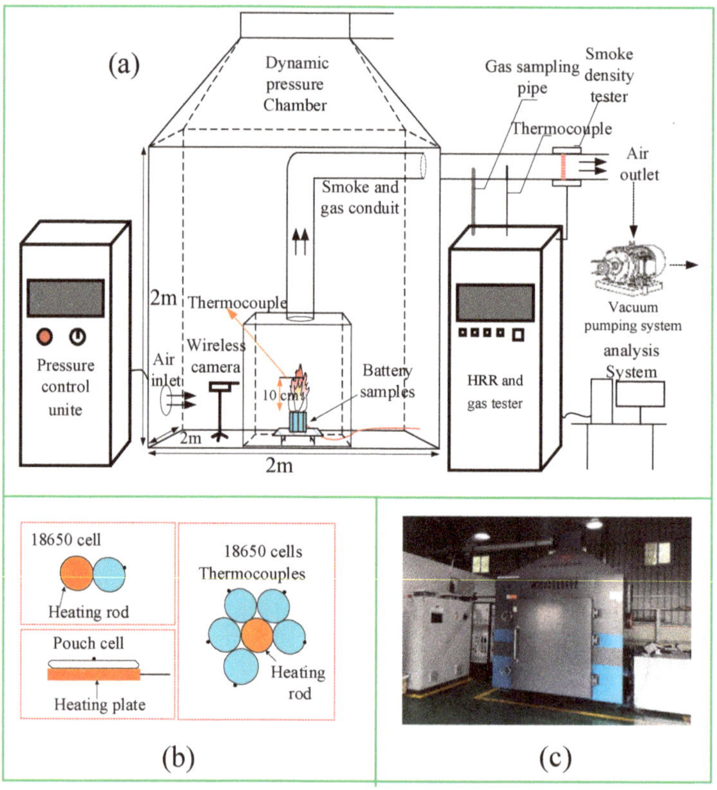

Figure 3. Diagrammatic diagram of experimental platform and setup: (**a**) experimental platform, (**b**) experimental setup in dynamic cabin, and (**c**) dynamic pressure cabin and test systems.

The dynamic pressure cabin and test systems were calibrated and carried out according to the procedures in the ISO 9705 standard [42]. Combustion emission products of LIBs were collected by the exhaust hood and transported out in the conduit through the air outlet system (vacuum pump). Smoke and gas product samples were extracted to the HRR and gas tester by a vacuum pump through the gas sampling pipe. CO_2, O_2, CO, and THC were measured and recorded by the tester, with a detection limit of 1%. A smoke density tester was used to detect the density of emissions by laser transmittance. The temperature of smoke and gas emissions was obtained by the thermocouple placed in the conduit. A digital camera was placed in front of the LIB samples to record the TR fire behaviors.

Four groups of experiments were designed and conducted at the pressures of 95 kPa (the atmospheric pressure of Guanghan, Deyang, China) and 20 kPa (the pressure of the cruise altitude of civil aviation aircraft).

Group A: Single 18650 LIB of 100% SOC.

Group B: Multiple 18650 LIBs at 100% SOC. The number of LIBs is 5.

Group C: Single 10 Ah pouch cell with 100% SOC.

Group D: Single 15 Ah pouch cell with 100% SOC.

As shown in Figure 3b, in the single-cell test, the heating rod with 18 mm in diameter and 65 mm in length is next to a battery with a heating power of 240 W. And the arrangement of multiple cylindrical cells with a heating rod is shown. In the pouch cell TR tests, a heating aluminum plate (200 mm × 100 mm × 20 mm), with a heating power of 400 W, was placed under the pouch cell. All LIB TR experiments were performed in the reaction chamber, with the ambient temperature around 25 °C. In order to reduce the experimental error, each test was repeated at least 3 times, and the average values were taken. The temperature of the TR fire and battery surface temperature were recorded.

2.3. Procedures

The dynamic pressure cabin and test systems are shown in Figure 3c. When the expected values of pressure and airflow rate inside the dynamic chamber were set by the touch screen of the pressure control system, the air inside was extracted to reduce the pressure to the target value by the vacuum pump, while the electromagnetic valve was used for dynamical control. After the chamber pressure was kept constant, the heating source was started, and the experiment began. Meanwhile, the data were recorded. The air inside flows with a certain value controlled and stabilized by the inlet and outlet systems to keep the pressure inside the chamber stable at the target value, even if certain amounts of gases are released from batteries to the chamber nonlinearly during the TR process of batteries.

3. Results and Discussion

3.1. Thermal Runaway Behaviors

3.1.1. Cylindrical Cells

As shown in Figure 4, the typical TR processes of cylindrical single-cell and multiple-cell experiments under the pressures of 95 kPa and 20 kPa are presented, respectively.

When the heating rate of the battery surface exceeds 10 °C/s, the occurrence of TR is determined. At this time, the spark injection time t_2 is almost the same as the trigger time of the TR. Because of the phenomenon of triggering the TR, the heating plate is closed. As shown in Figure 4a, during the TR process of 95 kPa, a single cylindrical cell of 18650 went through five stages, including (1) first venting (the opening of the safety valve), (2) spark ejections (the trigger of the TR), (3) jet fire/explosion, (4) combustion, and (5) extinguishment. However, at a pressure of 20 kPa, there are no combustion flames in the whole TR process. This is probably because large amounts of combustible gases and ejections, released rapidly after the sparks (safety valve opened completely), are not ignited under 20 kPa for the reason that oxygen mass concentration per volume is less than the stoichiometric ratio. Oxygen mass concentration has a functional relationship with pressure. And the ambient pressure would change the oxygen mass concentration and combustion reaction rates, as shown in Equation (10). As shown in Figure 4b, the early behaviors of the battery before the TR play a crucial role in the early detection and warning of TR hazards, e.g., the gases and volatiles are emitted out and can be detected after the first vent. Therefore, the first vent time is defined as t_1, and the first vent time is earlier than that of low AP. The parameter of t_2 is the average trigger time of spark ejections, i.e., bright solid sparks with gas products, indicating the activation of the TR. The criterion for the TR is the temperature rise rate in this work. It indicates that the TR of the battery was triggered when the battery temperature rise rate was over 10 °C/s, as presented in Equation (1), where k indicates the time index of the data, which was recorded with a sampling time of 1 s (Figure S1). Since the spark injection time t_2 and the TR trigger time are almost the same, this is regarded as the phenomenon of triggering the TR, and it is found that the trigger time of the TR is earlier at low AP.

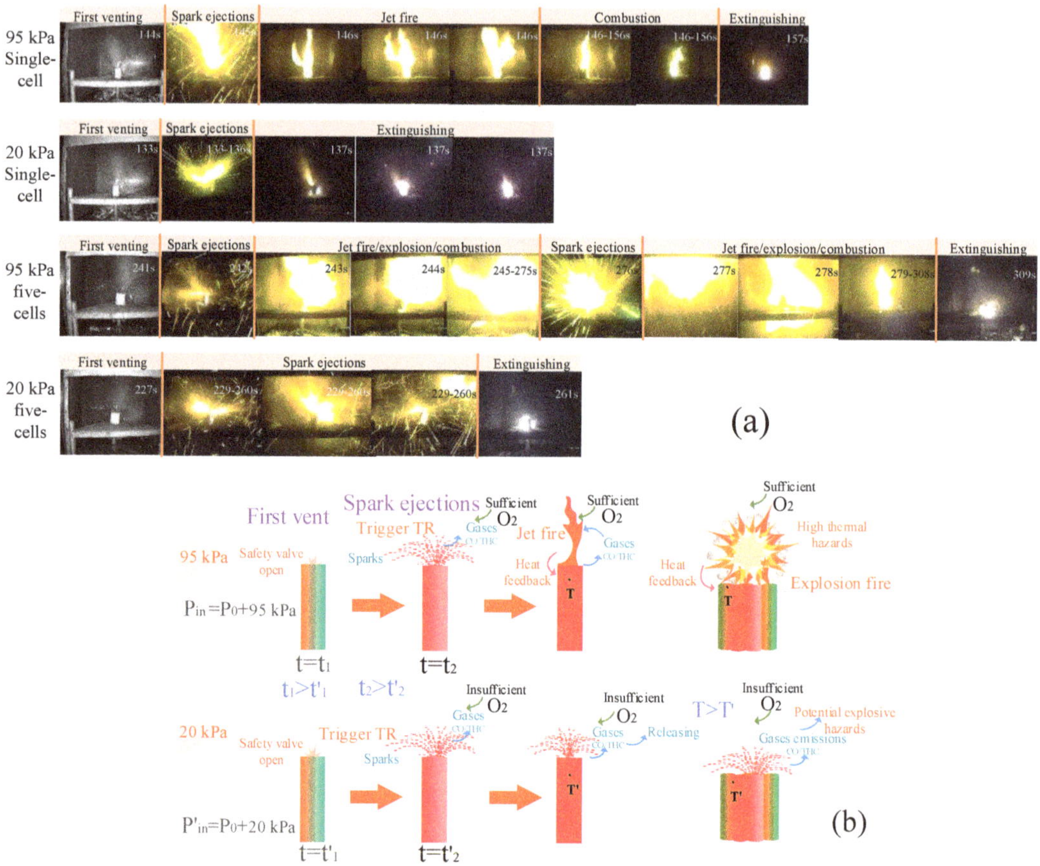

Figure 4. (**a**) Thermal runaway phenomenon of cylindrical LIBs under 95 kPa and 20 kPa, (**b**) Schematic diagram of Thermal runaway processes of cylindrical LIBs under 95 kPa and 20 kPa.

$$t_2 = \{k : T_{k+1} - T_k > 10\ ^\circ C\} \quad (1)$$

It is obvious that the jet fire and the combustion process, with large flames, are more vigorous for multiple cells than a single cell in 95 kPa. The visible explosion fire and audible bang are more common, asynchronous, and violent. It can be attributed to the mutual interference and the variation in ignition time, which can accelerate the nearby cells and ignite the gases released swiftly from the cell. The increasing number of batteries leads to significantly larger-scale fires and fiercer ejection. Comparing Figure 5a,b, it can be found that the battery structure after combustion almost remains intact in the single-cylindrical cell experiments of 95 kPa. However, for multiple-cell tests under 95 kPa, the shell and safety valve were broken severely, and this can be seen in Figure 6c The TR process for multiple cells has a relatively higher risk of fire and explosion than for single cells, and this risk also increases with the number of batteries.

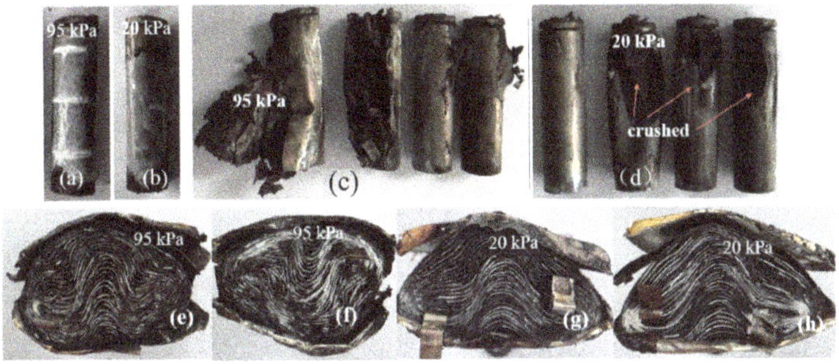

Figure 5. Typical phenomena of LIBs after thermal runaway: (**a**) group A—single cell in 95 kPa, (**b**) group A—single cell in 20 kPa, (**c**) group B—multiple cells in 95 kPa, (**d**) group B—multiple cells in 20 kPa, (**e**) 10 Ah pouch cell under 95 kPa, (**f**) 15 Ah pouch cell under 95 kPa, (**g**) 10 Ah pouch cell under 20 kPa, and (**h**) 15 Ah pouch cell under 20 kPa.

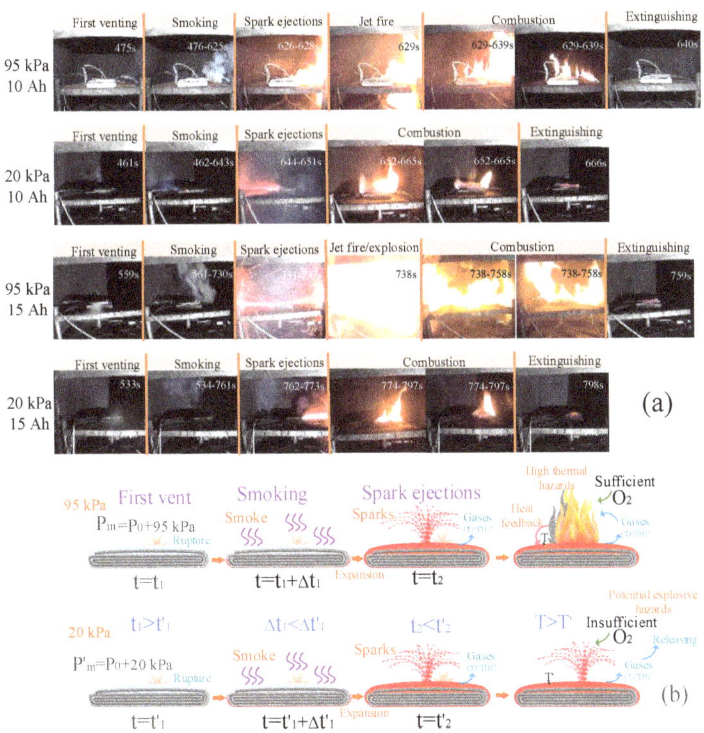

Figure 6. (**a**) Thermal runaway phenomenon of pouch LIBs under 95 kPa and 20 kPa, (**b**) Schematic diagram of Typical thermal runaway process of pouch LIBs under 95 kPa and 20 kPa.

It can be seen that the sparks and combustion flames are also obviously weaker for multiple cylindrical cells at 20 kPa than for the tests under 95 kPa. In addition, there is no fierce gas explosion and intense flame ejection, except for the ejections and sparks. As can be seen from the carbonized batteries after tests, the batteries after tests under 20 kPa are also broken or crushed, which is different from the single-cell tests under 20 kPa, as seen in Figure 5d. Those phenomena indicate that mutual interference still exists under 20 kPa, but this process is not as vigorous as the multiple cells under 95 kPa, without the violent fire behaviors of explosion and jet flames. This mutual interference may be ascribed to the heat transfer through the pathway of direct contact among the cells, which is not affected by the variation in pressure. The number of LIBs that were broken or crushed after the TR in each test was about three-fourths of all the batteries both under 95 kPa and 20 kPa. It can be seen that as the ambient pressure rises, the rate of heat propagation between battery cells may accelerate, and the whole system produces more energy per unit time, which is more destructive to the surrounding environment. Under different pressures, the battery pack should consider a reasonable space design.

3.1.2. Pouch Cells

It can be observed that the pouch cell under 95 kPa undergoes six stages during the TR process: (1) first venting (the rupture of the aluminum–polymer bag), (2) smoking (gas releasing), (3) sparks (huge irritant gases and products jet out, and the TR is triggered), (4) jet fire with flames and sparks, (5) combustion, and (6) extinguishing. The pouch cell firstly expands slowly in volume with the temperature increasing, indicating the beginning of the thermal decomposition reactions. Then, the space between the layers increases and is filled with gas products, as can be seen in Figure 5e. As the accumulation of gases, the pressure in the battery increases, and the bag is ruptured. Irritant gases and smoke gradually release from the side of the battery's external terminal, and the decomposition reactions are further accelerated. Subsequently, huge flammable gases along with solids and liquid droplets are ejected out swiftly and ignited, forming intense jet flames and white sparks. Lastly, the combustion fire gradually becomes weak and stable, and finally extinct.

As shown in Figure 6a, the typical TR phenomenon of pouch cells with 10 Ah and 15 Ah under 95 kPa and 20 kPa are presented, respectively. Under 95 kPa, with thick smoke and large flames, the TR of 10 Ah is particularly fiercer than 20 kPa. With the increase in cell capacity, the intensity of the TR increases. Fiercer explosions and larger fire flames occurred under 95 kPa for 15 Ah. Under the effects of hypobaric hypoxia in 20 kPa, there are no jet fire flames, and the sparks are not typical white but dark red for 10 Ah and 15 Ah. Only small fire flames were observed after the ejection of sparks for 15 Ah under 20 kPa. As shown in Figure 6b, Δt_1 represents the average duration of smoking after the first venting. The results show that the t_1 time is shortened at low AP, while Δt_1 is prolonged and the TR trigger time t_2 is delayed.

The typical pouch cells after the tests are shown in Figure 5e–h. The aluminum–polymer bag burned and melted into pieces, and the whole battery was severely carbonized, both under 95 kPa and 20 kPa. Obviously, there is an increase in thickness, while the gaps between the layers are expanded, especially for the batteries after a TR under 20 kPa. Compared with cylindrical batteries, pouch cells are less bound and show an overall fluffy enlargement after a TR. And the size of deformation under different environmental pressures may be related to the intensity of the whole combustion process.

3.2. First Venting and Trigger of TR

Figure 7 depicts the time values during the TR processes under two ambient pressures of 95 kPa and 20 kPa. As the pressure decreases from 95 kPa to 20 kPa, the average values of t_1 (the average open time of the safety valve or the rupture time of the bag, which was counted by the video recorded in the tests) decrease from 144 s to 133 s, 475 s to 461 s, and 559 s to 533 s, for 18650 cylindrical cells, 10 Ah pouch cells, and 15 Ah pouch cells, respectively (Table S1). A low-pressure atmosphere shortens the time of first venting (rupture of the safety valve/bag). It is mainly due to the competition between the internal pressure and atmospheric pressure, as shown in Equation (2) [33].

$$P_{in} = P_0 + P_{out} \quad (2)$$

Here, P_0 (kPa) is the rupture pressure of the safety valve/aluminum–polymer bag, which is a fixed value. P_{in} (kPa) refers to the internal pressure of the battery due to the accumulation of gases and heat released from the thermal decomposition reactions, and P_{out} (kPa) is the external pressure (atmospheric pressure). The first venting occurred in a case where internal pressure exceeds the sum of the atmospheric pressure and rupture pressure of the safety valve/bag. P_{in} decreases as P_{out} from 95 kPa to 20 kPa; therefore, the time for the rupture of the safety valve/bag is shortened. The earlier behaviors of the safety valve opening/rupturing of the bag are probably helpful for the safety warning of LIBs, especially under low-pressure environments.

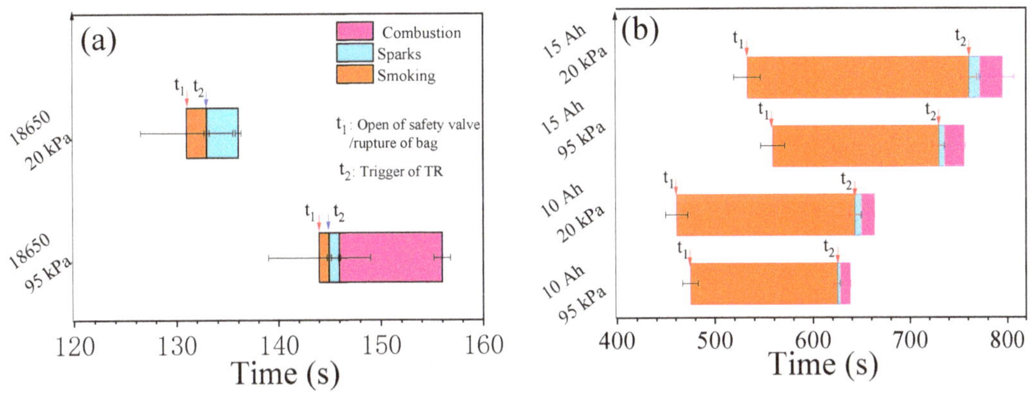

Figure 7. The average values of time for different stages during the TR process: (**a**) cylindrical cells and (**b**) pouch cells.

The average trigger time (t_2) of spark ejections for cylindrical cells is 145 s in 95 kPa and 133 s in 20 kPa. However, t_2 increases from 625 s to 643 s, and from 730 s to 761 s for 10 Ah and 15 Ah pouch cells, when the pressure decreases from 95 kPa to 20 kPa. The average trigger time of spark ejections is delayed for pouch cells and shortened for cylindrical cells, under a low-pressure environment of 20 kPa. This is attributed to the difference in smoking duration between the cylindrical and pouch cells. It can be seen from the smoking duration (Δt_1) between the first venting and the TR. Δt_1 increases from 1 s to 2 s, when the pressure decreases from 95 kPa to 20 kPa, for cylindrical cells. The influence of low pressure on smoking duration is minimal for cylindrical cells, which can be neglected. However, this average smoking duration is greatly extended, from 150 s to 182 s and from 171 s to 228 s for 10 Ah and 15 Ah pouch cells, as the pressure varies from 95 kPa to 20 kPa, leading to a delay of the average trigger time of the TR at 20 kPa. These phenomena for pouch cells can be explained by the acceleration effect of exothermal reactions with the oxygen from the environment in 20 kPa which is not as obvious as 95 kPa, after the rupture of the bag. In addition, the timely release of gases under low ambient pressure is not beneficial for the accumulation of heat, during the exothermal reactions in cells. Thus, it takes more time for the battery to reach the threshold of the TR under 20 kPa. However, this effect for cylindrical cells is not significant because of the structural characteristics of steel shells.

3.3. Heat Release Rates and Temperatures

As the most critical factor, HRR is used to evaluate thermal hazards and determine whether the neighbor cells will be ignited in the battery pack. Generally, the values of HRR and THR were calculated by the oxygen consumption principle based on Thornton [43]. According to previous studies [44], HRR, based on the oxygen consumption principle, cannot reflect the heat that does not consume any oxygen from the atmosphere, but it can better present the heat released from the oxidation with oxygen in the air, considering that the decomposition energy in a TR is determined by stored electric energy in LIBs [45]. Therefore, HRR and THR are better used in thermal hazard assessments to reflect the heat from the oxidation or combustion of TR product emissions with the oxygen from the atmosphere, e.g., flammable gases, combustible battery materials, and electrolytes.

$$q = E\left(m_{O_2}^0 - m_{O_2}\right) \tag{3}$$

$$m_{O_2} = 26.54 \cdot \frac{A \cdot k_c}{f(R_e)} \cdot \sqrt{\frac{\Delta p}{T}} \tag{4}$$

$$q = 1.1E \cdot 0.21 \cdot m_e \left[\frac{\Phi - 0.172(1-\Phi)\frac{X_{CO}^A}{X_{O_2}^A}}{(1-\Phi) + 1.105\Phi} \right] \tag{5}$$

$$\Phi = \frac{X_{O_2}^0\left(1 - X_{CO_2}^A - X_{CO}^A\right) - X_{O_2}^A\left(1 - X_{CO_2}^0\right)}{\left(1 - X_{O_2}^A - X_{CO_2}^A - X_{CO}^A\right)X_{O_2}^0} \tag{6}$$

$$Q = \int_0^t q(\tau)d\tau \tag{7}$$

Here, q represents the HRR (kW), $m_{O_2}^0$ and m_{O_2} are the flow rates of oxygen from the incoming and outgoing air ($Kg \cdot s^{-1}$), and E is the energy release per mass unit of O_2 consumed for a given fuel ($13.1 \pm 5\%$ $MJ \cdot Kg^{-1}$); the mass flow rate (m_{O_2}) can be calculated by the formula in Equation (4), where A represents the cross-sectional area (m^2), kc represents the wind velocity distribution coefficient, f(Re) represents the Reynolds number correction function, $\triangle P$ represents the pitot tube centerline pressure difference (Pa), and T represents the catheter centerline temperature (K). q is calculated by changes in O_2, CO, and CO_2 because it is actually an incomplete combustion, where $X_{O_2}^0$ and $X_{O_2}^A$ are the initial and final oxygen volume percentage concentrations (%), $X_{CO_2}^0$ and $X_{CO_2}^A$ are the initial and final carbon dioxide volume percentage concentrations (%), and X_{CO}^A is the final CO volume percentage concentration (%). Q is the THR (kJ).

Figure 8a depicts the HRR curves at two different atmospheric pressures. The peak HRR decreases from 1.2 kW to 0.65 kW, 14.2 kW to 1.2 kW, 26.3 kW to 6.1 kW, and 41.1 kW to 8.7 kW for 1 cylindrical cell, 5 cylindrical cells, a 10 Ah pouch cell, and a 15 Ah pouch cell, respectively. It can be seen that the peak HRR values under 95 kPa are higher than the values under 20 kPa for the external heating tests. These results indicated that the heat from the oxidation combustion reactions in the atmosphere outside the cell decreases greatly for low oxygen mass concentration per volume under 20 kPa. It ought to be mainly ascribed to the reason that the fierce fire behaviors are not formed, e.g., visible explosion fire flames and jet fires, for the insufficient oxygen in 20 kPa.

Moreover, it was confirmed that fire developed slowly as the ambient pressure declined for LIBs and common fuel. The empirical formula [30] of burning rate ($\dot{m}''$) on ambient pressure (P_{out}) can be expressed as follows:

$$\dot{m}'' \sim {P_{out}}^{0.57} \tag{8}$$

The mass burning rate slows as the atmospheric pressure decreases. As presented in this empirical formula, the burning rate reduces with the drop in ambient pressure. It can be derived that the combustion accompanied by a TR would be inhibited by ambient low pressure. The burning rate of ejected combustible products reduces when the pressure reduces from 95 kPa to 20 kPa. Consequently, the values of HRR and THR decrease with the reduction in atmospheric pressure, and the thermal hazards of TR fire for both cylindrical and pouch cells are lower than the tests under 95 kPa, as shown in Figure 8a.

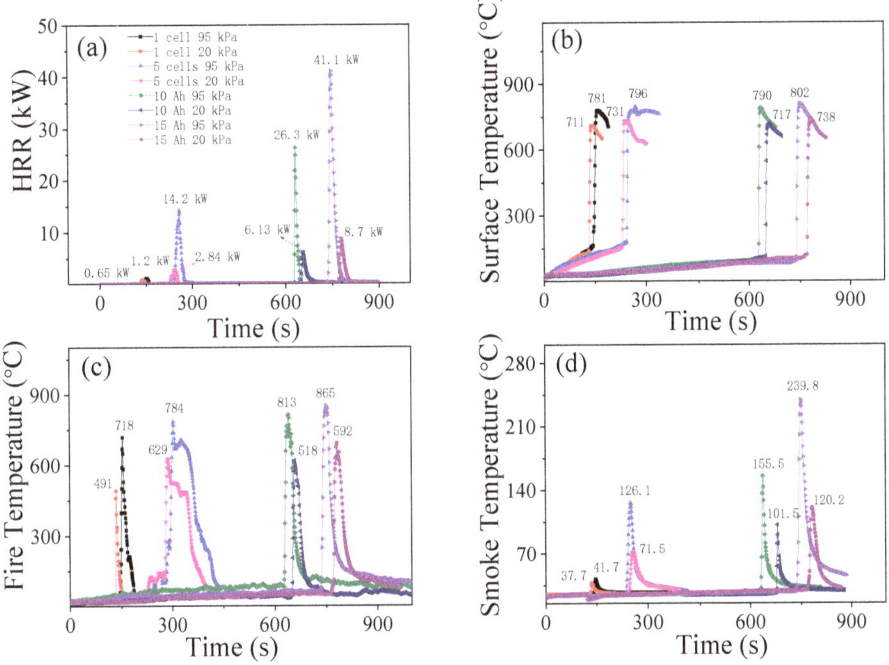

Figure 8. Typical curves and average values: (**a**) HRR; (**b**) battery surface temperatures; (**c**) thermal runaway fire temperatures; and (**d**) thermal runaway smoke temperatures.

It also can be seen from Figure 9 that the THR increases with atmospheric pressure and the number of cylindrical batteries in a power function. The data are fitted with regression lines of the following form:

$$Q = an^b \tag{9}$$

where Q represents the total heat release measured through the oxygen consumption principle, and a and b are the coefficients. The coefficients a and b decrease with the ambient pressure from 95 kPa to 20 kPa.

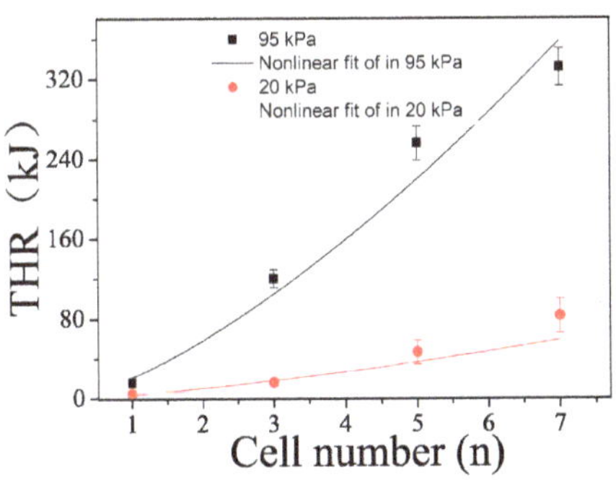

Figure 9. The average values of THR versus the number of cylindrical cells.

Figure 10a shows the data graph of THR and peak HHR, the variation in THR and HRR may account for the variations in fire temperature and smoke temperature, in Figure 9c,d. The peak fire temperature decreased from 718 °C to 491 °C, 784 °C to 629 °C, 813 °C to 518 °C, and 865 °C to 592 °C, for four groups of tests, when the pressure declined from 95 kPa to 20 kPa. The peak smoke temperatures of 41.7 °C, 126.1 °C, 155.5 °C, and 239.8 °C under 95 kPa are higher than 37.7 °C, 71.5 °C, 101.5 °C, and 120.5 °C under 20 kPa, for four groups of tests, independently. The curves of battery surface temperature can be seen in Figures 9b and 10b. The average initial temperature of the TR decreases from 161 °C to 139 °C for cylindrical cells and increases from 104 °C to 118 °C, and 112 °C to 124 °C, for two types of pouch cells, when the pressure decreases from 95 kPa to 20 kPa. This is consistent with the change in the average trigger time (t_2) of spark ejections discussed before. Due to the difference between the locally measured surface temperature and the actual TR temperature, the low pressure has different effects on the cylindrical and soft-pack batteries, and the structure of the soft-pack battery is greatly expanded by heat and produces a large amount of smoke and is released, which is not conducive to the accumulation of heat, resulting in inconsistent times to reach a TR, thus affecting the value of the initial TR temperature. The results show consistent patterns in different quantities and capacity states; the cell average peak surface temperature declines with the decrease in pressure from 95 kPa to 20 kPa, e.g., from 781 °C to 711 °C, from 796 °C to 731 °C, from 790 °C to 717 °C, and from 802 °C to 738 °C. It can be better understood that the heat feedback to the battery reduces greatly under 20 kPa, for the absence of fire flames, e.g., jet fire, explosion, and fierce combustion. The low combustion efficiency and burning rate under low-pressure environments also prevent the generation of heat, leading to a decrease in heat feedback from combustion to the battery. It has been proved that the combustion energy released in a TR is as high as 2–~6 times its electric capacity [46]. This is probably the dominant factor for the decline in surface temperature. In addition, the early release of gas products is in contrast to the accumulation of heat, for a certain amount of gas before releasing out of the battery, as shown in Equation (11).

$$\rho_{O_2} = \frac{M_{O_2}P}{TR} \tag{10}$$

$$T = \frac{P_{in}V}{nR} = \frac{(P_{out} + P_0)V}{nR} \tag{11}$$

Here, T is the temperature (°C), n is the number of moles of gas, R is the ideal gas constant, ρ_{O_2} is the mass concentration of O_2 in volume, and M_{O_2} is the molar mass of O_2. In theory, the temperature (T) of gases decreases with the decrease in atmospheric pressure (P_{out}) from 95 kPa to 20 kPa, at the moment of gas release. Therefore, the cells' peak surface temperatures are lower at 20 kPa than those under 95 kPa.

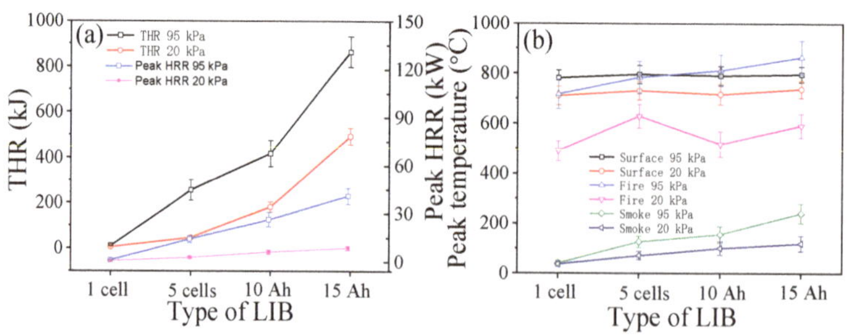

Figure 10. (**a**) The average values of THR and peak HRR, and (**b**) the average peak values of temperature.

3.4. Vent Gas Density and O_2

As presented in Figure 11b, the valley values of cylindrical cells and pouch cells increase from 20.71 vol.% to 20.83 vol.%, 19.28 vol.% to 20.42 vol.%, 17.57 vol.% to 20.18 vol.%, and 14.19 vol.% to 19.69 vol.%, with the decrease in pressure from 95 kPa to 20 kPa, respectively. It indicates that the consumption of oxygen under 20 kPa decreases during the TR. These consequences are due to the difficulty of ignition for emitted combustible substances, as well as the decrease in further thermal decomposition and combustion during a TR under 20 kPa. This also can be seen from the smoke light transmittance curves in Figure 11a,c, which represents the smoke density during the TR. It is obvious that the values of light transmittance under 20 kPa are higher than the values of 95 kPa, corresponding to the fierce TR fire behaviors, e.g., jet fire and explosion under 95 kPa. Under the effect of fire plume and violent behaviors in 95 kPa, more ejections flow with the smoke, and the smoke density is higher than that under 20 kPa.

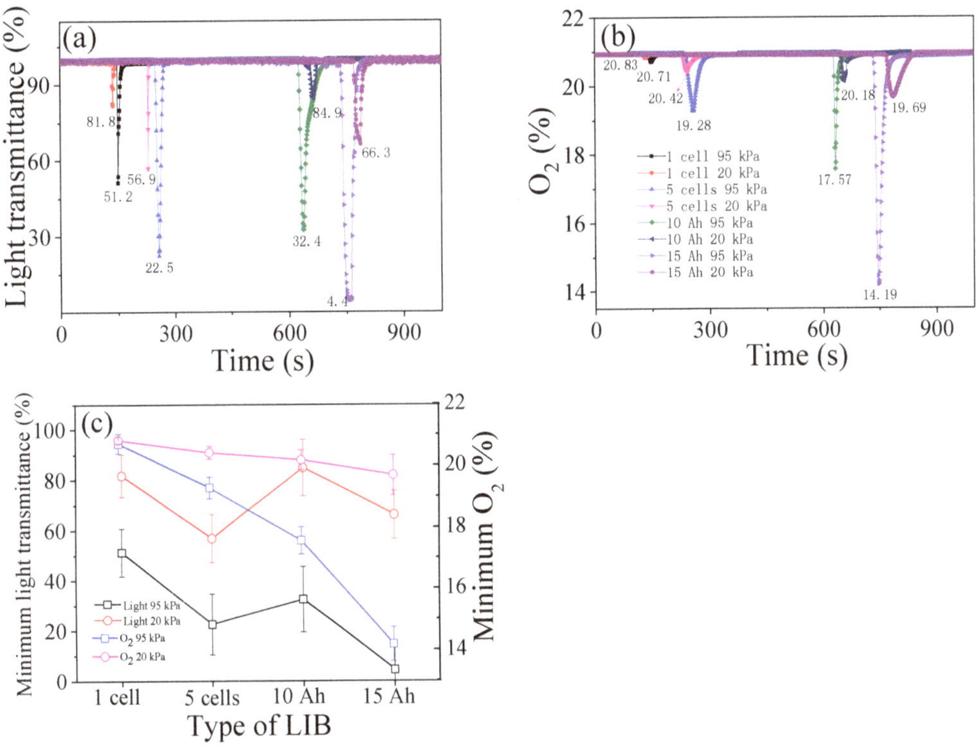

Figure 11. (**a**) The typical curves of light transmittance, (**b**) the mass concentration of oxygen in volume, (**c**) the average peak values of minimum light transmittance and O_2.

3.5. CO_2 and CO/CO_2

Figure 12a shows the curves of CO_2 concentration in volume. It is well known that the CO_2 generation is tightly related to the combustion efficiency in fire reactions. As the external pressure increases from 20 kPa to 95 kPa, the peak values of CO_2 for groups A, B, C, and D increase from 0.11 vol.%, 0.52 vol.%, 0.69 vol.%, and 1.81 vol.% to 0.25 vol.%, 1.10 vol.%, 2.02 vol.%, and 5.34 vol.%, respectively. One reason for the huge proportion of CO_2 is believed to be due to the decomposition of $LiPF_6$ and electrolyte solvents, and the further reduction in hydrocarbons inside the cell. Another source is the external combustion reaction products, which are affected by the oxygen mass concentration per volume in the atmosphere, combustion efficiency, and TR fire behaviors. The CO/CO_2 ratio can present fire hazards. The ratios of CO/CO_2 are 1.55, 0.96, 0.45, and 0.21, for the four groups of tests under 95 kPa, respectively, and are lower than the ratios of 2.05, 1.99, 1.87, and 2.55 under 20 kPa, as can be seen in Figure 12c,d. This indicates that the combustion efficiency of combustible ejections decreases with the reduction in the external pressure from 95 kPa to 20 kPa.

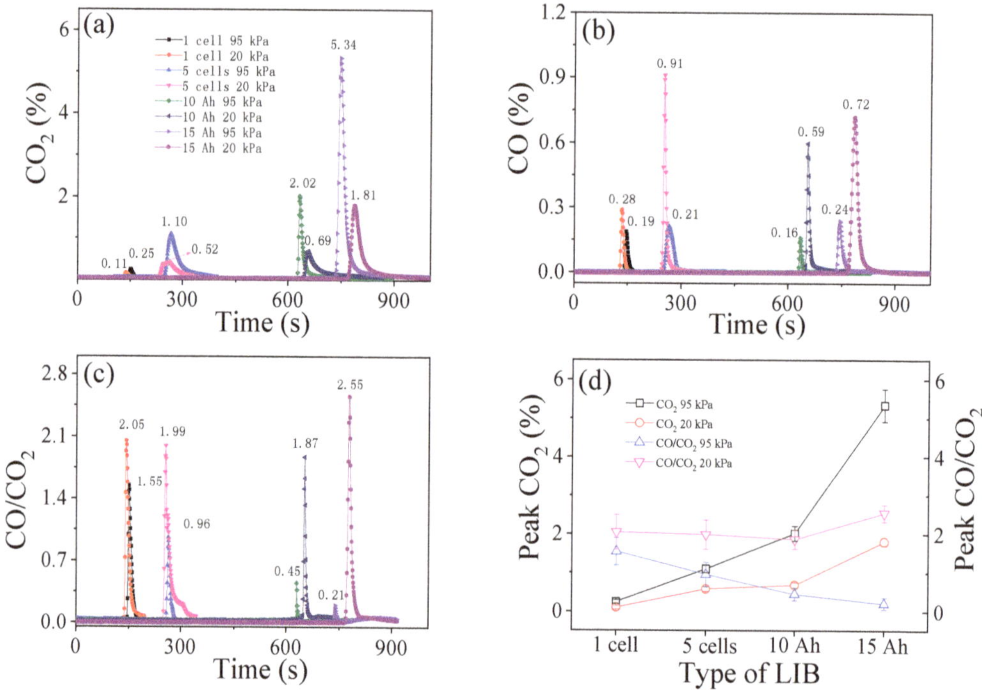

Figure 12. (**a**) The percentage volume concentration of CO_2, (**b**) CO, and (**c**) CO/CO_2, and (**d**) the average peak values of CO_2 and CO/CO_2.

For cylindrical cells, huge gases are released instantaneously with high velocity after the opening of the pressure valve. Thus, most parts of combustible gases are not oxidized sufficiently during the TR. But for the pouch cell, with the melting of the aluminum–polymer bag and thermal decomposition inside, the layers inside are exposed to air, and the gas products are released in time, and then oxidized with oxygen. Therefore, the oxidation efficiency of emissions released from pouch cells is higher than the cylindrical cells, resulting in more combustible emissions being oxidized into CO_2. Hence, the TR behaviors of various formats of batteries seem to have a distinct difference in the gas emission characteristics.

3.6. THC and CO

Figure 13a displays the typical concentration curves of hydrocarbon under two ambient pressures. The average peak THC values in 20 kPa are about 1275 ppm, 3474 ppm, 819 ppm, and 1641 ppm for four groups of tests, which can be seen in Figure 13b, which are higher than 1019 ppm, 256 ppm, 127 ppm, and 559 ppm under 95 kPa, respectively. From the CO curves in Figure 12b, the average peak values of CO increase from 0.19 vol.% to 0.28 vol.%, 0.21 vol.% to 0.91 vol.%, 0.16 vol.% to 0.59 vol.%, and 0.24 vol.% to 0.72 vol.%, with the pressure drop from 95 kPa to 20 kPa. The intensity of CO under 95 kPa is lower than that of 20 kPa. This phenomenon is consistent with the flammable emission gas of

THC, demonstrating a great decrease in oxidation between the combustible emitted gas products and oxygen in air under the pressure of 20 kPa. In other words, the proportion of flammable gas in total emitted gases increases because of the difficulty of ignition and insufficient oxidation.

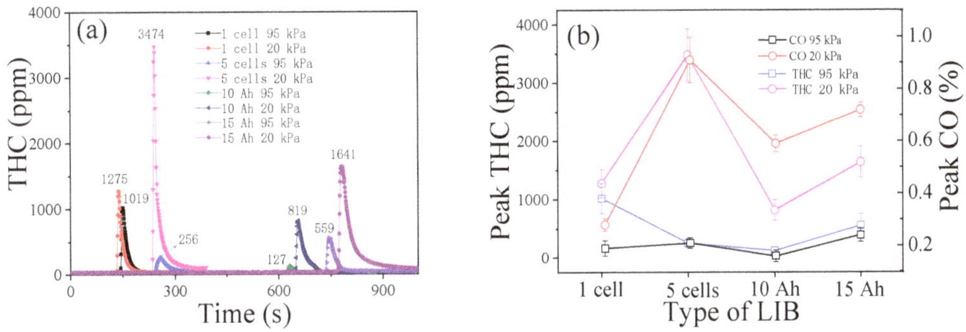

Figure 13. (**a**) The typical curves of THC, and (**b**) the average peak values of THC and CO.

Huge products, e.g., battery materials, series of intermediates and small molecule products, release the battery out in the form of solid particles, liquids, and gases. Those products undergo complex chemical reactions, such as degradation and oxidation. As the main reactions during the TR process, the reactions of degradation can be summarized as in Equations (12)–(16), and the reactions of oxidation are presented by Equations (17)–(22).

$$2Li + C_3H_4O_3(EC) \rightarrow Li_2CO_3 + C_2H_4 \quad (12)$$

$$2Li + C_3H_6O_3(DMC) \rightarrow Li_2CO_3 + C_2H_6 \quad (13)$$

$$2Li + C_4H_8O_3(EMC) \rightarrow Li_2CO_3 + C_3H_8 \quad (14)$$

$$2LiPF_6 + C_2H_5OCOOC_2H_5 \rightarrow C_2H_5OCOOPF_4 + HF + LiF + C_2H_4 \quad (15)$$

$$C_2H_5OCOOPF_4 \rightarrow HF + CO_2 + POF_3 + C_2H_4 \quad (16)$$

$$2O_2 + C_3H_4O_3(EC) \rightarrow 3CO_2 + H_2O \quad (17)$$

$$2O_2 + C_3H_6O_3(DMC) \rightarrow 3CO_2 + H_2O \quad (18)$$

$$2O_2 + C_4H_8O_3(EMC) \rightarrow 4CO_2 + H_2O \quad (19)$$

$$-CH_2 - CF_2 - +2H_2 + 2O_2 \rightarrow CO_2 + H_2O + HF \quad (20)$$

$$C_nH_x + \left(n + \frac{x}{4}\right)O_2 \rightarrow nCO_2 + \left(\frac{x}{2}\right)H_2O \quad (21)$$

$$C + O_2 \rightarrow CO_2 \quad (22)$$

The oxidation of products is affected by the density of oxygen in the atmosphere, as well as the degradation, which can be accelerated by high temperatures. Under a low-pressure environment of 20 kPa, the low density of oxygen does not contribute to the oxidation of the TR emissions, leading to higher values of combustible gases.

It can also be better understood from the theory of the carbon triangle. More carbon, hydrocarbon, carbon monoxide, and other organic products are oxidized with oxygen to carbon dioxide under 95 kPa. However, the combustible gas emissions released from the cell could not be ignited easily under a low-pressure environment of 20 kPa, with thin air and low oxidizer density, as can be seen from the TR behaviors above. Therefore, the incomplete oxidation/combustion and the difficulty of ignition under the anoxia environment of 20 kPa lead to higher values of CO and THC and lower values of CO_2, compared with the tests of 95 kPa. It can be concluded that the hazards of toxic/potentially explosive gases are higher than the tests under 95 kPa, during the TR process of LIBs. Hence, it is crucial to maintain good ventilation or take feasible measures to suppress gas explosions.

3.7. Hazards Analysis

Figure 14 illustrates the regional distribution of the peak values of HRR, THC, and CO during the TR tests of two types of LIBs with 100% SOC under the ambient pressures of 95 kPa and 20 kPa. Obviously, the thermal parameter of peak HRR varies inversely with the inflammable and potentially explosive gas emissions of THC and CO in the TR process. More combustion energy is released due to the oxidation combustion reactions of combustible gas products with oxygen in the air, resulting in the decrease in flammable/potentially explosive gases. On the contrary, the concentration of combustible/toxic gases would be high in the TR tests when the TR behaviors of fierce combustion are not formed for low oxygen mass concentration under 20 kPa. Although the energy released is reduced, with the rising concentration of combustible gases per unit volume, once supplemented by the entry of oxygen, and pre-mixed to reach the explosion limit, it will have a potential risk of thermal hazards.

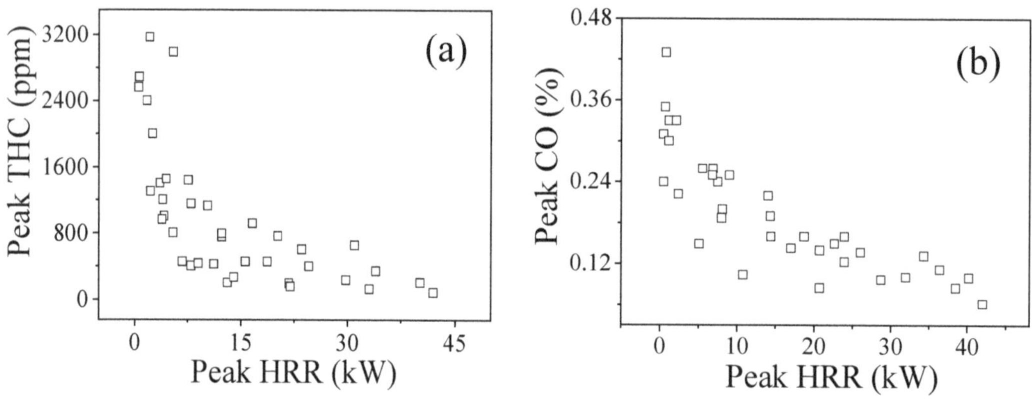

Figure 14. Average peak values of THC (**a**) and CO (**b**) versus average peak HRR for different LIBs under 95 kPa and 20 kPa.

As the pressure decreases from 95 kPa to 20 kPa, the average peak values of the parameters of HRR, THR, and CO_2 decrease, while the peak values of THC and CO increase, indicating a lower thermal hazard of TR fire and a higher concentration of toxic/flammable gas emission in the tests of 20 kPa, compared with the tests under 95 kPa. These results can also be characterized by the distribution of average values per unit capacity under two atmospheric pressures, as shown in Figure 15a,b. The average peak values per capacity of THR, HRR, and CO_2 in 20 kPa are lower, while the gas peak concentration values per capacity of CO/CO_2, THC, and CO are higher, compared with the tests under 95 kPa. It can be found that the low atmospheric pressure has a significant influence on the LIB parameters of TR hazards. In other words, the thermal hazards of TR fire, e.g., fierce jet fire, deflagration, or explosion, under 95 kPa are more severe than the TR in 20 kPa, but the hazards of potentially explosive and toxic gas emissions, e.g., the THC and CO, need be paid more attention in 20 kPa, particularly in a confined space with life, such as the cabin or cargo compartment of aircraft.

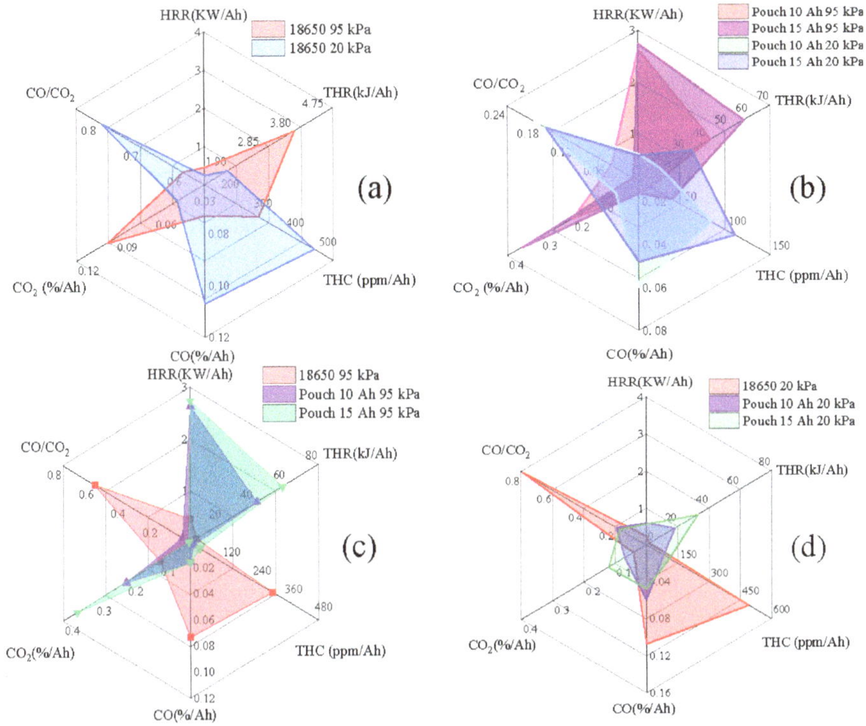

Figure 15. The distribution of average peak value per capacity for the TR tests of 18650 and pouch LIBs under 95 kPa and 20 kPa: (**a**) 18650 LIBs under 95 kPa and 20 kPa, (**b**) 10 Ah and 15 Ah pouch LIBs under 95 kPa and 20 kPa, (**c**) 18650 and pouch LIBs under 95 kPa, and (**d**) 18650 and pouch LIBs under 20 kPa.

With high values of peak HRR and THR, the pouch cell with a soft can has a higher thermal risk of fire than the hard-can cell of 18650, both at 95 kPa and 20 kPa. Meanwhile, the cylindrical cells have a higher concentration of THC and CO at two ambient pressures than the pouch cells. Also, as seen in Figure 15c,d, the values of peak HRR, THR, and peak CO per unit capacity for the pouch cell are higher than the values of the cylindrical cell, but the peak values of CO/CO_2, THC and CO per capacity are lower than in cylindrical cells. It indicates that the pouch cell with a soft aluminum–polymer bag has higher thermal hazards of TR fire, but a lower concentration of toxic/combustible gases than cylindrical cells with hard cans. The cylindrical cell has a higher risk of toxicity and potential explosion than the pouch cell in the TR process. In addition to a high concentration of toxic/potentially explosive gases, the TR fire behaviors are more aggressive for cylindrical cells, due to the structure of the steel shell. It can be concluded that the type of cell packaging also has an obvious impact on the characteristics of the TR.

4. Conclusions

In this study, two types of LIBs of cylindrical cells (18650 with 2.6 Ah) and pouch cells (10 Ah and 15 Ah) were tested under the pressures of 95 kPa and 20 kPa, in a dynamic pressure cabin. The major conclusions can be summarized as follows:

(1) With the pressure drop from 95 kPa to 20 kPa, the average open time of the safety valve (the rupture of the bag for pouch cells) is shortened. The average trigger time of the TR is shortened for cylindrical cells, but delayed for pouch cells under 20 kPa. The

average peak values of HRR, THR, surface/fire/smoke temperature, smoke density, and CO_2 decrease, but the values of THC and CO increase.

(2) The proportion of CO and THC varies inversely with the thermal hazards of TR fire. TR behaviors under 20 kPa are not as vigorous as the tests in 95 kPa, without violent fire, explosion, and huge jet flames. The thermal hazards of TR fire under 20 kPa are lower than that in 95 kPa, but the proportion of flammable/potentially explosive gas emissions increases greatly which poses a huge risk, especially in a confined space.

(3) The number of batteries and type of packaging also have an obvious effect on the performance of the TR. Pouch cells have a higher thermal hazard of fire, but lower combustible/toxic emitted gases, than the cylindrical cells both under 95 kPa and 20 kPa. With fierce TR behaviors, the combustible gas products are released at high speeds, leading to a higher risk of toxicity and potential explosion of cylindrical cells than pouch cells in the TR process.

Under low-pressure environments of cruise altitude, the thermal hazards of TR fire for pouch cells and the toxic/potentially explosive hazards of gas emissions of cylindrical cells need more attention, especially in the cabin or compartments of aircraft. The measures of ventilation or gas explosion suppression and the feasible design of extinguishing and fire protection systems should be considered, for the use of cylindrical and pouch LIBs. This discovery may be helpful to the safety design and protection of LIBs used in low-pressure environments, such as the installed Li battery on aircraft.

Supplementary Materials: The following supporting information can be downloaded at: https://www.mdpi.com/article/10.3390/batteries10090298/s1, Figure S1: A dT-T curve of the ARC test; Table S1: The average values of time for different stages during the TR process: (a) a cylindrical cell and (b) a pouch cell.

Author Contributions: Conceptualization, Q.S. and H.L.; methodology, Q.S.; software, Q.S.; validation, Q.S., H.L. and Z.W.; formal analysis, Y.M.; investigation, Q.S.; resources, Z.W.; data curation, Q.S.; writing—original draft preparation, Q.S. and H.L.; writing—review and editing, Q.S., H.L. and Z.W.; visualization, Y.W.; supervision, C.X.; project administration, Q.W.; funding acquisition, Q.S. and H.L. All authors have read and agreed to the published version of the manuscript.

Funding: This work was financially supported by the Opening Foundation of Civil Aircraft Fire Science and Safety Engineering Key Laboratory of Sichuan Province (MZ2023KF06), the General Program of Civil Aviation Flight University of China (QJ2022-171), and the General Program of Sichuan Vocational College of Chemical Technology (SCHYB-2024-08).

Data Availability Statement: The data presented in this study.

Conflicts of Interest: The authors declare no conflict of interest.

References

1. Sripad, S.; Viswanathan, V. Performance Metrics Required of Next-Generation Batteries to Make a Practical Electric Semi Truck. *ACS Energy Lett.* **2017**, *2*, 1669–1673. [CrossRef]
2. Bills, A.; Sripad, S.; Fredericks, W.L.; Singh, M.; Viswanathan, V. Performance Metrics Required of Next-Generation Batteries to Electrify Commercial Aircraft. *ACS Energy Lett.* **2020**, *5*, 663–668. [CrossRef]
3. Feng, X.; Sun, J.; Ouyang, M.; He, X.; Lu, L.; Han, X.; Fang, M.; Peng, H. Characterization of large format lithium ion battery exposed to extremely high temperature. *J. Power Sources* **2014**, *272*, 457–467. [CrossRef]
4. Zhu, X.; Wang, H.; Wang, X.; Gao, Y.; Allu, S.; Cakmak, E.; Wang, Z. Internal short circuit and failure mechanisms of lithium-ion pouch cells under mechanical indentation abuse conditions: An experimental study. *J. Power Sources* **2020**, *455*, 227939. [CrossRef]
5. Xu, J.; Wang, L.; Guan, J.; Yin, S. Coupled effect of strain rate and solvent on dynamic mechanical behaviors of separators in lithium ion batteries. *Mater. Des.* **2016**, *95*, 319–328. [CrossRef]
6. Wang, Q.; Ping, P.; Zhao, X.; Chu, G.; Sun, J.; Chen, C. Thermal runaway caused fire and explosion of lithium ion battery. *J. Power Sources* **2012**, *208*, 210–224. [CrossRef]
7. Li, W.; Wang, H.; Zhang, Y.; Ouyang, M. Flammability characteristics of the battery vent gas: A case of NCA and LFP lithium-ion batteries during external heating abuse. *J. Energy Storage* **2019**, *24*, 10077–10086. [CrossRef]
8. Larsson, F.; Andersson, P.; Blomqvist, P.; Lorén, A.; Mellander, B.-E. Characteristics of lithium-ion batteries during fire tests. *J. Power Sources* **2014**, *271*, 414–420. [CrossRef]

9. Ping, P.; Wang, Q.; Huang, P.; Sun, J.; Chen, C. Thermal behaviour analysis of lithium-ion battery at elevated temperature using deconvolution method. *Appl. Energy* **2014**, *129*, 261–273. [CrossRef]
10. Fredericks, W.L.; Sripad, S.; Bower, G.C.; Viswanathan, V. Performance Metrics Required of Next Generation Batteries to Electrify Vertical Takeoff and Landing (VTOL) Aircraft. *ACS Energy Lett.* **2018**, *3*, 2989–2994. [CrossRef]
11. Feng, X.; Sun, J.; Ouyang, M.; Wang, F.; He, X.; Lu, L.; Peng, H. Characterization of penetration induced thermal runaway propagation process within a large format lithium ion battery module. *J. Power Sources* **2015**, *275*, 261–273. [CrossRef]
12. Lamb, J.; Orendorff, C.J.; Steele, L.A.M.; Spangler, S.W. Failure propagation in multi-cell lithium ion batteries. *J. Power Sources* **2015**, *283*, 517–523. [CrossRef]
13. Ping, P.; Kong, D.; Zhang, J.; Wen, R.; Wen, J. Characterization of behaviour and hazards of fire and deflagration for high-energy Li-ion cells by over-heating. *J. Power Sources* **2018**, *398*, 55–66. [CrossRef]
14. Feng, X.; Ouyang, M.; Liu, X.; Lu, L.; Xia, Y.; He, X. Thermal runaway mechanism of lithium ion battery for electric vehicles: A review. *Energy Storage Mater.* **2018**, *10*, 246–267. [CrossRef]
15. Xu, C.; Feng, X.; Huang, W.; Duan, Y.; Chen, T.; Gao, S.; Lu, L.; Jiang, F.; Ouyang, M. Internal temperature detection of thermal runaway in lithium-ion cells tested by extended-volume accelerating rate calorimetry. *J. Energy Storage* **2020**, *31*, 101670. [CrossRef]
16. Chen, Y.; Liu, N.; Jie, Y.; Hu, F.; Li, Y.; Wilson, B.P.; Xi, Y.; Lai, Y.; Yang, S. Toxicity Identification and Evolution Mechanism of Thermolysis-Driven Gas Emissions from Cathodes of Spent Lithium-Ion Batteries. *ACS Sustain. Chem. Eng.* **2019**, *7*, 18228–18235. [CrossRef]
17. Ouyang, D.; Weng, J.; Chen, M.; Wang, J. Impact of high-temperature environment on the optimal cycle rate of lithium-ion battery. *J. Energy Storage* **2020**, *28*, 101242. [CrossRef]
18. Guo, L.S.; Wang, Z.R.; Wang, J.H.; Luo, Q.K.; Liu, J.J. Effects of the environmental temperature and heat dissipation condition on the thermal runaway of lithium ion batteries during the charge-discharge process. *J. Loss Prev. Process Ind.* **2017**, *49*, 953–960. [CrossRef]
19. Sun, J.; Li, J.; Zhou, T.; Yang, K.; Wei, S.; Tang, N.; Dang, N.; Li, H.; Qiu, X.; Chen, L. Toxicity, a serious concern of thermal runaway from commercial Li-ion battery. *Nano Energy* **2016**, *27*, 313–319. [CrossRef]
20. Koch, S.; Fill, A.; Birke, K.P. Comprehensive gas analysis on large scale automotive lithium-ion cells in thermal runaway. *J. Power Sources* **2018**, *398*, 106–112. [CrossRef]
21. Golubkov, A.W.; Fuchs, D.; Wagner, J.; Wiltsche, H.; Stangl, C.; Fauler, G.; Voitic, G.; Thaler, A.; Hacker, V. Thermal-runaway experiments on consumer Li-ion batteries with metal-oxide and olivin-type cathodes. *RSC Adv.* **2014**, *4*, 3633–3642. [CrossRef]
22. Larsson, F.; Andersson, P.; Blomqvist, P.; Mellander, B.E. Toxic fluoride gas emissions from lithium-ion battery fires. *Sci. Rep.* **2017**, *7*, 10018. [CrossRef]
23. Baird, A.R.; Archibald, E.J.; Marr, K.C.; Ezekoye, O.A. Explosion hazards from lithium-ion battery vent gas. *J. Power Sources* **2020**, *446*, 227257. [CrossRef]
24. Zhang, C.; Santhanagopalan, S.; Sprague, M.A.; Pesaran, A.A. Coupled mechanical-electrical-thermal modeling for short-circuit prediction in a lithium-ion cell under mechanical abuse. *J. Power Sources* **2015**, *290*, 102–113. [CrossRef]
25. Larsson, F.; Bertilsson, S.; Furlani, M.; Albinsson, I.; Mellander, B.-E. Gas explosions and thermal runaways during external heating abuse of commercial lithium-ion graphite-$LiCoO_2$ cells at different levels of ageing. *J. Power Sources* **2018**, *373*, 220–231. [CrossRef]
26. Gao, T.; Wang, Z.; Chen, S.; Guo, L. Hazardous characteristics of charge and discharge of lithium-ion batteries under adiabatic environment and hot environment. *Int. J. Heat Mass Transf.* **2019**, *141*, 419–431. [CrossRef]
27. Zhu, J.; Koch, M.M.; Lian, J.; Li, W.; Wierzbicki, T. Mechanical Deformation of Lithium-Ion Pouch Cells under In-Plane Loads—Part I: Experimental Investigation. *J. Electrochem. Soc.* **2020**, *167*, 090533. [CrossRef]
28. Peng, Y.; Yang, L.; Ju, X.; Liao, B.; Ye, K.; Li, L.; Cao, B.; Ni, Y. A comprehensive investigation on the thermal and toxic hazards of large format lithium-ion batteries with $LiFePO_4$ cathode. *J. Hazard. Mater.* **2020**, *381*, 120916. [CrossRef]
29. Kriston, A.; Kersys, A.; Antonelli, A.; Ripplinger, S.; Holmstrom, S.; Trischler, S.; Döring, H.; Pfrang, A. Initiation of thermal runaway in Lithium-ion cells by inductive heating. *J. Power Sources* **2020**, *454*, 227914. [CrossRef]
30. Ma, Q.; Liu, Q.; Zhang, H.; Tian, R.; Ye, J.; Yang, R. Experimental study of the mass burning rate in n-Heptane pool fire under dynamic pressure. *Appl. Therm. Eng.* **2017**, *113*, 1004–1010. [CrossRef]
31. Li, C.; Yang, R.; Yao, Y.; Tao, Z.; Liu, Q. Factors affecting the burning rate of pool fire in a depressurization aircraft cargo compartment. *Appl. Therm. Eng.* **2018**, *135*, 350–355. [CrossRef]
32. Chen, M.; Liu, J.; He, Y.; Yuen, R.; Wang, J. Study of the fire hazards of lithium-ion batteries at different pressures. *Appl. Therm. Eng.* **2017**, *125*, 1061–1074. [CrossRef]
33. Chen, M.; Ouyang, D.; Weng, J.; Liu, J.; Wang, J. Environmental pressure effects on thermal runaway and fire behaviors of lithium-ion battery with different cathodes and state of charge. *Process Saf. Environ. Prot.* **2019**, *130*, 250–256. [CrossRef]
34. Fu, Y.; Lu, S.; Shi, L.; Cheng, X.; Zhang, H. Ignition and combustion characteristics of lithium ion batteries under low atmospheric pressure. *Energy* **2018**, *161*, 38–45. [CrossRef]
35. Xie, S.; Ren, L.; Yang, X.; Wang, H.; Sun, Q.; Chen, X.; He, Y. Influence of cycling aging and ambient pressure on the thermal safety features of lithium-ion battery. *J. Power Sources* **2020**, *448*, 1181–1189. [CrossRef]
36. Xie, S.; Ren, L.; Gong, Y.; Li, M.; Chen, X. Effect of charging/discharging rate on the thermal runaway characteristics of lithium-ion batteries in low pressure. *J. Electrochem. Soc.* **2020**, *167*, 140503. [CrossRef]

37. Xie, S.; Gong, Y.; Li, G.; Ping, X. Effect of low-pressure environment on thermal runaway behavior of NCM523/graphite pouch cells with different overcharge cycles. *J. Energy Storage* **2022**, *55*, 105444. [CrossRef]
38. Sun, Q.; Liu, H.; Zhi, M.; Chen, X.; Lv, P.; He, Y. Thermal characteristics of thermal runaway for pouch lithium-ion battery with different state of charges under various ambient pressures. *J. Power Sources* **2022**, *527*, 231175. [CrossRef]
39. Sun, Q.; Liu, H.; Zhi, M.; Zhao, C.; Jia, J.; Lv, P.; Xie, S.; He, Y.; Chen, X. Lithium-ion battery thermal runaway propagation characteristics under 20 kPa with different airflow rates. *Fire Technol.* **2023**, *59*, 1157–1179. [CrossRef]
40. Pandian, G.; Pecht, M.; Zio, E.; Hodkiewicz, M. Data-driven reliability analysis of Boeing 787 Dreamliner. *Chin. J. Aeronaut.* **2020**, *33*, 1969–1979. [CrossRef]
41. Yayathi, S.; Walker, W.; Doughty, D.; Ardebili, H. Energy distributions exhibited during thermal runaway of commercial lithium ion batteries used for human spaceflight applications. *J. Power Sources* **2016**, *329*, 197–206. [CrossRef]
42. *ISO 9705:1993*; Fire Test-Full-Scale Room Fire Test for Surface Products. International Standards Organization: Geneva, Switzerland, 1993.
43. Chow, W.K.; Han, S.S. Heat release rate calculation in oxygen consumption calorimetry. *Appl. Therm. Eng.* **2011**, *31*, 304–310. [CrossRef]
44. Lyon, R.E.; Walters, R.N. Energetics of lithium ion battery failure. *J Hazard Mater* **2016**, *318*, 164–172. [CrossRef] [PubMed]
45. Fu, Y.; Lu, S.; Li, K.; Liu, C.; Cheng, X.; Zhang, H. An experimental study on burning behaviors of 18650 lithium ion batteries using a cone calorimeter. *J. Power Sources* **2015**, *273*, 216–222. [CrossRef]
46. Quintiere, J.G. More on methods to measure the energetics of lithium ion batteries in thermal runaway. *Fire Saf. J.* **2021**, *124*, 103382. [CrossRef]

batteries

MDPI

Review

Review of Flame Behavior and Its Suppression during Thermal Runaway in Lithium-Ion Batteries

Yikai Mao [1,2], Yin Chen [1,2,*] and Mingyi Chen [1,2,*]

[1] School of Emergency Management, Jiangsu University, Zhenjiang 212013, China; myk_0727@163.com
[2] School of the Environment and Safety Engineering, Jiangsu University, Zhenjiang 212013, China
* Correspondence: chenyin@ujs.edu.cn (Y.C.); chenmy@ujs.edu.cn (M.C.)

Abstract: Lithium-ion batteries (LIBs) are extensively utilized in electric vehicles (EVs), energy storage systems, and related fields due to their superior performance and high energy density. However, battery-related incidents, particularly fires, are increasingly common. This paper aims to first summarize the flame behavior of LIBs and then thoroughly examine the factors influencing this behavior. Based on these factors, methods for suppressing LIB flames are identified. The factors affecting flame behavior are categorized into two groups: internal and external. The paper then reviews the flame behavior within battery modules, particularly in confined spaces, from both experimental and simulation perspectives. Furthermore, methods for suppressing battery flames are classified into active and passive techniques, allowing for a more comprehensive analysis of their effectiveness. The paper concludes with a summary and outlook, offering new insights for future research and contributing to the development of safer and more efficient battery systems.

Keywords: lithium-ion battery; flame behavior; thermal runaway; suppression

1. Introduction

In 1976, Whittingham et al. [1] proposed a new battery operating principle, intercalation, which laid a solid foundation for the subsequent commercialization of lithium-ion batteries (LIBs). Nowadays, LIBs have been widely used in various fields, especially in electric vehicles (EVs), because of their long cycle life, high energy density, and high charging speed [2–4]. However, battery modules with higher energy densities tend to have lower thermal stability, which has led to a global rise in fire accidents in EVs, posing a serious threat to the personal safety of passengers [5,6]. Whereas most of the battery thermal runaway (TR) occur in confined structures, such as the top plate of an electric vehicle (EV) [7–9], this battery flame behavior, which occurs in confined spaces, tends to be more violent and has the potential to explode, leading to the structural failure of the roof, ultimately posing a significant hazard to the environment and nearby personnel [10]. Flames can affect the electrochemical processes within the cell by raising its temperature and ultimately exacerbate thermal runaway propagation (TRP), especially in confined space where this behavior is more pronounced [11]. Therefore, the study of the flame behavior of batteries and how to suppress the flame of batteries is particularly important.

There are many ways to trigger TR, including thermal abuse, mechanical abuse, and electrical abuse [12]. Many scholars have studied battery flame behavior under each of these three scenarios, including flame behavior under mechanical abuse triggered by penetration [13,14], flame behavior under thermal abuse in superheated conditions [15,16], and flame behavior under electrical abuse under overcharged conditions [17]. Regardless of the type of battery fire initiated by the triggering method, the factors affecting the behavior of these flames are similar, and they all exacerbate the TR behavior of the battery to varying degrees. Therefore, it is important to understand the factors affecting the flame behavior of batteries in order to better suppress flames during TR. These factors are categorized into internal and external factors. The internal factors include the battery state of charge (SOC) [18,19], battery healthiness (SOH), etc. [20]. External factors include ambient temperature [21], environmental pressure [22,23], ambient oxygen concentration, etc. [24,25].

According to the factors affecting the flame behavior of batteries, scholars have investigated methods of suppressing battery flames from different perspectives. These methods can be categorized into two directions at a large level: one is active suppression [26,27], and the other is passive suppression [28,29]. Active suppression can achieve the suppression effect more directly compared to passive suppression, but it cannot achieve suppression in the first moment and is more costly. Passive suppression is less costly, but the suppression is not as efficient as that of active cooling. There are considerable papers that mention active flame suppression versus passive flame suppression, but there is a limited overview of them. Therefore, this paper discusses them in detail in the Battery Flame Suppression Section to clarify their relationship more clearly.

The novelty of this work lies in the systematic summary of the behavior of battery flames and the mechanisms that influence flame behavior. Based on these mechanisms, the methods to suppress battery flames are logically summarized. A developmental perspective provides a good idea for future research on battery flame behavior and the suppression of battery flames. In the past, studies on TR process flames have focused on flame behavior. Currently, there are more and more studies on flame behavior affecting TR, both from experimental and simulation perspectives. There are many existing methods for TRP suppression, and a large portion of these studies start from suppressing the flames during TR. Based on practical factors, many scholars have started to study the characteristics of TR flame behavior in confined spaces at the experimental as well as simulation levels. However, there are limited reviews currently that discuss the behavioral mechanisms of battery flames during TR in correspondence with the methods for suppressing battery flames. Hence, this paper provides a comprehensive account of the behavior of flames during TR, as well as the methods to suppress this behavior and the existing challenges through a systematic review of the existing studies.

2. Thermal Safety of Lithium-Ion Batteries

2.1. *Flame Disaster of Lithium-Ion Batteries*

Whether it is thermal, mechanical, or electrical abuse, TR is triggered whenever a critical condition is reached that triggers TR in LIBs. Prior to TR in LIBs, batteries conform to the following three-dimensional self-ignition model [30]:

$$\rho C_{p,b}\frac{dT}{dt} = Q + \frac{\partial}{\partial x}\left(k_x\frac{\partial T}{\partial x}\right) + \frac{\partial}{\partial y}\left(k_y\frac{\partial T}{\partial y}\right) + \frac{\partial}{\partial z}\left(k_z\frac{\partial T}{\partial z}\right) \tag{1}$$

where ρ is the cell density, $C_{p,b}$ is the average heat capacity of the cell, T is the temperature of the cell, Q (W/m^3) is the heat generation term as reflected by the side reactions, and k_x, k_y, and k_z are the thermal conductivity of the cell in the x, y, and z axes. And once the battery temperature rises abnormally under abusive conditions, a chain of chemical reactions within it occurs [31], from the collapse of the SEI membrane to the reaction between the electrolyte and the anode; ultimately, the short-circuit inside the battery releases a large amount of heat and flammable gases, causing a flame disaster when the fire triangle is satisfied [12]. This potential flame hazard increases dramatically as the SOC of the battery increases [22]. This is attributed to the fact that the SOC greatly influences the Joule heat released at the moment of TR of the battery, which can be expressed by the following equation [32]:

$$Q_J = \int_0^{t_{end}} I_{ISC,t}^2 R_{ISC,t}dt = \int_0^{t_{end}} \frac{U_t^2}{R_{ISC,t}}dt \tag{2}$$

where U_t denotes the terminal voltage of the cell, $R_{ISC,t}$ denotes the ISC resistance, and t denotes the end time. The more heat and the more gases that are released from combustion, the greater the flame hazard that can be caused by LIBs. And for the existing popularized packaged battery structure in EVs, once TR occurs, the flame disaster caused by TR in a confined space may be even more serious [10]. Therefore, it is necessary to suppress the flame disaster of lithium batteries.

2.2. The Necessity to Suppress TR Flames

When searching for the keyword "battery fire accident" in Web of science and CNKI, we can find the trend of the number of items in the literature on this topic from 2013 to 2023, as shown in Figure 1. It can be observed that the research trends of Chinese scholars and international scholars are more or less the same. At the international level, the number of articles studying battery fire accidents has risen sharply since 2017. These indirectly reflect the frequent occurrence of battery fire safety problems in recent years. TR-induced battery fires are on a growing trend. This makes it necessary to study the flame behavior during TR in order to explore its mechanism and to suppress TR flames according to the mechanism of battery flames. Therefore, this paper explores the flame mechanism of TR in detail in Section 3 and summarizes the existing methods for suppressing TR flames in detail in Section 4.

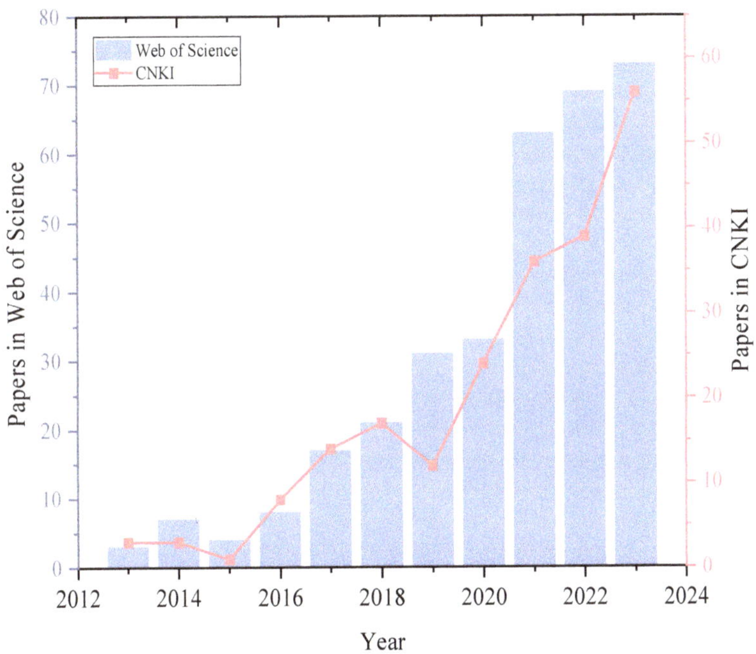

Figure 1. Trend chart of the number of related literature entries in the last ten years.

2.3. Literature Search

The literature search for this review paper followed the PRISMA guidelines and this was conducted to make the paper more scientifically credible. The detailed steps are shown in the Figure 2. The search for this paper was completed in August 2024. The following figure shows the workflow based on the PRISMA 2020 statement. In Web of science, the following phrases were used as a search term: lithium-ion battery, flame, and thermal runaway. Then, the language of the review was limited to English. The search for this review was restricted to 2018 and beyond. The types of articles for the review included research articles and review articles to simplify the review process. The articles were further screened based on their titles, abstracts, and related elements. Sixty-nine articles were finally selected for in-depth analysis.

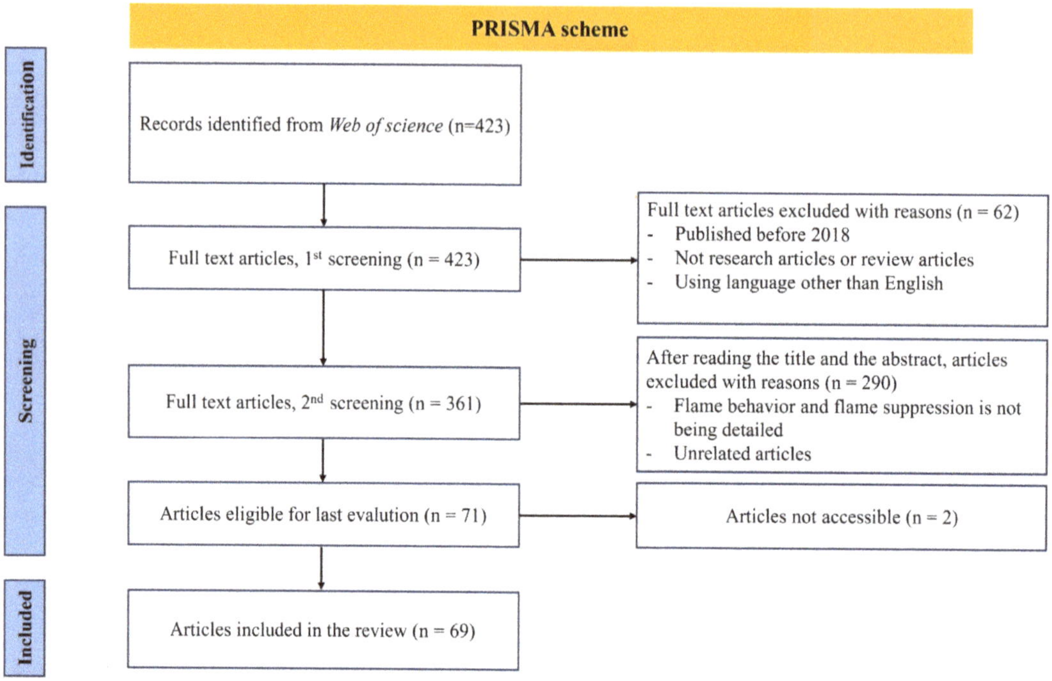

Figure 2. The PRISMA scheme for literature search.

3. Flame Behavior during TR

3.1. Flame Behavior and TR in Single Batteries

3.1.1. TR Progress

There are many scholars who have divided the combustion process of battery TR into several stages in the course of their research. According to Liu et al. [33], the burning behavior is divided into four stages: (I) ignition and venting, (II) steady combustion, (III) jet fire, and (IV) abatement. They found that, for 0% SOC cells, there were no jet flames, and the results are shown in Figure 3a. At stage I, under continuous heating, flammable substances collected in the combustion chamber and were ignited by an external igniter, forming a deflagration fireball. The behavior of fire at stage II was observed by a video camera; after stable combustion, a high-speed jet flame immediately appeared vertically. Subsequently, the flame decayed until extinguishment. Some investigations interpreted the fire behavior in a battery module as a three-stage process: (I) venting/igniting, (II) combustion, and (III) extinguishment [15,34,35]. On the one hand, unlike the above study, they combined steady combustion and jet flame into the combustion module, and, on the other hand, just like in the previous study, they mentioned the influence of the SOC and gas generated from the battery on the behavior of fire in the battery module. And these phenomena are shown in Figure 3b–d. They concluded that, as the SOC rises, the more violent the battery's flame behavior is.

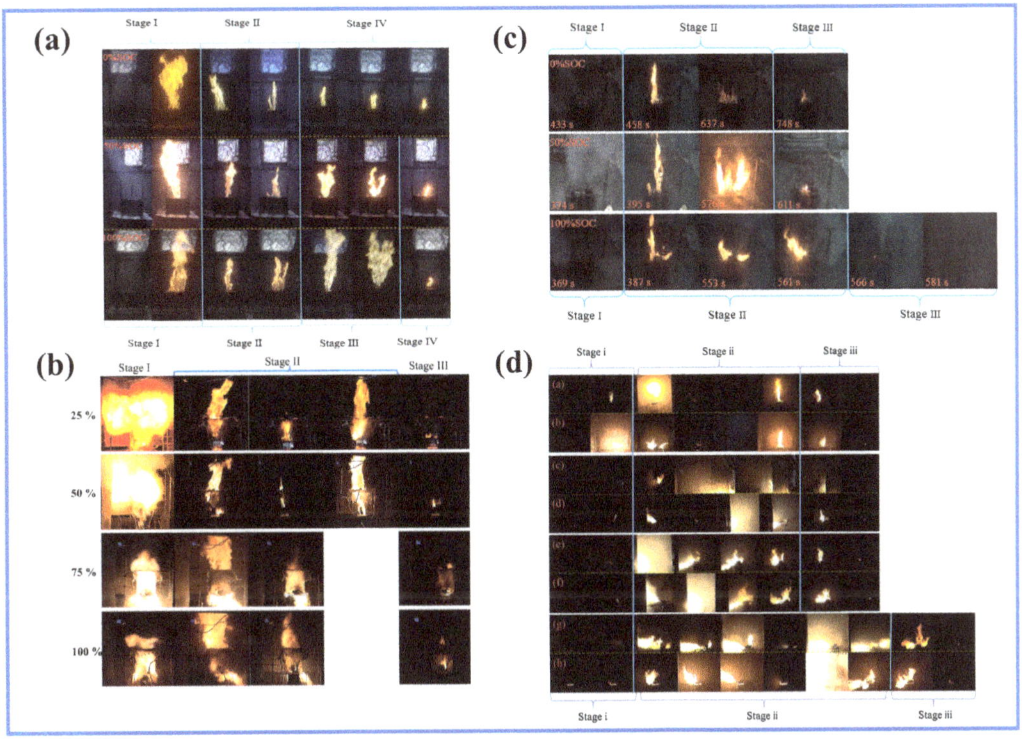

Figure 3. Diagram of phenomena at different stages of TR: (**a**) Burning behavior of LFP LIBs of different SOC [33], (**b**) Flame behavior of NCM pouch cells of different SOC [34], (**c**) Combustion behavior of LFP LIBs of different SOC [15], and (**d**) Combustion behavior of NCM prismatic cells under different heating methods [35].

3.1.2. Influence of Internal Factors

With the increase in the SOC, the mass loss ratio and highest temperature of lithium-ion cells are higher and the time when TR occurs decreased [36]. Wang et al. [34] studied the flame behavior of batteries under external ignition. The results revealed that the trigger time strongly declined with an increase in the SOC. However, there is not a monotonically increasing relationship between the total heat release and SOC. The residual mass of the cell decreases with the increase in the SOC, indicating a very violent ejection of materials, some of which are ignited extremely rapidly. Although the SOC is not relevant to the total heat release, it has an impact on the heat release rate (HRR). According to Chen et al. [22], the HRR is often used to quantify thermal hazards. It was discovered that the SOC of the cell has a significant influence on the peak HRR and the production of CO and HF increase with the increase in the SOC, indicating that the SOC plays an important role in causing thermal hazard. The main reason for this is that, on the one hand, the higher the SOC of the battery, the more reactive its positive and negative electrodes and the lower its thermal stability. Therefore, the lower the trigger temperature of the battery TR, the lower the safety. On the other hand, a high SOC battery releases more energy during TR, which is more destructive to the surrounding environment. Liu et al. [15] investigated TR and the fire behaviors of LIBs induced by overheating, and it was found that, compared to 0% SOC and 50% SOC cells, the jet flow was more violent in 100% SOC battery, which aggravates thermal hazard. According to Zou et al. [37], the thermal ejection duration and flame burning duration of a 38 Ah battery are affected by the incident heat flux and the severity of the TR, both of which increase with the increase in the SOC. Similarly, the ignition time and extinguishing time of a 78 Ah battery also decrease with the increase in the incident heat flux, while the duration of the sustained burning flame remains approximately constant.

Due to many of the gases generated by batteries being flammable, another key risk factor of LIB fire is flammable gas emissions, which are toxic [33]. As the SOC of the battery grows, the battery TR phenomenon advances and even becomes more intensive, and the types of exhaust components released increase, with unsaturated hydrocarbons being the majority of them [38]. As shown in Figure 4a, three types of batteries have been studied; the gases generated from them were H_2, CH_4, C_2H_4, and C_2H_6 [39]. Most components of the gases generated from batteries are flammable [40], which are beneficial to the behavior of jet fire, and the main components of thermally induced runaway gas in LIBs are not influenced by the SOC [41]. The details are shown in Figure 4b. Although there is little difference in the main components, the concentration of each gas increases significantly with the increase in the SOC, especially CO and H_2, which poses a potential risk of fire and explosion in the battery system [41]. Wang et al. [42] investigated the gas production behavior and flame behavior of 50% SOC and 100% SOC LIBs when TR occurred. It was found that, under the conditions of gas production and flame, the maximum temperature of the 100% SOC battery was much higher than that of the 50% SOC battery. What is more, as the gas was ignited, the accumulation of large quantities of gas will cause a risk of toxicity and explosion. The gas temperature, concentration, and typical components are shown in Figure 4c. It is apparent from the information supplied that the battery's SOC has a considerable influence on the generation of gases, and both flammable gases and the SOC are relevant to the behavior of flames.

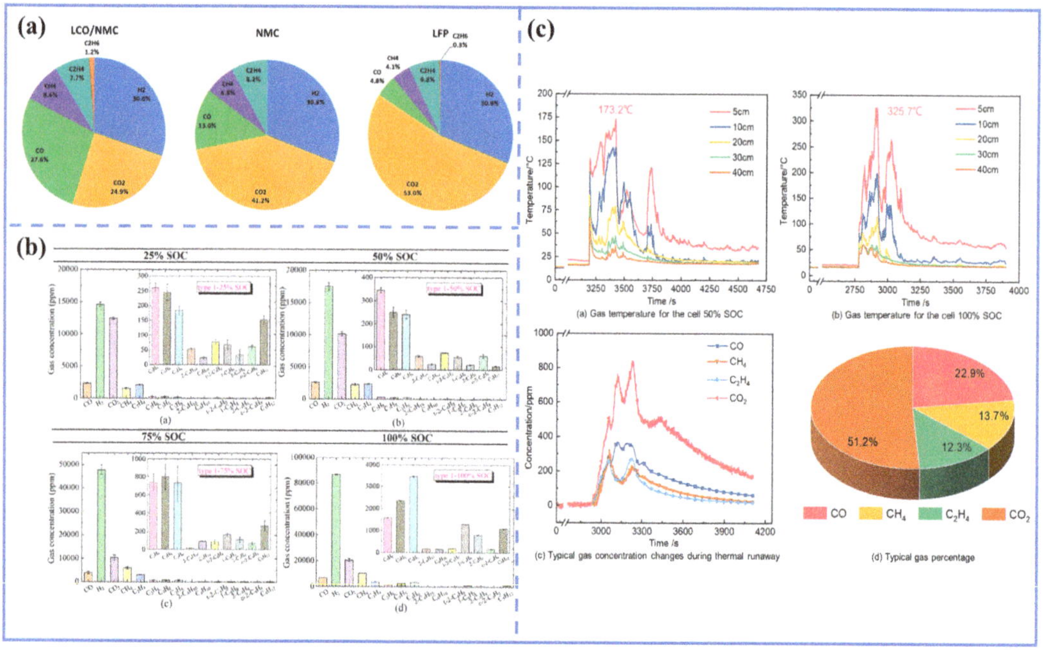

Figure 4. (**a**) Composition of gases from three different batteries [39]. (**b**) Main components of thermally induced runaway gas in LIBs with different SOCs [41]. (**c**) Gas temperature of different SOC LIBs and the concentration and typical components of a battery's gas [42].

In addition to the impact of the battery capacity on the gas produced during TR of the battery, the health of the battery also has an impact on it, which indirectly affects the flame behavior of the battery during TR. Some studies had shown that higher SOC and SOH cause more intensive reactions with a greater gas release and heat energy [43,44]. Liu et al. [20]

studied the factors that have an impact on LIBs in a confined space, and they found that the cumulative exothermic side reactions increased with the decrease in SOH, and that a lower SOH increased the risk of danger in LIBs to a certain extent, because batteries with lower SOH values were more prone to fire accidents. Wu et al. [45] explained the reason for this phenomenon in more detail, because at the same temperature, the self-heating rate of aging batteries is higher, which shows that the lithium deposited on the anode during low-temperature aging will greatly reduce the thermal stability of LIBs. The following three chemical Equations (3)–(5) [46] are the reaction of lithium metal with the electrolyte to produce gas, which corresponds to the peak of the reaction when the battery bursts. The difference in cell self-heating rate is most pronounced at this time for different battery health values, which explains why batteries with a low battery health exhibit lower rupture temperatures.

$$2Li + C_3H_4O_3(EC) \rightarrow LiCO_3 + C_2H_4 \tag{3}$$

$$2Li + C_4H_6O_3(PC) \rightarrow LiCO_3 + C_3H_6 \tag{4}$$

$$2Li + C_3H_6O_3(DMC) \rightarrow LiCO_3 + C_2H_6 \tag{5}$$

During the TR of a battery, the battery produces gases as well as some fumes, which are not only harmful to the human body [47], but also have an impact on the flame behavior of the battery [48,49]. It is worth noting that the influence of flue gas on the flame during the TR of the battery has not been paid attention to until recently; so, the current research in this aspect is relatively limited, especially to experimental study.

Wang et al. [50] studied the TR hazard of LIBs, and details are shown in Figure 5. The own hazardous factor of the battery (B1) is mainly concerned with its own dangerous state when the battery experiences TR. B2 included the factor of vented gases, which were characterized by their explosion hazard, toxicity, and pressure. B3 was jet fire and high temperature, and B4 consisted of ejected powder. According to Liu et al. [51], the results were the fact that the charging rate, ambient temperature, and battery aging have an influence on TR, and the conclusion of the orthogonal tests indicated that the rank of the influence on the thermal runaway risk was charging rate > ambient temperature > aging. It is apparent that, except for the charging rate and ambient temperature, all of the factors mentioned above are produced from an internal cause, which have an impact on TR and fire behaviors at some scale. As a matter of fact, there are some external factors of the environment that have an influence on flame behavior as well.

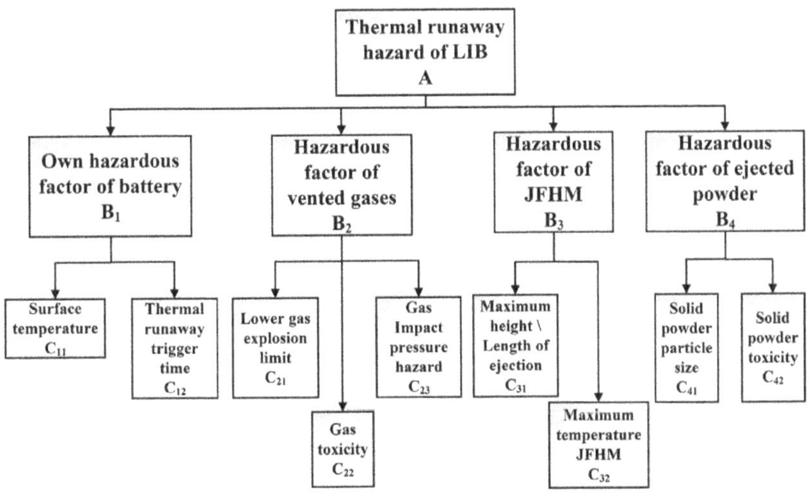

Figure 5. Hierarchical index system for the multi-parameter assessment of TR in lithium-ion batteries [50].

3.1.3. Impact of External Factors

Previous studies have shown that TR and the battery's flame behavior are related to the concentration of oxygen [52–54], and the flame is almost not able to exist under 12 precent oxygen concentration for common hydrogen fuel gases [55,56]. What is more, inert gas dilution has a profound effect on the flame propagation and extinction [57], the reason for which being that inert gases can lower the oxygen concentration to smother the flame [58]. Weng et al. [24] investigated the effects of the oxygen level and dilution gas on alleviating on battery TRP. By mainly measuring the temperature of jet fires of different heights and observing the flame duration, it is obvious that lowering the oxygen concentration can weaken the flame behavior when TR occurs, and details are shown in Figure 6a. Additionally, by comparing different inert gases, the results showed that both argon and nitrogen dilution can slow down TRP, one essential reason for which being that decreasing the oxygen concentration through extra inert gas dilution can reduce the value of the flame heating (q_f) and total heating (q_t), indicating that the fire behavior was also suppressed because of the inert gases. The above conclusions were also verified in an experiment that better reflects the characteristics of the battery module. According to Yan et al. [59], it has been found that a completely confined space suppresses the maximum temperature of the battery due to the insufficient combustion of the gases emitted from the battery. Therefore, reducing the amount of oxygen in the battery module may be beneficial in preventing TRP. The central reason for the diminished flame behavior of the battery described above is that the oxygen in the battery's burning fire triangle is weakened. There are current studies on how the oxygen concentration affects the flame behavior and how inert gases can be used to suppress TRP. In the future, methods for suppressing cell flames by attenuating the oxygen concentration need to be further refined. How to make the operation easier and to improve the environmental friendliness of the method are current issues that need to be addressed.

Except for the impact of the battery' SOC and ambient concentration of oxygen, the environment's pressure also has influence on the flame behavior of the battery [60]. Chen et al. [22] found that the ambient pressure has a significant effect on the heat of combustion, and the (total heat release) THR value at a high pressure is relatively larger than that at a low pressure. Also, the unit growth rate of the heat of combustion at both pressures increases with the increase in the SOC. The mechanism is depicted in Figure 6b, where a simplified diagram of the pressure by competition between the internal pressure and environmental pressure and the crack pressure is defined as: $P_{crack} = P_{in} - P_{out}$. What is more, the difference in the relief crack pressure also has an effect on the combustion behavior of LIBs. Wang et al. [61] found that, as the ambient pressure decreases, the TR trigger time becomes longer and the maximum surface temperature decreases. Ding et al. [23] investigated the environmental pressure effects on the TR properties of 21,700 LIBs with high energy density, and they found that the pressure had an impact on the height of the jet flame, which reached the maximum at 50 kPa. Nowadays, there are many studies on the effect of the pressure on the flame behavior of the battery TR, while studies on suppressing the flame behavior of the battery in this regard are limited. Some scholars have studied the effect of different safety valve types on battery disasters. And the application of a safety valve is closely related to the environmental pressure around the battery. Therefore, in the future, it is important to analyze the design of safety valves in combination with the environmental pressure where the battery is located. The study of a safer safety valve is very promising for the reduction in battery flame disasters.

According to Wang et al. [62], the combustion and explosion of outgassing in the event of a battery failure can be catastrophic for an electrochemical energy storage system. The fire hazard posed by outgassing from lithium iron phosphate batteries is greater than that from NCM batteries. The spray intensity at 30 kPa is significantly stronger than that at 95 kPa due to the higher pressure difference at which the combustible material is released [63]. Wang et al. [64] also studied the TR behavior of lithium-ion phosphate batteries. And they studied the effect of the charging rate on the battery exhaust behavior, which in turn affected the flame behavior of the battery during TR. They divided the TR behavior of lithium-ion phosphate batteries overcharged at different rates into the following stages: electrolyte vapor discharge, jet ignition, stable combustion, attenuated combustion, secondary injection, combustion and smoke, and gradual extinguishment. The results showed that the high-rate battery exhibited a more severe ignition and ejection than the low-rate one (0.5 C), not only evolving faster, but also appearing twice. At 0.5 C, the electrolyte vapor injection is more violent, which may be due to the prolonged accumulation of internal gases generated after a long period of charging. Some relevant details are shown in Figure 6c. Also, under charging conditions, Meng et al. [21] studied the effect of the ambient temperature on TR during the charging of LIBs, and the results showed that the battery exhibited greater thermal hazards at high charging rates and high ambient temperatures. Talele et al. [65] came to a similar conclusion through numerical simulations; the higher the temperature, the shorter the time the battery took from being heated to TR, as shown in Figure 6d.

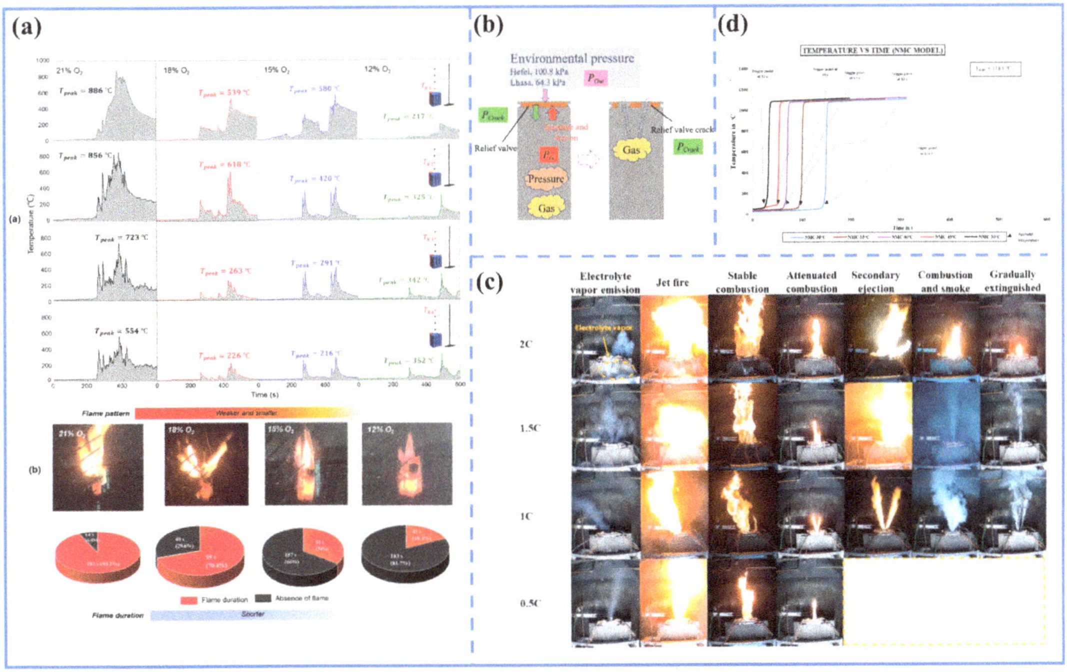

Figure 6. (**a**) Temperature of the battery module and flame pattern at different oxygen concentrations [24]. (**b**) Schematic representation of the effect of the ambient pressure on a battery [22]. (**c**) TR evolution under overcharge at different rates [64]. (**d**) Electrochemical model TR trigger point at different ambient temperatures [65].

To conclude, when TR occurs in batteries, the battery flame behavior caused by it is influenced by many elements, which consist of two large scales: external factors and internal factors, including battery health, state of charge, gas and particles, ambient temperature, charging rate, ambient concentration of oxygen, and environment pressure. It is obvious that, when a battery undergoes TR, the jet fire becomes a more and more important factor worthy of study, which can be influenced by plenty of factors. The jet flame is a special behavior when a battery TR occurs, and it causes great harm to surrounding objects [66–68]. And ambient batteries are inevitably influenced by it; as a consequence, the flame behavior in the battery module is even more worthy of study.

spread of heat to the surrounding cells, such as materials like aerogel felt. Suppressing the absorption of flame radiation with qualified radiation protection materials is also an idea to effectively suppress TRP. How these methods can be better synthesized and applied in practice in the future is something that requires further research.

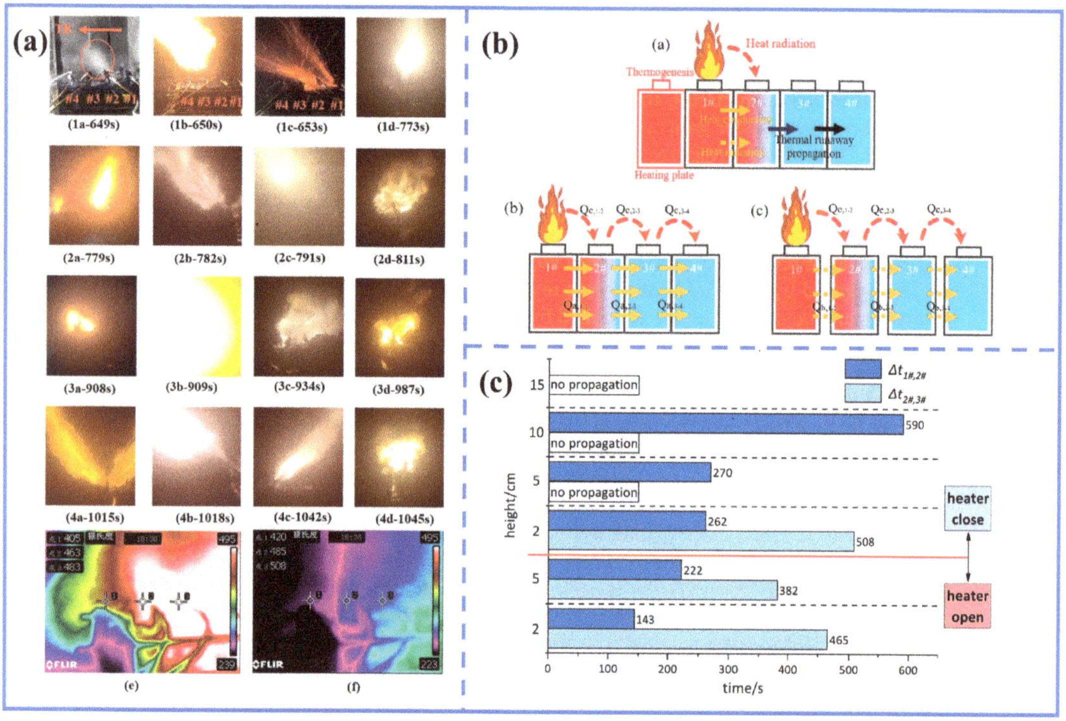

Figure 7. (**a**) The flame behavior of NCM battery and its thermal image during TRP [78]. (**b**) Two main ways of TRP path in NCM battey module: heat conduction and heat radiation [69]. (**c**) The TRP time between adjacent batteries for all experimental conditions [11].

3.4. Flame Behavior and TPR in Confined Space

3.4.1. Difference between CS and OS

The combustion performance of cells in a confined space is very different from that in open space [7,59], and so it does in the behavior of TRP. According to Wang et al. [81], when TR occurred in a confined space, combustible gases were ejected and then spread to the upper surfaces of neighboring cells, blocked by the ceiling. The spread of the ejected fire enlarged the heat transfer area and enhanced the radiative and convective heat flux between the cell and the fire. At the same time, the total amount of combustible gases was greater in a confined space than in open space. Therefore, it possesses larger potential deflagration risks, which will cause the speed of TR to be accelerated. Zhang et al. [82] investigated the effect of a plate obstacle on the flame behavior of 18,650 LIBs, and the phenomenon of the different performance of TR in a confined space and open space was described. Comparing the heat flux in a confined space (CS) and open space (OS), it was found that the radiation heat from the flame was affected by the plate obstacle with larger values, and the maximum heat flux was about four times that of OS. As shown in Figure 8b, the maximum length of the fireball before the formation of the jet flame at different plate heights is different from the case of OS, where the fireball length is larger for heights up to 10 cm and increases linearly with the decrease in height. This phenomenon may lead to more serious consequences of TR, such as a stronger thermal radiation to the surrounding area. What is more, as depicted in Figure 8a, with the decrease in the height of obstacle, the radiant heat flux rose more and more rapidly than that of OS, illustrating the difference between OS and CS. Liu et al. [83] compared the TRP interval, battery voltage, and the maximum temperature of the environment in a confined space and open space during the TRP period of batteries, and found that the confined space has a great influence on the ambient temperature, and the maximum temperature of environment in a confined space is much higher than that in open space, as shown in Figure 8c. Due to space constraints, the high ambient temperature and flame radiation accelerate the TRP between battery modules. This phenomenon can be more intuitively reflected in the energy storage station. For instance, Wang et al. [84] studied the propagation of the battery flame in the energy storage station in the half-reduced order model of the multi-scale simulation of battery fire propagation in the energy storage system. What they found was that the fire spread upward rapidly after the gases were released from the battery modules. The upward flame almost impacted all the upper battery module, which promoted the upward TRP by heating the packs' side surfaces through thermal convection and radiation.

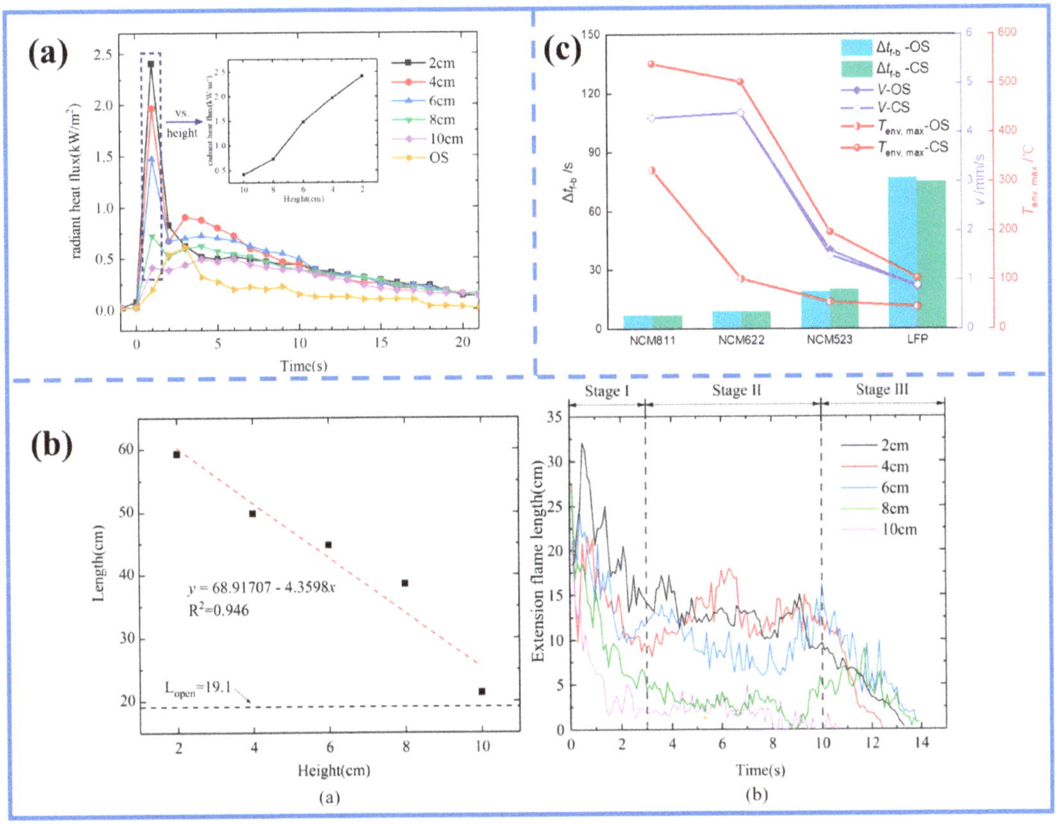

Figure 8. (**a**) Heat flux in a confined space at different heights and in an open space [82]. (**b**) The length of fire balls and extension flame at the plate at various plate heights for 18650 cylindrical NCA batteries [82]. (**c**) The maximum temperature and TRP time in a confined space and open space [83].

3.4.2. Flame Function

In a confined space, a large amount of heat is generated by the flame, which directly makes the TR behavior more severe, and the HRR is significant in quantitatively evaluating the flame development process and subsequent flame intensity [83]. The flame's behavior in a confined space has been widely studied in other fields. For instance, You et al. [85] conducted an experiment about the heat transfer characteristics of turbulent plumes and fire behavior on a horizontal ceiling, and it was found that, as the degree of confinement grew, the heat fluxes around the confined space tended to increase. Zhang et al. [86] used a three-regime lay to interpret the change in the maximum temperature at the impingement zone. What is more, the term "effective ceiling HRR" was defined to characterize the confined ceiling jet properties, by which the ceiling flame length was physically correlated [87]. According to Wang et al. [88], a systematic investigation on the vertical plate impinged by horizontal jet fires was conducted, and the flame extension area as well as the temperature profile in the thermal impinging flow were studied. Additionally, their corresponding correlations were proposed. All investigations mentioned above are important to understand the mechanism of flames on the TR behavior of batteries in a confined space. Zhang et al. [11] investigated the effect of flame heating on the TRP of LIBs in a confined space, as depicted in Table 2. The TRP of the battery module mainly contributed to the ceiling flame in a confined space and the influence of the flame increased as the height decreased, indicating the energy from the flame was the most dominant heat source to trigger the TRP in a confined space.

Table 2. The proportion of heat to trigger TR during combustion [11].

Heating Mode	Height (cm)	$Q_{transfer}$	Q_{tr}	Q_{flame}
Non-continuous heating	2	2.7%	22.1%	75.2%
	5	5.2%	20.2%	74.7%
	10	29.4%	17.1%	53.5%
	15	/	/	/
Continuous heating	2	3.7%	20.6%	75.7%
	5	16.5%	17.4%	66.1%

As shown in Figure 9a, compared to the situation of an open space, which did not have a ceiling plate, the smoke temperature in a confined space was higher, and with the growth in the SOC, the temperature increased [8]. When it comes to the angle of the plate in a confined space, the results indicated the TRP only happened at a lower ceiling angle and the horizontal ceiling structure was the most dangerous. And some details are shown in Figure 9b,c [89]. To conclude, in a confined space, the ceiling plate blocks the flame emitted from the combustion zone, which makes the flame play an important role in the TPR of batteries in a confined space. Obviously, in a confined space, the height of the plate is an essential factor that cannot be neglected; the height of the roof affects the tightness of the confined space to a certain extent and also affects the behavior of the flame. For instance, with the decline in the height of the plate, the flame heating power to batteries 2 and 3 grew. What is more, no matter whether the heater was opened or closed, the TRP time was shorter. Some relevant details are shown in Figure 9d,e [11]. In essence, it is because the radiant heat flux increases greatly as the height decreases [82].

At this stage, the study of flame behavior during the TR of batteries in a confined space is relatively rare, especially for fully enclosed battery spaces. The difficulty lies in the fact that the experimental process is more difficult to observe and the experiment is more difficult to operate. How to suppress TRP in a confined space is a focus of the research needed nowadays, and the current research on this is very limited. In the future, we can combine simulation and experiment more often to verify the experimental results while reducing the complexity of operation. Meanwhile, the rise in AI technology also provides a new direction for predicting the flame behavior during the TRP of batteries in a confined space in the future.

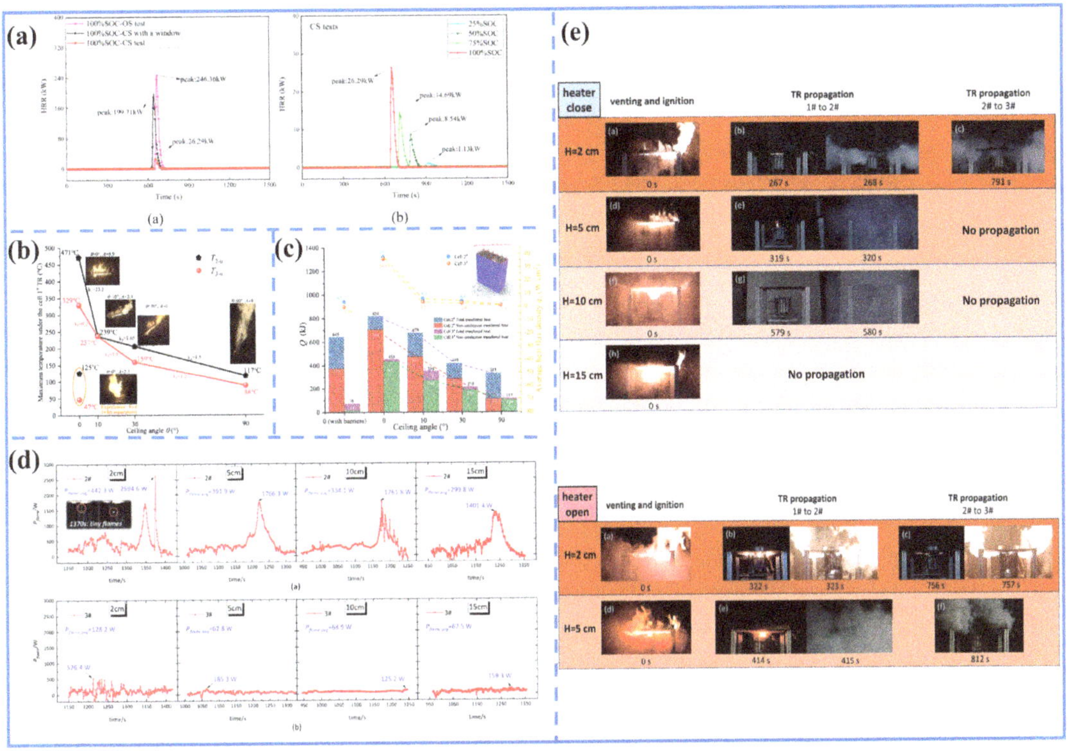

Figure 9. (**a**) HRR variations in cells under open space and confined space; HRR variations for cells of different SOC in confined space [8]. (**b**) Maximum temperature of the cells' upper surface for different ceiling angles [89]. (**c**) The average heat flux density in the z direction for different ceiling angles [89]. (**d**) Flame heating power to batteries at different plate heights in confined space [11]. (**e**) The evolution of the TR behavior for the battery module with non-continuous heating and continuous heating [11].

3.5. Relevant Flame Modeling

3.5.1. Development of Flame Modeling

A number of models of the TR behavior can be found in many papers. Firstly, Dahn et al. [90] formulated a mathematical model on the basis of the SEI layer and the decomposition rates of the anode. Later, Hatchard et al. [91] investigated the oven exposure testing of a standard benchmark and developed an one-dimensional predicting model for it. The main aim of this article was to predict the response of a new battery. Considering Hatchard did not illustrate abuse behaviors, Kim et al. [92] used Fluent to developed a three-dimensional model on the basis of Hatchard's model. All models introduced above have a lack of venting and electrolyte vaporization, as a consequence of which they obviously cannot predict the exact TR behavior. To improve the accuracy of the model, Coman et al. [93] built a lumped model with a series of ODES representing the decomposition rates, the energy balance, and the ideal gas flow equations. The model showed a good agreement when compared in terms of the temperature and vented gas tendencies. Almost at the same time, paying attention to the battery itself, Feng et al. [94] built a 3D TRP model to find solutions for the prevention of TRP and the simulation they used was the same as that of Comen. Not long after that, Coman et al. [95] improved their model to predict the temperature–pressure behavior and gas generation in 18,650 Graphite/LCO cells. In this study, they firstly illustrated the jet flow dynamics by isentropic flow equations. Bugryniec et al. [96] also used COMSOL to build an advanced abuse model. Due to its improved temperature predictions, it considerably enhanced the LIB TR field of study. Other scholars used computational fluid dynamics (CFD) techniques, which overcame the limitations of the lumped model to provide flow field evolution and distribution [97], and many people used it to establish fire modeling in the field of battery.

According to Kong et al. [98], they developed a numerical model, which combined the thermal decomposition reactions, pressure build-up and venting mechanisms, and combustion process. By coupling conjugate heat transfer with CFD to capture the cell temperature and internal pressure evolution under thermal abuse, they indicated that the radiative heat loss from solid particles and soot was neglected in their model, as a consequence of which it possibly caused 20–40 percent overprediction in jet flame temperatures. To predict the hazards to a wider environment induced by the heat and gas release of the battery flame, one model used the CFD simulation of a battery fire test, which was based on the three main heat-generating TR mechanisms [99]. However, the transport of the ejected particles was not considered in any of the above models, and few studies have considered the effects of solid particle radiation in a prediction model of TRP [49,97]. Based on the above-mentioned situation, some scholars have studied the influence of the particles generated during the TR of the battery on the TR behavior and the influence of the battery flame by establishing the corresponding model on the basis of previous studies.

To solve the problems above, Wang et al. [97] investigated how particles are generated and how particles and gases interact in their new model. They successfully simulated the particles' ejection process during TR; however, their model lacked a combustion process. Aiming to fill this gap, a model integrating the effects of the conduction, convection, and radiation of particles was established by Zhang et al. [49]. This multi-physics model based on ANSYS Fluent can be used to predict the TR process, which included many reactions. Not long after that, as the effect of solid particles on the jet flame generated during TR had not been thoroughly studied, Wang et al. [48] used the discrete particle model (DPM) to calculate particle–fluid coupling. And they demonstrated that the particles had a significant impact on the jet flame characteristics and dynamic behavior during TR with their model. The development of battery fire modeling is as shown in Table 3.

Table 3. The development of battery fire modeling.

Year	Aim	Methods	Simulations	Highlights	Ref.
1999	Calculate self-heating rate profiles	ARC	Mathematical model	A mathematical model was used to study the thermal stability	Dahn [90]
2001	Predict the response of a new battery	ARC Fortran	One-dimensional predictive model	A one-dimensional predictive model was established	Hatchard [91]
2007	Illustrate abuse behaviors	Fluent	Thermal–fluid dynamics	A three-dimensional thermal abuse model was established	Kim [92]
2016	Analyze thermal runaway during venting	Comsol	Mathematical and multi-physics model	Flow equations were used to better understand thermal runaway	Coman [93]
2016	Find solutions for preventing TPR	Comsol	Mathematical and multi-physics model	A 3D TPR model for a battery pack was built	Feng [94]
2017	Predict the temperature–pressure behavior in battery packs	Comsol	Mathematical and multi-physics model	The pressure inside the battery was predicted	Coman [95]
2020	Investigate cell pressurization and abuse reaction of LIBS	Comsol	AAM (advanced abuse model)	Capable of predicting pressure accumulation	Bugryniec [96]
2022	Predict the thermal abuse reactions and jet dynamics	CFD	A coupled numerical model	Combined many sub-models that are necessary for prediction	Kong [98]
2022	Predict the hazards induced by heat and gas release of battery fire	CFD	Thermal–fluid dynamics	It can determine the release of heat and gases from the battery	Voigt [99]
2022	Simulate the particle ejection process	CFD	Multi-scale and multiphase modeling framework	The complete evolution of the ejected particles was simulated	Wang [97]
2023	Analyze the effect of particles' radiation	CFD	A coupled numerical model	Combined the effect of particle radiation into TR innovatively	Zhang [49]
2023	Analyze the influence of particles on the jet flame	CFD	A coupled numerical model	The connection of the particles and characteristics of the jet flame was revealed	Wang [48]

It is not difficult to see from the above that, with the passage of time, the model was developed from the simplest one-dimensional model to the current multi-scale and multi-stage modeling framework. And the goals achieved by the model are becoming more and more accurate, and the tools are gradually refined. In the field of fluids, the CFD module in COMSOL is very popular. Especially in the study of battery flame behavior, CFD is the most used, because CFD is especially suitable for the fluid field.

3.5.2. Flame Modeling in a Confined Space

At this stage, the simulation of batteries in a confined space is still relatively scarce. In the research of battery TR simulation, initially, people only focused on the TR itself inside the battery. Subsequently, the gas emission during the TR of the battery and the influence of the subsequent combustion behavior on the TR behavior were considered. With the refinement of the model, many scholars modeled the jet flame and combustion behavior during the TR of the battery. Finally, the research object was expanded from the TR of a single battery to TRP of a battery module.

Kim et al. [80] developed an exhaust model, and to improve the accuracy of the exhaust gas combustion heat, they considered the ratio of hydrogen and volatile organic compounds in terms of the battery temperature. This model was built in a confined space and could predict the flame combustion heat release rate of the battery, but it lacked the simulation of the flame morphology and the interaction between the flame and the confined space. Wang et al. [81] further studied the TRP behavior of batteries under the influence of ceiling jet fires on the basis of their previous flame model. The model was based on the conjugate propagation modeling framework to explore the interaction between the jet flame and TRP behavior, highlighting the influence of the jet flame on the TRP behavior. The idea of the full text is shown in Figure 10.

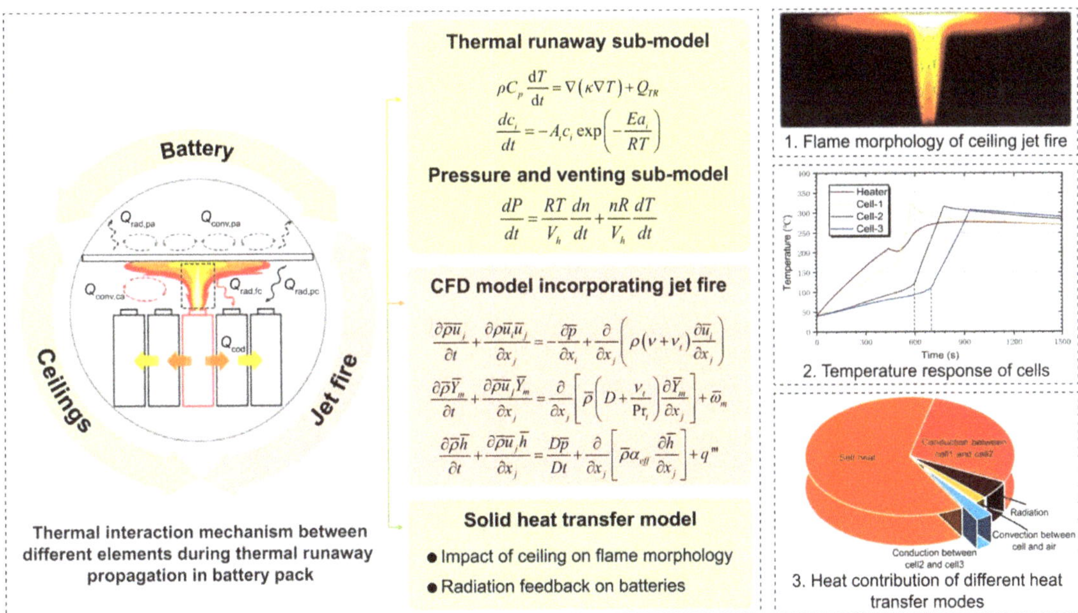

Figure 10. Modeling of flames in a confined space and its mechanism [81].

To sum up, at this stage, the simulation in a confined space is very limited, and some factors have not been fully considered, such as the lack of the radiation effect of particles in some models, and TR in a confined space triggered by mechanical abuse is not yet available, which are the directions of future research.

4. Methods of Suppressing Lithium-Ion Battery Flames

In recent years, the frequency of fire accidents in LIBs has been high, and it has become important for the safety management of batteries in electric vehicles. Whether in energy storage or in electric vehicles, the safety of batteries is no longer limited to individual batteries, but rather to the TRP between the battery module, which can exacerbate the fire hazard of batteries [12,100]. This section categorizes the existing methods of flame suppression for LIBs into active and passive methods of suppression, highlighting their superiority and limitations. Finally, a comprehensive discussion is conducted to provide certain directions for future research.

4.1. Active Suppression

For a flame to continue to burn, the "fire triangle" consists of fuel, oxygen, and heat. In the case of a battery, its safety characteristics can also be characterized by the "fire triangle". When one or more elements of the "fire triangle" are removed, battery flames can be extinguished and even prevented [101]. Based on current articles on active battery flame suppression, they can all be analyzed from a "fire triangle" perspective.

Perfluorohexanone ($C_6F_{12}O$) is a liquid that evaporates readily at room temperature and pressure [102]. It has been studied and applied by a wide range of scholars in the fire suppression of LIBs. Wang et al. [103] applied perfluorohexanone to suppress a battery fire during TR, and they effectively prolonged the TRP interval. Zhang et al. [104] investigated the suppression effects of substances such as perfluorohexanone and CO_2 on LIBs fires. They found that perfluorohexanone could quickly extinguish battery flames, while CO_2 took longer. Perfluorohexanone has a good extinguishing efficiency direct rapid cooling so that the battery cannot reach the point of ignition and played a certain role in the isolation of combustion aids. However, its operation is not very convenient, especially for the battery flame in a confined space.

Some scholars have also approached the issue from the perspective of reducing the oxygen concentration in order to inhibit the combustion of LIBs. Inert gas dilution reduces the temperature of the flame while lowering the concentration of combustion aids, and this approach is considered a promising way to suppress battery fires [24,105,106]. Previously, Lin et al. [107] and Golubkov et al. [39] investigated the suppression effect of CO_2 and N_2 on the combustion explosion of gases such as CH_4. They both found that both gases play a certain role in the suppression of combustion and explosion. The main components of the combustion gas generated during the battery TR are CH_4, H_2, and other gases. Zhou et al. [29] investigated the suppression effects of CO_2 and N_2 on battery combustion and explosion during TR. They concluded that both CO_2 and N_2 had inhibitory effects on the battery flame and explosion, and CO_2 had better effects than N_2. Weng et al. [24] investigated the effects of inert gas and oxygen concentration on the flame behavior and TRP of LIBs. They concluded that both inert gas addition and oxygen concentration reduction can suppress the battery combustion and TRP. In general, gases such as CO_2 and N_2 can be used to reduce the concentration of combustion aids by diluting oxygen and to achieve a certain cooling effect. And its operation is easier than perfluorohexanone, but the cost is relatively higher. And the fire extinguishing efficiency is not as good as that of perfluorohexanone.

In recent years, there has been an increasing number of studies utilizing fine water mist (WM) and wind methods to suppress battery fires. Miao et al. [108] investigated the suppression effect of WM on the battery flame at different stages of TR. The results showed that releasing the WM at the moment of safety valve cracking could achieve a fire extinguishing effect within a few seconds, whereas by releasing WM after the battery TR, the extinguishing time was longer and took about three minutes. Hong et al. [109] studied the suppression effect of flood cooling on fire and TRP for different types of batteries. They found that flood cooling is effective in suppressing both TRP and fire for LMO and LFP batteries, but less effective for NCM batteries. On the other hand, for the case of a confined space, Wang et al. [110] successfully suppressed the TR of the cell in a confined channel with the flame of the cell using longitudinal winds, which acted as an effective cooling agent. Jiang et al. [111] utilized WM and wind for coupling to suppress fire in a confined space, the battery enclosure. The results showed that a 4.5 m/s wind and 0.5 MPa WM had the best suppression effect, which can achieve a high heat dissipation capacity and reduce the concentration of toxic and hazardous gases. In summary, wind and WM suppress battery fires mainly by keeping the battery from reaching the ignition point. It is better for LIB fire suppression in a confined space, especially in tunnels. However, it is more demanding on the operator and has many subjective uncertainties.

By analyzing previous studies on active flame suppression, it is obvious that they all start from the fire triangle and the factors affecting the flame behavior of the battery as a way to achieve the purpose of suppressing the flame during the suppression of the battery TR. The specific battery fire triangle mechanism and the main existing active methods for suppressing the battery TR flame are shown in Figure 11. Whether it is perfluorohexanone, inert gas, or WM and wind, they each have advantages and disadvantages. They also share the common advantage of active suppression, which is the relatively high efficiency of cooling.

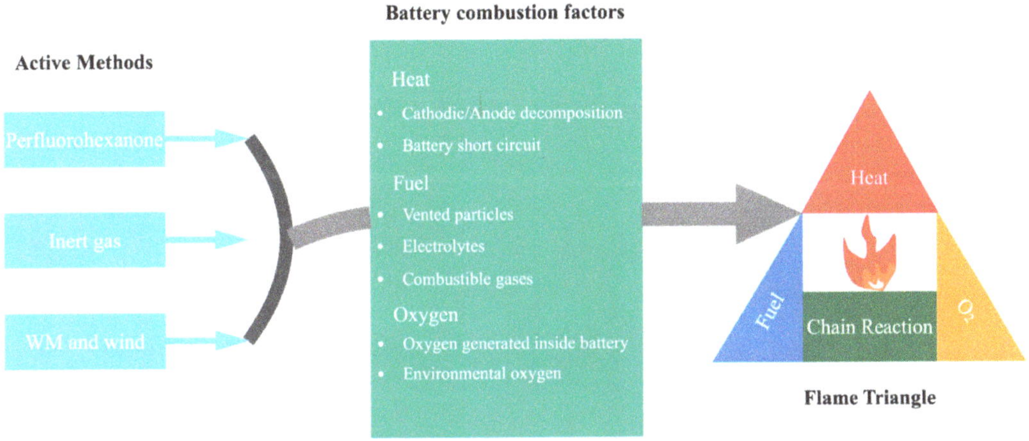

Figure 11. The specific battery fire triangle mechanism and the main existing active methods for suppressing the battery TR flame.

4.2. Passive Suppression

The opposite of the active suppression of a battery flame is the passive suppression of a battery flame. A number of studies have shown that the temperature of a battery without an effective battery thermal management system (BTMS) can rise dramatically under extreme conditions, leading to the risk of fire and explosion [112–114]. This highlights the importance of effective BTMS in the passive suppression of battery flames. Also, the intrinsic safety design of the battery is an important component of the passive suppression of a battery flame. This intrinsic safety design includes flame-retardant electrolytes [115], improved electrode materials [116], and ceramic separators [117]. Here, we can refer to the internal safety design of the battery as internal passive suppression (IPS) and the BTMS of the battery as external passive suppression (EPS).

Relevant researchers have conducted considerable research on the internal safety of batteries, especially for the design of fire protection aspects. Liu et al. [118] designed a non-flammable electrolyte by introducing low-density and low-cost fluorobenzene (FB) as a co-solvent and bridging solvent in a diluted high-Concentration electrolyte (DHCE) system through the introduction of low-density and low-cost fluoro-benzene. Shao et al. [119] designed and fabricated thermally polymerized composite gel polymer electrolytes (CPEs) for quasi-solid-state batteries. They verified that CPEs exhibited a better flame retardancy and excellent thermal stability compared to commercial PP separators (Figure 12a). Lv et al. [120] started to study IPS by improving the electrode materials for batteries.

They designed and synthesized self-extinguishing phenothiazine-based polymers using 4-bromobenzene as a flame-retardant device and 1,4-divinylbenzene as a cross-linking agent. The results showed that this designed polymer significantly shortened the self-extinguishing time without deteriorating its intrinsic thermodynamic and electrochemical properties (Figure 12b). Huang et al. [121] combined flame-retardant electrolyte additives with commercially available PE separators to improve the flame resistance of battery ceramic separators. The results showed that the improved separator helped to contribute to the safety and flame retardancy of the battery at high temperatures (Figure 12c).

On the other hand, compared to IPS, EPS improves more with the phase change material (PCM). In order to ensure a good heat dissipation at the same time, it has a certain degree of flame retardancy, which is used to suppress the battery flame [122,123]. Weng et al. [124] added the flame-retardant Al(OH)3 to PCM to improve the flame retardancy of PCM. The results showed that the flame-retardant additive played a positive role in delaying the ignition time and reducing the HRR as well as toxic gases (Figure 12d). Chen et al. [125] developed a flame-retardant phase-change material (FRPCM) and combined it with an aerogel felt. They found that the combined use of FRPCM and the aerogel felt effectively blocked TRP and suppressed the battery flame to a certain extent. Some scholars have also utilized flame retardants directly for EPS. Mei et al. [126] utilized TRRH (phenoxycyclic phospholipids), TRRM (magnesium hydroxide), and TRRS (sodium acetate trihydrate) to suppress the battery TR. They found that TRRH (phenoxycyclic phospholipids), which has a free-trapping effect, suppresses the battery flame best. The TRRS with a high heat-absorbing capacity could better control the battery surface temperature (Figure 12e).

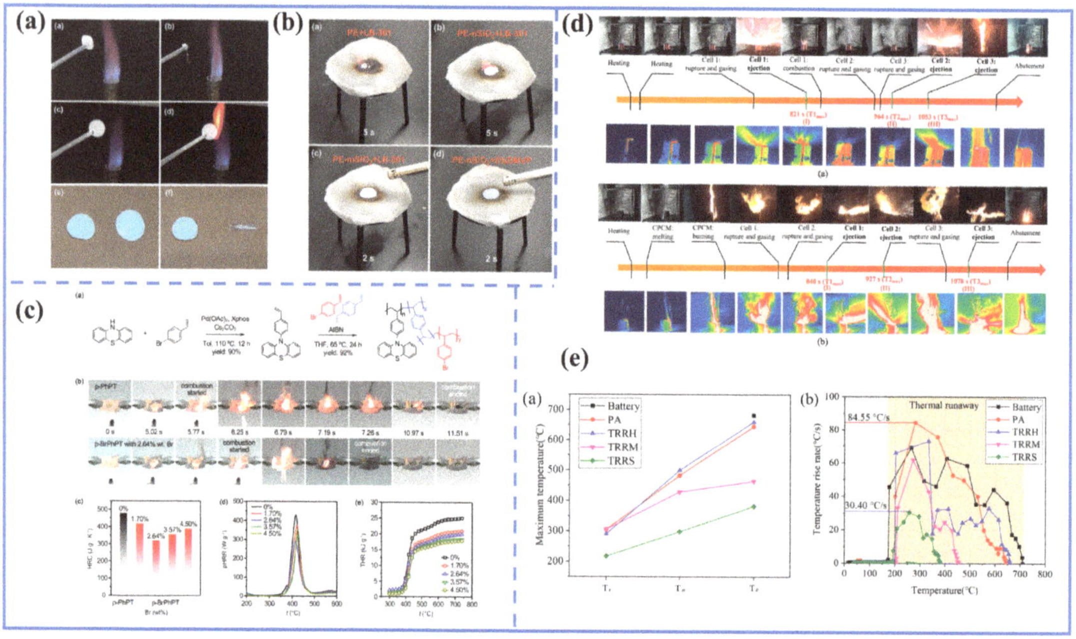

Figure 12. (**a**) Comparison of combustion experiments before and after electrolyte material modification [119]. (**b**) Comparison of combustion experiments before and after the improvement in the electrode material [120]. (**c**) Improved ceramic separator flame-retardant test [121]. (**d**) Photos and infrared images of the typical moments in the heating tests: cuboid module and tubular module [124]. (**e**) The maximum temperature of the shell, the suppression material, and the battery under different suppression conditions; The rate of temperature rise rate during TR of the battery under different suppression materials [126].

The above is a summary of passive battery flame suppression methods. For the passive suppression of battery flame, some methods start from the safety of the inside of the battery, while others start from the safety of the outside of the battery. What they have in common is that they can suppress the battery flame when the battery fire first occurs, without subjective human factors. Compared to active suppression, their disadvantage is also obvious, that is, the efficiency of cooling and flame suppression is not as high as active suppression. Therefore, in the future, the suppression of the battery flame should be strengthened in two aspects. On the one hand, the operability of active suppression should be strengthened, and on the other hand, the efficiency of passive suppression should be strengthened. Through in-depth development and research in these areas, the safety of batteries will be elevated to a new dimension.

5. Conclusions and Outlook

5.1. Conclusions

This work provides an in-depth discussion summarizing the flame behavior mechanism during the TR of LIBs and a consolidation of the current state of the methods to suppress the flame behavior of LIBs. The following conclusions are drawn:

(1) The flame intensity of LIBs during TR is influenced by external factors, such as ambient pressure and temperature, as well as internal factors like battery health and the state of charge.

(2) In a battery module, the heat contribution from the flame primarily arises from its radiant heat. The flame accelerates the TRP within the battery module, particularly in confined spaces where the heat radiation from the flame is further amplified. Currently, CFD models are the primary tools used for simulating battery flames.

(3) LIB flame suppression methods are broadly categorized into active and passive approaches. Both methods can be analyzed based on the internal and external factors affecting the battery flame and operate from the perspective of the flame triangle. Active suppression is more efficient than passive suppression but offers less intrinsic safety. A balanced approach combining both methods could enhance battery safety in the future.

5.2. Outlook

Through a comprehensive analysis of the relevant literature, the limitations were identified, and the following recommendations were made:

(1) Research on the flame behavior of batteries and the exploration of their intrinsic mechanisms is very necessary. The current research on the flame behavior in a confined space is relatively limited. On the one hand, the TRP and flame characteristics of a fully enclosed confined space in practical applications cannot be fully studied due to its spatial structure. On the other hand, for flame simulation in a confined space, multiple cell types must be considered and multiple physical fields must be coupled in order to come closer to the actual situation. Therefore, in the future, combining experiments and simulations to comprehensively verify the behavior of battery flames will be a worthwhile scheme to try.

(2) The current methods for suppressing the battery flame behavior can be divided into two main categories: active and passive suppression. At the level of active flame suppression, how to minimize human error is a challenging issue. For passive flame suppression, the development of green, energy-saving, and highly efficient safety systems will be the trend of the future, whether we start from the inside of the battery or from the environment in which the battery is located.

Further research in the field of battery flames is essential to improve the efficiency and safety of the battery module.

Author Contributions: Author Contributions: Conceptualization, M.C.; Investigation, Y.C. and Y.M.; Writing—original draft preparation, Y.M.; Writing—review and editing, Y.C. and M.C.; Supervision, Y.C. and M.C.; Funding acquisition, M.C. All authors have read and agreed to the published version of the manuscript.

Funding: This research was funded by the National Natural Science Foundation of China (52204213).

Conflicts of Interest: The authors declare that they have no known competing financial interests or personal relationships that could have appeared to influence the work reported in this paper.

References

1. Whittingham, M.S. Electrical Energy Storage and Intercalation Chemistry. *Science* **1976**, *192*, 1126–1127. [CrossRef] [PubMed]
2. Wang, F.; Wu, X.; Li, C.; Zhu, Y.; Fu, L.; Wu, Y.; Liu, X. Nanostructured positive electrode materials for post-lithium ion batteries. *Energy Environ. Sci.* **2016**, *9*, 3570–3611. [CrossRef]
3. May, G.J.; Davidson, A.; Monahov, B. Lead batteries for utility energy storage: A review. *J. Energy Storage* **2018**, *15*, 145–157. [CrossRef]
4. Zeng, X.; Li, M.; Abd El-Hady, D.; Alshitari, W.; Al-Bogami, A.S.; Lu, J.; Amine, K. Commercialization of Lithium Battery Technologies for Electric Vehicles. *Adv. Energy Mater.* **2019**, *9*, 1900161. [CrossRef]
5. Noh, H.-J.; Youn, S.; Yoon, C.S.; Sun, Y.-K. Comparison of the structural and electrochemical properties of layered Li[NixCoyMnz]O_2 (x = 1/3, 0.5, 0.6, 0.7, 0.8 and 0.85) cathode material for lithium-ion batteries. *J. Power Sources* **2013**, *233*, 121–130. [CrossRef]
6. Wang, Q.; Mao, B.; Stoliarov, S.I.; Sun, J. A review of lithium ion battery failure mechanisms and fire prevention strategies. *Prog. Energy Combust. Sci.* **2019**, *73*, 95–131. [CrossRef]
7. Mao, B.; Chen, H.; Jiang, L.; Zhao, C.; Sun, J.; Wang, Q. Refined study on lithium ion battery combustion in open space and a combustion chamber. *Process Saf. Environ. Prot.* **2020**, *139*, 133–146. [CrossRef]
8. Liu, P.; Sun, H.; Qiao, Y.; Sun, S.; Wang, C.; Jin, K.; Mao, B.; Wang, Q. Experimental study on the thermal runaway and fire behavior of LiNi0.8Co0.1Mn0.1O_2 battery in open and confined spaces. *Process Saf. Environ. Prot.* **2022**, *158*, 711–726. [CrossRef]
9. Qin, P.; Jia, Z.; Wu, J.; Jin, K.; Duan, Q.; Jiang, L.; Sun, J.; Ding, J.; Shi, C.; Wang, Q. The thermal runaway analysis on LiFePO4 electrical energy storage packs with different venting areas and void volumes. *Appl. Energy* **2022**, *313*, 118767. [CrossRef]
10. Li, J.; Gao, P.; Tong, B.; Cheng, Z.; Cao, M.; Mei, W.; Wang, Q.; Sun, J.; Qin, P. Revealing the mechanism of pack ceiling failure induced by thermal runaway in NCM batteries: A coupled multiphase fluid-structure interaction model for electric vehicles. *eTransportation* **2024**, *20*, 100335. [CrossRef]
11. Zhang, Y.; Zhao, H.; Wang, G.; Gao, X.; Ping, P.; Kong, D.; Yin, X. Effect of flame heating on thermal runaway propagation of lithium-ion batteries in confined space. *J. Energy Storage* **2024**, *78*, 110052. [CrossRef]
12. Feng, X.; Ren, D.; He, X.; Ouyang, M. Mitigating Thermal Runaway of Lithium-Ion Batteries. *Joule* **2020**, *4*, 743–770. [CrossRef]
13. Wang, H.; Yang, Z.; Jiang, C.; Ji, Z.; Zhu, Z. Internal temperature and flame jet characteristics during thermal runaway triggered by nail penetration for NCM811 lithium-ion battery. *J. Therm. Anal. Calorim.* **2022**, *147*, 14925–14938. [CrossRef]
14. Mao, B.; Chen, H.; Cui, Z.; Wu, T.; Wang, Q. Failure mechanism of the lithium ion battery during nail penetration. *Int. J. Heat Mass Transf.* **2018**, *122*, 1103–1115. [CrossRef]
15. Liu, P.; Liu, C.; Yang, K.; Zhang, M.; Gao, F.; Mao, B.; Li, H.; Duan, Q.; Wang, Q. Thermal runaway and fire behaviors of lithium iron phosphate battery induced by over heating. *J. Energy Storage* **2020**, *31*, 101714. [CrossRef]
16. Zou, K.; Li, Q.; Lu, S. An experimental study on thermal runaway and fire behavior of large-format LiNi0.8Co0.1Mn0.1O_2 pouch power cell. *J. Energy Storage* **2022**, *49*, 104138. [CrossRef]
17. Liu, P.; Li, S.; Jin, K.; Fu, W.; Wang, C.; Jia, Z.; Jiang, L.; Wang, Q. Thermal Runaway and Fire Behaviors of Lithium Iron Phosphate Battery Induced by Overheating and Overcharging. *Fire Technol.* **2023**, *59*, 1051–1072. [CrossRef]
18. Kim, S.W.; Park, S.G.; Lee, E.J. Assessment of the explosion risk during lithium-ion battery fires. *J. Loss Prev. Process Ind.* **2022**, *80*, 104851. [CrossRef]
19. Tao, C.; Ye, Q.; Wang, C.; Qian, Y.; Wang, C.; Zhou, T.; Tang, Z. An experimental investigation on the burning behaviors of lithium ion batteries after different immersion times. *J. Clean. Prod.* **2020**, *242*, 118539. [CrossRef]
20. Liu, J.; Zhang, Y.; Zhou, L.; Han, C.; He, T.; Wang, Z. Influencing factors of lithium-ion battery thermal runaway in confined space. *J. Energy Storage* **2023**, *73*, 109125. [CrossRef]
21. Meng, D.; Wang, X.; Chen, M.; Wang, J. Effects of environmental temperature on the thermal runaway of lithium-ion batteries during charging process. *J. Loss Prev. Process Ind.* **2023**, *83*, 105084. [CrossRef]
22. Chen, M.; Ouyang, D.; Weng, J.; Liu, J.; Wang, J. Environmental pressure effects on thermal runaway and fire behaviors of lithium-ion battery with different cathodes and state of charge. *Process Saf. Environ. Prot.* **2019**, *130*, 250–256. [CrossRef]
23. Ding, C.; Zhu, N.; Yu, J.; Li, Y.; Sun, X.; Liu, C.; Huang, Q.; Wang, J. Experimental investigation of environmental pressure effects on thermal runaway properties of 21700 lithium-ion batteries with high energy density. *Case Stud. Therm. Eng.* **2022**, *38*, 102349. [CrossRef]
24. Weng, J.; Ouyang, D.; Liu, Y.; Chen, M.; Li, Y.; Huang, X.; Wang, J. Alleviation on battery thermal runaway propagation: Effects of oxygen level and dilution gas. *J. Power Sources* **2021**, *509*, 230340. [CrossRef]
25. Tao, C.; Zhu, Y.; Liu, Z.; Li, R.; Chen, Z.; Gong, L.; Liu, J. The experimental investigation of thermal runaway characteristics of lithium battery under different nitrogen concentrations. *J. Therm. Anal. Calorim.* **2023**, *148*, 12097–12107. [CrossRef]
26. Ün, Ç.; Aydin, K.; Garo, J.P.; Coudour, B.; Gentilleau, M. Experimental study of fire suppression for li-ion electric batteries with H_2O. *Fresenius Environ. Bull.* **2021**, *30*, 790–801.

27. Meng, X.; Li, S.; Fu, W.; Chen, Y.; Duan, Q.; Wang, Q. Experimental study of intermittent spray cooling on suppression for lithium iron phosphate battery fires. *eTransportation* **2022**, *11*, 100142. [CrossRef]
28. Huang, Q.; Li, X.; Zhang, G.; Weng, J.; Wang, Y.; Deng, J. Innovative thermal management and thermal runaway suppression for battery module with flame retardant flexible composite phase change material. *J. Clean. Prod.* **2022**, *330*, 129718. [CrossRef]
29. Zhou, W.; Zhao, H.; Wu, D.; Yang, Y.; Yuan, C.; Li, G. Inhibition effect of inert gas jet on gas and hybrid explosions caused by thermal runaway of lithium-ion battery. *J. Loss Prev. Process Ind.* **2024**, *90*, 105336. [CrossRef]
30. Zhao, L.; Li, W.; Luo, W.; Zheng, M.; Chen, M. Numerical study of critical conditions for thermal runaway of lithium-ion battery pack during storage. *J. Energy Storage* **2024**, *84*, 110901. [CrossRef]
31. Feng, X.; Ouyang, M.; Liu, X.; Lu, L.; Xia, Y.; He, X. Thermal runaway mechanism of lithium ion battery for electric vehicles: A review. *Energy Storage Mater.* **2018**, *10*, 246–267. [CrossRef]
32. Li, Y.; Zhang, N.; Jiang, L.; Wei, Z.; Zhang, Y.; Yu, Y.; Song, L.; Wang, L.; Duan, Q.; Sun, J.; et al. Assessment of the complete chain evolution process of LIBs from micro internal short circuit failure to thermal runaway under mechanical abuse conditions. *Process Saf. Environ. Prot.* **2024**, *185*, 296–306. [CrossRef]
33. Liu, P.; Li, Y.; Mao, B.; Chen, M.; Huang, Z.; Wang, Q. Experimental study on thermal runaway and fire behaviors of large format lithium iron phosphate battery. *Appl. Therm. Eng.* **2021**, *192*, 116949. [CrossRef]
34. Wang, Z.; Zhao, Q.; Yin, B.; Zhai, H.; Wang, J.; An, W. Characterization of fire behaviors associated with a thermal runaway in large-scale commercial LiNi0.8Co0.1Mn0.1O_2/graphite cells under external ignition. *Case Stud. Therm. Eng.* **2023**, *47*, 103126. [CrossRef]
35. Wang, Z.; Yang, H.; Li, Y.; Wang, G.; Wang, J. Thermal runaway and fire behaviors of large-scale lithium ion batteries with different heating methods. *J. Hazard. Mater.* **2019**, *379*, 120730. [CrossRef]
36. Zhu, X.; Sun, Z.; Wang, Z.; Wang, H.; Lin, N.; Shan, C. Thermal runaway in commercial lithium-ion cells under overheating condition and the safety assessment method: Effects of SoCs, cathode materials and packaging forms. *J. Energy Storage* **2023**, *68*, 107768. [CrossRef]
37. Zou, K.; Lu, S. Comparative study on the influence of incident heat flux on thermal runaway fire development of large-format lithium-ion batteries. *Process Saf. Environ. Prot.* **2023**, *176*, 831–840. [CrossRef]
38. Zhang, Q.; Niu, J.; Zhao, Z.; Wang, Q. Research on the effect of thermal runaway gas components and explosion limits of lithium-ion batteries under different charge states. *J. Energy Storage* **2022**, *45*, 103759. [CrossRef]
39. Golubkov, A.W.; Fuchs, D.; Wagner, J.; Wiltsche, H.; Stangl, C.; Fauler, G.; Voitic, G.; Thaler, A.; Hacker, V. Thermal-runaway experiments on consumer Li-ion batteries with metal-oxide and olivin-type cathodes. *RSC Adv.* **2014**, *4*, 3633–3642. [CrossRef]
40. Yuan, L.; Dubaniewicz, T.; Zlochower, I.; Thomas, R.; Rayyan, N. Experimental study on thermal runaway and vented gases of lithium-ion cells. *Process Saf. Environ. Prot.* **2020**, *144*, 186–192. [CrossRef]
41. Yang, M.; Rong, M.; Pan, J.; Ye, Y.; Yang, A.; Chu, J.; Yuan, H.; Wang, X. Thermal runaway behavior analysis during overheating for commercial LiFePO4 batteries under various state of charges. *Appl. Therm. Eng.* **2023**, *230*, 120816. [CrossRef]
42. Wang, S.; Song, L.; Li, C.; Tian, J.; Jin, K.; Duan, Q.; Wang, Q. Experimental study of gas production and flame behavior induced by the thermal runaway of 280 Ah lithium iron phosphate battery. *J. Energy Storage* **2023**, *74*, 109368. [CrossRef]
43. Perea, A.; Paolella, A.; Dubé, J.; Champagne, D.; Mauger, A.; Zaghib, K. State of charge influence on thermal reactions and abuse tests in commercial lithium-ion cells. *J. Power Sources* **2018**, *399*, 392–397. [CrossRef]
44. Doose, S.; Hahn, A.; Fischer, S.; Müller, J.; Haselrieder, W.; Kwade, A. Comparison of the consequences of state of charge and state of health on the thermal runaway behavior of lithium ion batteries. *J. Energy Storage* **2023**, *62*, 106837. [CrossRef]
45. Wu, S.; Wang, C.; Luan, W.; Zhang, Y.; Chen, Y.; Chen, H. Thermal runaway behaviors of Li-ion batteries after low temperature aging: Experimental study and predictive modeling. *J. Energy Storage* **2023**, *66*, 107451. [CrossRef]
46. Spotnitz, R.; Franklin, J. Abuse behavior of high-power, lithium-ion cells. *J. Power Sources* **2003**, *113*, 81–100. [CrossRef]
47. Zhang, Y.; Wang, H.; Li, W.; Li, C.; Ouyang, M. Size distribution and elemental composition of vent particles from abused prismatic Ni-rich automotive lithium-ion batteries. *J. Energy Storage* **2019**, *26*, 100991. [CrossRef]
48. Wang, J.-h.; Jiang, Z.; Mei, M.; Qiu, H.; Wang, Y. Numerical simulation study on two-phase flow of thermal runaway evolution and jet fire of 18650 lithium-ion battery under thermal abuse. *Case Stud. Therm. Eng.* **2024**, *53*, 103726. [CrossRef]
49. Zhang, P.; Lu, J.; Yang, K.; Chen, H.; Huang, Y. A 3D simulation model of thermal runaway in Li-ion batteries coupled particles ejection and jet flow. *J. Power Sources* **2023**, *580*, 233357. [CrossRef]
50. Wang, Z.; Chen, S.; He, X.; Wang, C.; Zhao, D. A multi-factor evaluation method for the thermal runaway risk of lithium-ion batteries. *J. Energy Storage* **2022**, *45*, 103767. [CrossRef]
51. Liu, J.; Wang, Z.; Bai, J. Influences of multi factors on thermal runaway induced by overcharging of lithium-ion battery. *J. Energy Chem.* **2022**, *70*, 531–541. [CrossRef]
52. Said, A.O.; Lee, C.; Stoliarov, S.I. Experimental investigation of cascading failure in 18650 lithium ion cell arrays: Impact of cathode chemistry. *J. Power Sources* **2020**, *446*, 227347. [CrossRef]
53. Liu, X.; Stoliarov, S.I.; Denlinger, M.; Masias, A.; Snyder, K. Comprehensive calorimetry of the thermally-induced failure of a lithium ion battery. *J. Power Sources* **2015**, *280*, 516–525. [CrossRef]
54. Liu, X.; Wu, Z.; Stoliarov, S.I.; Denlinger, M.; Masias, A.; Snyder, K. Heat release during thermally-induced failure of a lithium ion battery: Impact of cathode composition. *Fire Saf. J.* **2016**, *85*, 10–22. [CrossRef]

55. Natarajan, J.; Lieuwen, T.; Seitzman, J. Laminar flame speeds of H_2/CO mixtures: Effect of CO_2 dilution, preheat temperature, and pressure. *Combust. Flame* **2007**, *151*, 104–119. [CrossRef]
56. Qi, S.; Du, Y.; Zhang, P.; Li, G.; Zhou, Y.; Wang, B. Effects of concentration, temperature, humidity, and nitrogen inert dilution on the gasoline vapor explosion. *J. Hazard. Mater.* **2017**, *323*, 593–601. [CrossRef]
57. Zhang, Y.; Shen, W.; Zhang, H.; Wu, Y.; Lu, J. Effects of inert dilution on the propagation and extinction of lean premixed syngas/air flames. *Fuel* **2015**, *157*, 115–121. [CrossRef]
58. Hafiz, N.M.; Mansor, M.R.A.; Wan Mahmood, W.M.F. Simulation of the combustion process for a CI hydrogen engine in an argon-oxygen atmosphere. *Int. J. Hydrogen Energy* **2018**, *43*, 11286–11297. [CrossRef]
59. Yan, W.; Wang, Z.; Chen, S. Quantitative analysis on the heat transfer modes in the process of thermal runaway propagation in lithium-ion battery pack under confined and semi-confined space. *Int. J. Heat Mass Transf.* **2021**, *176*, 121483. [CrossRef]
60. Chen, M.; Liu, J.; He, Y.; Yuen, R.; Wang, J. Study of the fire hazards of lithium-ion batteries at different pressures. *Appl. Therm. Eng.* **2017**, *125*, 1061–1074. [CrossRef]
61. Wang, H.; Du, Z.; Liu, L.; Zhang, Z.; Hao, J.; Wang, Q.; Wang, S. Study on the Thermal Runaway and Its Propagation of Lithium-Ion Batteries Under Low Pressure. *Fire Technol.* **2020**, *56*, 2427–2440. [CrossRef]
62. Wang, H.; Xu, H.; Zhang, Z.; Wang, Q.; Jin, C.; Wu, C.; Xu, C.; Hao, J.; Sun, L.; Du, Z.; et al. Fire and explosion characteristics of vent gas from lithium-ion batteries after thermal runaway: A comparative study. *eTransportation* **2022**, *13*, 100190. [CrossRef]
63. Xie, S.; Gong, Y.; Li, G.; Ping, X. Effect of low-pressure environment on thermal runaway behavior of NCM523/graphite pouch cells with different overcharge cycles. *J. Energy Storage* **2022**, *55*, 105444. [CrossRef]
64. Wang, K.; Wu, D.; Chang, C.; Zhang, J.; Ouyang, D.; Qian, X. Charging rate effect on overcharge-induced thermal runaway characteristics and gas venting behaviors for commercial lithium iron phosphate batteries. *J. Clean. Prod.* **2024**, *434*, 139992. [CrossRef]
65. Talele, V.; Moralı, U.; Patil, M.S.; Panchal, S.; Fraser, R.; Fowler, M.; Thorat, P.; Gokhale, Y.P. Computational modelling and statistical evaluation of thermal runaway safety regime response on lithium-ion battery with different cathodic chemistry and varying ambient condition. *Int. Commun. Heat Mass Transf.* **2023**, *146*, 106907. [CrossRef]
66. Huang, P.; Ping, P.; Li, K.; Chen, H.; Wang, Q.; Wen, J.; Sun, J. Experimental and modeling analysis of thermal runaway propagation over the large format energy storage battery module with Li4Ti5O12 anode. *Appl. Energy* **2016**, *183*, 659–673. [CrossRef]
67. Ping, P.; Wang, Q.; Huang, P.; Li, K.; Sun, J.; Kong, D.; Chen, C. Study of the fire behavior of high-energy lithium-ion batteries with full-scale burning test. *J. Power Sources* **2015**, *285*, 80–89. [CrossRef]
68. Huang, P.; Wang, Q.; Li, K.; Ping, P.; Sun, J. The combustion behavior of large scale lithium titanate battery. *Sci. Rep.* **2015**, *5*, 7788. [CrossRef]
69. Zhu, M.; Zhang, S.; Chen, Y.; Zhao, L.; Chen, M. Experimental and analytical investigation on the thermal runaway propagation characteristics of lithium-ion battery module with NCM pouch cells under various state of charge and spacing. *J. Energy Storage* **2023**, *72*, 108380. [CrossRef]
70. An, Z.; Li, W.; Du, X.; Jia, L.; Li, Q.; Zhang, D. Experimental study on behaviors of lithium-ion cells experiencing internal short circuit and thermal runaway under nail penetration abuse condition. *Appl. Therm. Eng.* **2024**, *247*, 123058. [CrossRef]
71. Kong, D.; Zhao, H.; Ping, P.; Zhang, Y.; Wang, G. Effect of low temperature on thermal runaway and fire behaviors of 18650 lithium-ion battery: A comprehensive experimental study. *Process Saf. Environ. Prot.* **2023**, *174*, 448–459. [CrossRef]
72. Shahid, S.; Agelin-Chaab, M. A review of thermal runaway prevention and mitigation strategies for lithium-ion batteries. *Energy Convers. Manag. X* **2022**, *16*, 100310. [CrossRef]
73. Song, L.; Huang, Z.; Mei, W.; Jia, Z.; Yu, Y.; Wang, Q.; Jin, K. Thermal runaway propagation behavior and energy flow distribution analysis of 280 Ah LiFePO4 battery. *Process Saf. Environ. Prot.* **2023**, *170*, 1066–1078. [CrossRef]
74. Mishra, D.; Tummala, R.; Jain, A. Investigation of propagation of thermal runaway during large-scale storage and transportation of Li-ion batteries. *J. Energy Storage* **2023**, *72*, 108315. [CrossRef]
75. Wang, Z.; He, T.; Bian, H.; Jiang, F.; Yang, Y. Characteristics of and factors influencing thermal runaway propagation in lithium-ion battery packs. *J. Energy Storage* **2021**, *41*, 102956. [CrossRef]
76. Zhou, Z.; Li, M.; Zhou, X.; Ju, X.; Yang, L. Investigating thermal runaway characteristics and trigger mechanism of the parallel lithium-ion battery. *Appl. Energy* **2023**, *349*, 121690. [CrossRef]
77. Jin, C.; Sun, Y.; Wang, H.; Zheng, Y.; Wang, S.; Rui, X.; Xu, C.; Feng, X.; Wang, H.; Ouyang, M. Heating power and heating energy effect on the thermal runaway propagation characteristics of lithium-ion battery module: Experiments and modeling. *Appl. Energy* **2022**, *312*, 118760. [CrossRef]
78. Wang, H.; Xu, H.; Zhao, Z.; Wang, Q.; Jin, C.; Li, Y.; Sheng, J.; Li, K.; Du, Z.; Xu, C.; et al. An experimental analysis on thermal runaway and its propagation in Cell-to-Pack lithium-ion batteries. *Appl. Therm. Eng.* **2022**, *211*, 118418. [CrossRef]
79. Said, A.O.; Lee, C.; Stoliarov, S.I.; Marshall, A.W. Comprehensive analysis of dynamics and hazards associated with cascading failure in 18650 lithium ion cell arrays. *Appl. Energy* **2019**, *248*, 415–428. [CrossRef]
80. Kim, J.T.; Choi, J.Y.; Kang, S.; Han, N.G.; Kim, D.K. Development of thermal runaway propagation model considering vent gas combustion for electric vehicles. *J. Energy Storage* **2023**, *60*, 106535. [CrossRef]
81. Wang, G.; Ping, P.; Zhang, Y.; Zhao, H.; Lv, H.; Gao, X.; Gao, W.; Kong, D. Modeling thermal runaway propagation of lithium-ion batteries under impacts of ceiling jet fire. *Process Saf. Environ. Prot.* **2023**, *175*, 524–540. [CrossRef]

82. Zhang, Y.; Kong, D.; Ping, P.; Zhao, H.; Dai, X.; Chen, X. Effect of a plate obstacle on fire behavior of 18650 lithium ion battery: An experimental study. *J. Energy Storage* **2022**, *54*, 105283. [CrossRef]
83. Liu, P.; Wang, C.; Sun, S.; Zhao, G.; Yu, X.; Hu, Y.; Mei, W.; Jin, K.; Wang, Q. Understanding the influence of the confined cabinet on thermal runaway of large format batteries with different chemistries: A comparison and safety assessment study. *J. Energy Storage* **2023**, *74*, 109337. [CrossRef]
84. Wang, G.; Ping, P.; Peng, R.; Lv, H.; Zhao, H.; Gao, W.; Kong, D. A semi reduced-order model for multi-scale simulation of fire propagation of lithium-ion batteries in energy storage system. *Renew. Sustain. Energy Rev.* **2023**, *186*, 113672. [CrossRef]
85. You, H.Z.; Faeth, G.M. Ceiling heat transfer during fire plume and fire impingement. *Fire Mater.* **1979**, *3*, 140–147. [CrossRef]
86. Zhang, X.; Hu, L.; Zhu, W.; Zhang, X.; Yang, L. Flame extension length and temperature profile in thermal impinging flow of buoyant round jet upon a horizontal plate. *Appl. Therm. Eng.* **2014**, *73*, 15–22. [CrossRef]
87. Gao, Z.; Jie, J.; Wan, H.; Zhu, J.; Sun, J. Experimental investigation on transverse ceiling flame length and temperature distribution of sidewall confined tunnel fire. *Fire Saf. J.* **2017**, *91*, 371–379. [CrossRef]
88. Wang, Z.; Zhou, K.; Zhang, L.; Nie, X.; Wu, Y.; Jiang, J.; Dederichs, A.S.; He, L. Flame extension area and temperature profile of horizontal jet fire impinging on a vertical plate. *Process Saf. Environ. Prot.* **2021**, *147*, 547–558. [CrossRef]
89. Zhai, H.; Chi, M.; Li, J.; Li, D.; Huang, Z.; Jia, Z.; Sun, J.; Wang, Q. Thermal runaway propagation in large format lithium ion battery modules under inclined ceilings. *J. Energy Storage* **2022**, *51*, 104477. [CrossRef]
90. Dahn, J.; Richard, M. Accelerating Rate Calorimetry Study on the Thermal Stability of Lithium Intercalated Graphite in Electrolyte. II. Modeling the Results and Predicting Differential Scanning Calorimeter Curves. *Fuel Energy Abstr.* **1999**, *146*, 2078–2084.
91. Hatchard, T.D.; Macneil, D.D.; Basu, A.; Dahn, J.R. Thermal Model of Cylindrical and Prismatic Lithium-Ion Cells. *J. Electrochem. Soc.* **2001**, *148*, A755–A761. [CrossRef]
92. Kim, G.H.; Pesaran, A.; Spotnitz, R. Three-Dimensional Thermal Abuse Model for Lithium-Ion Cells. *J. Power Sources* **2007**, *170*, 476–489. [CrossRef]
93. Coman, P.T.; Rayman, S.; White, R.E. A lumped model of venting during thermal runaway in a cylindrical Lithium Cobalt Oxide lithium-ion cell. *J. Power Sources* **2016**, *307*, 56–62. [CrossRef]
94. Feng, X.; Lu, L.; Ouyang, M.; Li, J.; He, X. A 3D thermal runaway propagation model for a large format lithium ion battery module. *Energy* **2016**, *115*, 194–208. [CrossRef]
95. Coman, P.; Matefi-Tempfli, S.; Veje, C.; White, R. Modeling Vaporization, Gas Generation and Venting in Li-Ion Battery Cells with a Dimethyl Carbonate Electrolyte. *J. Electrochem. Soc.* **2017**, *164*, A1858–A1865. [CrossRef]
96. Bugryniec, P.J.; Davidson, D.J.N.; Brown, D.S.F. Advanced abuse modelling of Li-ion cells—A novel description of cell pressurisation and simmering reactions. *J. Power Sources* **2020**, *474*, 228396. [CrossRef]
97. Wang, G.; Kong, D.; Ping, P.; Wen, J.; He, X.; Zhao, H.; He, X.; Peng, R.; Zhang, Y.; Dai, X. Revealing particle venting of lithium-ion batteries during thermal runaway: A multi-scale model toward multiphase process. *eTransportation* **2023**, *16*, 100237. [CrossRef]
98. Kong, D.; Wang, G.; Ping, P.; Wen, J. A coupled conjugate heat transfer and CFD model for the thermal runaway evolution and jet fire of 18650 lithium-ion battery under thermal abuse. *eTransportation* **2022**, *12*, 100157. [CrossRef]
99. Voigt, S.; Sträubig, F.; Kwade, A.; Zehfuß, J.; Knaust, C. An empirical model for lithium-ion battery fires for CFD applications. *Fire Saf. J.* **2023**, *135*, 103725. [CrossRef]
100. Azizighalehsari, S.; Venugopal, P.; Singh, D.; Batista, T.; Rietveld, G. Empowering Electric Vehicles Batteries: A Comprehensive Look at the Application and Challenges of Second-Life Batteries. *Batteries* **2024**, *10*, 161. [CrossRef]
101. Placke, T.; Heckmann, A.; Schmuch, R.; Meister, P.; Beltrop, K.; Winter, M. Perspective on Performance, Cost, and Technical Challenges for Practical Dual-Ion Batteries. *Joule* **2018**, *2*, 2528–2550. [CrossRef]
102. Zhang, J.; Gao, W.; Chen, L.; Li, Y. Inhibition and enhancement of hydrogen explosion by perfluorohexanone. *Int. J. Hydrogen Energy* **2024**, *53*, 522–534. [CrossRef]
103. Wang, Z.; He, C.; Geng, Z.; Li, G.; Zhang, Y.; Shi, X.; Yao, B. Experimental study of thermal runaway propagation suppression of lithium-ion battery module in electric vehicle power packs. *Process Saf. Environ. Prot.* **2024**, *182*, 692–702. [CrossRef]
104. Zhang, L.; Li, Y.; Duan, Q.; Chen, M.; Xu, J.; Zhao, C.; Sun, J.; Wang, Q. Experimental study on the synergistic effect of gas extinguishing agents and water mist on suppressing lithium-ion battery fires. *J. Energy Storage* **2020**, *32*, 101801. [CrossRef]
105. Prathap, C.; Ray, A.; Ravi, M.R. Investigation of nitrogen dilution effects on the laminar burning velocity and flame stability of syngas fuel at atmospheric condition. *Combust. Flame* **2008**, *155*, 145–160. [CrossRef]
106. Li, S.; Zhang, Y.; Qiu, X.; Li, B.; Zhang, H. Effects of Inert Dilution and Preheating Temperature on Lean Flammability Limit of Syngas. *Energy Fuels* **2014**, *28*, 3442–3452. [CrossRef]
107. Lin, C.; Yan, H.; Qi, C.; Liu, Z.; Liu, D.; Liu, X.; Lao, L.; Li, Z.; Sun, Y. Thermal runaway and gas production characteristics of semi-solid electrolyte and liquid electrolyte lithium-Ion batteries: A comparative study. *Process Saf. Environ. Prot.* **2024**, *189*, 577–586. [CrossRef]
108. Miao, B.; Lv, J.; Wang, Q.; Zhu, G.; Guo, C.; An, G.; Ou, J. The Suppression Effect of Water Mist Released at Different Stages on Lithium-Ion Battery Flame Temperature, Heat Release, and Heat Radiation. *Batteries* **2024**, *10*, 232. [CrossRef]
109. Hong, Y.; Jin, C.; Chen, S.; Xu, C.; Wang, H.; Wu, H.; Huang, S.; Wang, Q.; Li, H.; Zheng, Y.; et al. Experimental study of the suppressing effect of the primary fire and thermal runaway propagation for electric bicycle batteries using flood cooling. *J. Clean. Prod.* **2024**, *435*, 140392. [CrossRef]

110. Wang, Z.; Zhao, Q.; Yin, B.; Shi, B.; Wang, J.; An, W. Influence of longitudinal wind on thermal runaway and fire behaviors of 18650 lithium-ion batteries in a restricted channel. *J. Power Sources* **2023**, *567*, 232974. [CrossRef]
111. Jiang, X.; Han, H.; Liu, X.; Zhang, P. The synergistic effect of wind and two-phase flow water mist on thermal runaway and its propagation of lithium-ion battery module within battery case. *J. Energy Storage* **2024**, *89*, 111701. [CrossRef]
112. Bhavsar, S.; Kant, K.; Pitchumani, R. Robust model-predictive thermal control of lithium-ion batteries under drive cycle uncertainty. *J. Power Sources* **2023**, *557*, 232496. [CrossRef]
113. Vikram, S.; Vashisht, S.; Rakshit, D. Performance analysis of liquid-based battery thermal management system for Electric Vehicles during discharge under drive cycles. *J. Energy Storage* **2022**, *55*, 105737. [CrossRef]
114. Nicholls, R.A.; Moghimi, M.A.; Sehhat, S. Thermal performance analysis of battery modules with passive cooling under different cycling loads in electric vehicles. *J. Energy Storage* **2024**, *94*, 112349. [CrossRef]
115. Qin, M.; Zeng, Z.; Wu, Q.; Liu, X.; Liu, Q.; Cheng, S.; Xie, J. 1,3,5-Trifluorobenzene endorsed EC-free electrolyte for high-voltage and wide-temperature lithium-ion batteries. *J. Energy Chem.* **2023**, *85*, 49–57. [CrossRef]
116. Wu, F.; Shi, Q.; Chen, L.; Dong, J.; Zhao, J.; Wang, H.; Gao, F.; Liu, J.; Zhang, H.; Li, N.; et al. New insights into dry-coating-processed surface engineering enabling structurally and thermally stable high-performance Ni-rich cathode materials for lithium ion batteries. *Chem. Eng. J.* **2023**, *470*, 144045. [CrossRef]
117. Roh, Y.; Jin, D.; Kim, E.; Byun, S.; Lee, Y.-S.; Ryou, M.-H.; Lee, Y.M. Highly improved thermal stability of the ceramic coating layer on the polyethylene separator via chemical crosslinking between ceramic particles and polymeric binders. *Chem. Eng. J.* **2022**, *433*, 134501. [CrossRef]
118. Liu, M.; Zeng, Z.; Zhong, W.; Ge, Z.; Li, L.; Lei, S.; Wu, Q.; Zhang, H.; Cheng, S.; Xie, J. Non-flammable fluorobenzene-diluted highly concentrated electrolytes enable high-performance Li-metal and Li-ion batteries. *J. Colloid Interface Sci.* **2022**, *619*, 399–406. [CrossRef]
119. Shao, D.; Yang, L.; Luo, K.; Chen, M.; Zeng, P.; Liu, H.; Liu, L.; Chang, B.; Luo, Z.; Wang, X. Preparation and performances of the modified gel composite electrolyte for application of quasi-solid-state lithium sulfur battery. *Chem. Eng. J.* **2020**, *389*, 124300. [CrossRef]
120. Lv, J.; Ye, J.; Dai, G.; Niu, Z.; Sun, Y.; Zhang, X.; Zhao, Y. Flame-retarding battery cathode materials based on reversible multi-electron redox chemistry of phenothiazine-based polymer. *J. Energy Chem.* **2020**, *47*, 256–262. [CrossRef]
121. Huang, B.; Hua, H.; Peng, L.; Wang, X.; Shen, X.; Li, R.; Zhang, P.; Zhao, J. The functional separator for lithium-ion batteries based on phosphonate modified nano-scale silica ceramic particles. *J. Power Sources* **2021**, *498*, 229908. [CrossRef]
122. Du, X.; Jin, L.; Deng, S.; Zhou, M.; Du, Z.; Cheng, X.; Wang, H. Recyclable, Self-Healing, and Flame-Retardant Solid–Solid Phase Change Materials Based on Thermally Reversible Cross-Links for Sustainable Thermal Energy Storage. *ACS Appl. Mater. Interfaces* **2021**, *13*, 42991–43001. [CrossRef] [PubMed]
123. Xu, Z.; Chen, W.; Wu, T.; Wang, C.; Liang, Z. Thermal management system study of flame retardant solid–solid phase change material battery. *Surf. Interfaces* **2023**, *36*, 102558. [CrossRef]
124. Weng, J.; Xiao, C.; Ouyang, D.; Yang, X.; Chen, M.; Zhang, G.; Yuen, R.K.K.; Wang, J. Mitigation effects on thermal runaway propagation of structure-enhanced phase change material modules with flame retardant additives. *Energy* **2022**, *239*, 122087. [CrossRef]
125. Chen, M.; Zhu, M.; Zhao, L.; Chen, Y. Study on thermal runaway propagation inhibition of battery module by flame-retardant phase change material combined with aerogel felt. *Appl. Energy* **2024**, *367*, 123394. [CrossRef]
126. Mei, J.; Shi, G.; Liu, H.; Wang, Z.; Chen, M. Experimental study on the effect of passive retardation method for thermal runaway mitigation of lithium-ion battery. *Appl. Therm. Eng.* **2023**, *230*, 120861. [CrossRef]

batteries

MDPI

Article

Pyrrolidinium-Based Ionic Liquids as Advanced Non-Aqueous Electrolytes for Safer Next Generation Lithium Batteries

Antía Santiago-Alonso [1,2], José Manuel Sánchez-Pico [1], Raquel San Emeterio [1], María Villanueva [1], Josefa Salgado [1,*] and Juan José Parajó [1,*]

1 NaFoMAT Research Group, Physic Faculty and Materials Institute (iMATUS), Universidade de Santiago de Compostela, Rúa de José María Suárez Núñez, s/n, 15782 Santiago de Compostela, A Coruña, Spain; antia.santiago.alonso@usc.es (A.S.-A.); josemanuel.sanchez.pico@rai.usc.es (J.M.S.-P.); raquel.san@rai.usc.es (R.S.E.); maria.villanueva@usc.es (M.V.)

2 ABCR LABORATORIOS, Lg. Vilapouca (PG Industrial), 36550 Forcarei, Pontevedra, Spain

* Correspondence: j.salgado.carballo@usc.es (J.S.); juanjose.parajo@usc.es (J.J.P.)

Abstract: In the current context of increasing energy demand, ionic liquids (ILs) are presented as possible candidates to replace conventional electrolytes and to develop more efficient energy storage devices. The IL 1-Methyl-1-propylpyrrolidinium bis(trifluoromethanesulfonyl)imide has been selected for this work, due to the good thermal and chemical stabilities and good electrochemical performance of the pyrrolidinium cation based ILs. Binary mixtures of this IL and lithium salt with the same anion, [TFSI], have been prepared with the aim of assessing them, as possible electrolytes for lithium batteries. These mixtures were thermally and electrochemically characterised through DSC and dielectric spectroscopy studies. The ionic conductivity decreases as the salt concentration increases, finding values ranging between 0.4 S/m and 0.1 S/m at room temperature. Additionally, a wide liquid range was found for the mixtures, which would reduce or even eliminate some of the most common problems of current electrolytes, such as their crystallisation at low temperatures and flammability. Finally, the toxicity of pure IL and the intermediate salt concentration was also evaluated in terms of the bioluminescence inhibition of the Alivibrio Fischeri bacteria, observing that, although the toxicity increases with the salt addition, both samples can be classified as practically harmless.

Keywords: ionic liquids; battery safety; non-aqueous electrolyte; DSC; BBDS; Microtox®

1. Introduction

One of the biggest challenges for contemporary society is the transition to a new energy model based on renewable resources instead of fossil fuels. This transition involves many demanding tasks that must be taken up by science and the whole society, including sustainable consumption, material recycling, and the circular economy. An-other fundamental aspect that needs to be addressed is the primary issue posed by renewable sources: their intermittency. Their output depends on environmental factors, seasonal variations, weather conditions, and the day/night cycles. As a result, peak production frequently does not coincide with peak demand, underscoring the necessity for systems to store surplus energy. For this reason, the development and innovation of energy storage systems is currently a strategic line. The actual battery market is dominated by lithium-ion batteries (LIBs) due to their numerous advantages over other battery systems. These include high specific capacity and voltage, a wide operation temperature range, excellent cycling performance, minimal self-discharge, and the absence of memory effects [1].

Nowadays, most LIBs are composed of a carbonate(s) in combination with a lithium salt [2]. Nevertheless, common carbonates have a flash point near room temperature, making them highly inflammable if exposed to extreme thermal, electrical, or mechanical conditions, in which can lead to serious events such as fire or explosion [3]. While the risk of failure in a single cell is minimal, the risk increases significantly when thousands of cells are enclosed in a single battery pack.

Aqueous-based electrolytes present advantages over carbonate-based ones in terms of safety and cost. Nevertheless, they exhibit lower energy densities, narrower electrochemical stability windows, and limitations at temperatures below the crystallisation point of water [4].

Ionic liquids (ILs) have recently attracted considerable interest because of their exceptional properties and potential applications across diverse fields. The ILs are composed entirely of ions and exist in the liquid state at temperatures below 100 °C. The identifying characteristic of ILs is their low volatility, which makes them environmentally friendly alternatives, reducing the risk of air pollution and flammability [5,6]. One of the most intriguing aspects of ILs is their tunability. By selecting different combinations of cations and anions, it is possible to customize their physical and chemical properties to match a particular application. This adaptability extends to their thermal and electrochemical stability, solubility, viscosity, and ionic conductivity [4]. For instance, ILs can be designed to dissolve a large sort of substances, including organic, inorganic, and polymeric substances, making them highly effective solvents for chemical reactions and extractions [7].

Moreover, ILs have unique solvation properties with significant implications for green chemistry, where ILs are used to replace volatile organic solvents, minimizing the environmental impact [7]. So, to overcome the problem of common carbonates, ionic liquids (ILs) and their blends with electrochemically relevant salts emerge as a viable alternative. These mixtures offer enhanced safety, non-flammability [8] and thermal stability, without compromising electrochemical performance [9].

The effectiveness of ionic liquid–Li-ion electrolyte solutions is directly influenced by Li-ion mobility, which is intricately linked to their local environment, including solubility and solvation properties, [7,10].

Between the practically unlimited combinations of ILs, pyrrolidinium based ILs emerge as a clear alternative due to their exceptional thermal and electrochemical stability, versatility, and compatibility with Li-based battery materials [11,12]. N-methyl-N-propylpyrrolidinium bis(trifluoromethanesulfonyl)imide ([C_3C_1Pyrr][TFSI]) was selected due to the favourable characteristics of pyrrolidini-um-based ionic liquids [13]. Since ILs are not inherently electroactive, their use in LiBs requires mixing with lithium salts. Instead of the frequently used $LiPF_6$ for the electrolytes preparation, lithium bis(trifluoromethanesulfonyl) imide (LiTFSI) was selected, whose high conductivity, resistance to moisture, and extraordinary thermal and electrochemical stability makes it a good alternative to the $LiPF_6$ salt [14].

The aim of this study is to advance the properties knowledge of ILs and their mixtures and their subsequent potential use in energy storage systems. Therefore, we propose a comprehensive analysis of the thermal behaviour, electrical conductivity, and ecotoxicity of mixtures of N-methyl-N-propylpyrrolidinium bis(trifluoromethanesulfonyl) imide ([C_3C_1Pyrr][TFSI]) and lithium bis(trifluoromethanesulfonyl)imide ([Li][TFSI]) at different concentrations.

2. Materials and Methods

2.1. Chemicals

The IL, 1-Methyl-1-propylpyrrolidinium bis(trifluoromethylsulfonyl)imide ([C_3C_1Pyrr][TFSI]) provided by IoLiTec, was selected for this work due to its promising electrochemical properties as it is reported by different authors [15–17] and, as is also mentioned before, their non-flammable character. The salt lithium bis(trifluoromethylsulfonyl)imide ([Li][TFSI]) was provided by Sigma Aldrich, sharing the anion with the IL. Table 1 summarizes the key characteristics of the ionic liquid and the salt, being the stated purity of both higher than 0.99. To know the effect of salt concentrations on the physicochemical and biological properties, mixtures of four different concentrations (mole of salt per litre) ranging from 0.2 m to 1 m were analysed (Table 2).

Table 1. Reagent used in this study.

Name	Molecular Mass (g·mol^{-1})	Structure	Short Name CAS Number
1-Methyl-1-propylpyrrolidinium bis(trifluoromethylsulfonyl)imide	408.38	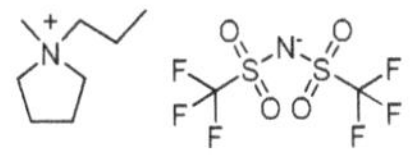	[C_3C_1Pyrr][TFSI] 223437-05-6
Lithium bis(trifluoromethylsulfonyl)imide	287.09	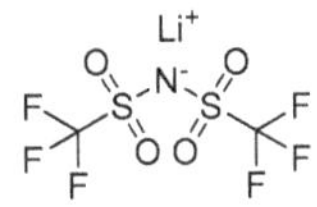	[Li][TFSI] 90076-65-6

Table 2. Amount of pure metal salts per kilogram of [C_3C_1Pyrr][TFSI] in the mixtures.

Molality/mol kg^{-1}	m_{LiTFSI}/g
0.2	57.42
0.5	143.55
0.75	215.32
1.0	287.10

2.2. *Differential Scanning Calorimetry*

The thermal behaviour and thermal transitions of the mixtures IL + salt were analysed using the Differential Scanning Calorimeter DSC Q2000 (Waters-TA Instruments. New Castle, DE, USA) with hermetically sealed aluminium pans, samples mass of (5–8) mg and nitrogen atmosphere. For the experimental part two heating-cooling cycles between −80 °C and 100 °C at two different rates, 10 °C min^{-1} and 5 °C min^{-1}, have been carried out, after an initial ramp from room temperature to 125 °C at 40 °C min^{-1} with an iso-thermal step at 125 °C for 45 min, to remove possible volatile impurities. Transition temperatures were determined as the onset point of the different peaks from the DSC curves during the reheating and re-cooling ramps. Transition temperatures were estimated with an uncertainty of 2 °C at 95% confidence [18]. Further details can be found in previously published papers [19,20].

2.3. *Ionic Conductivity*

Ionic conductivities were measured through broadband dielectric spectroscopy using an Agilent RLC precision meter HP 4284A (Agilent Technologies, Santa Clara, CA, USA) with a precision of 0.05% in the measurements. The applied frequency range was 20 Hz to 0.2 MHz to ensure a good fitting of the ohmic regime with a tolerance for the slope of the imaginary part of the dielectric constant of -1.00 ± 0.02 described thoroughly by J. Leys [21]. The mixtures were analysed using a symmetric two-electrode Swagelok configuration with stainless-steel electrodes, this setup ensures that the distance between electrodes remains constant. To control the temperature, the Swagelok cell was placed in a climate chamber (Memmert ICP400, Schwabahc, Germany); measurements were pursued in the range of 0 °C to 50 °C, starting with the lowest temperature. The experiment was done following a static isothermal method with an extra 15 min to assure the thermal equilibrium inside the chamber before performing the measurement.

2.4. *Toxicity*

The toxicity of the pure [C_3C_1Pyrr][TFSI] and the mixture of intermediate concentration (0.5 m) with [Li][TFSI], selected based on melting and conductivity results, was assessed using the Microtox® Toxicity Test kit (M500 Analyzer—Modern Water, Rema Tek LLC, Bonsall, CA, USA), following the procedure detailed in previous studies [22,23]. The Microtox® test measures the luminescence inhibition in the Gram-negative marine bacterium *Aliivibrio fischeri* (*A. fischeri*) by targeting a quorum sensing mechanism. When exposed to a toxic substance, bacterial respiration is disrupted, leading to a de-crease in bioluminescence, which correlates directly with the metabolic activity of the bacterial population. The results were compared with previous data on ILs of the same and different nature.

A range of diluted aqueous solutions (0 to 81.9%) of each compound was prepared for the acute toxicity test, with exposure times of 5, 15, and 30 min. The resulting data were used to estimate the effective concentrations causing 50%, 20%, and 10% (EC_{50}, EC_{20}, and EC_{10}, respectively) luminescence inhibition. The 95% confidence intervals were also calculated using non-linear regression and the least-squares method to fit the data to the logistic equation [22,23].

Passino and Smith [24] proposed six toxicity levels in their widely used toxicity classification based on EC50 values at 30 min, as follows: $EC_{50} > 1000$ mg/L: relatively harmless; 1000 mg/L > EC_{50} > 100 mg/L: practically harmless; 100 mg/L > EC_{50} > 1 mg/L: toxic; 1 mg/L > EC_{50} > 0.1 mg/L: highly toxic; 0.1 mg/L > EC_{50} > 0.01 mg/L: extremely toxic and EC_{50} < 0.01 mg/L: super toxic.

3. Results

3.1. DSC Results

The comparison of DSC curves registered on cooling and on subsequent heating at 5 C min^{-1} for mixtures of $[C_3C_1Pyrr][TFSI]$ + [Li][TFSI] and pure IL, are presented in Figure 1. As it can be observed, pure IL shows a unique peak on both, cooling and heating scans, which correspond to freezing and melting processes, respectively. The shift between the onset temperatures of these two peaks, around 20 °C; known as the super-cooling effect, is common in ILs due to the difficulty of crystallization that these com-pounds present where in addition to the Coulomb forces, Van der Waals interactions have great importance [20,25].

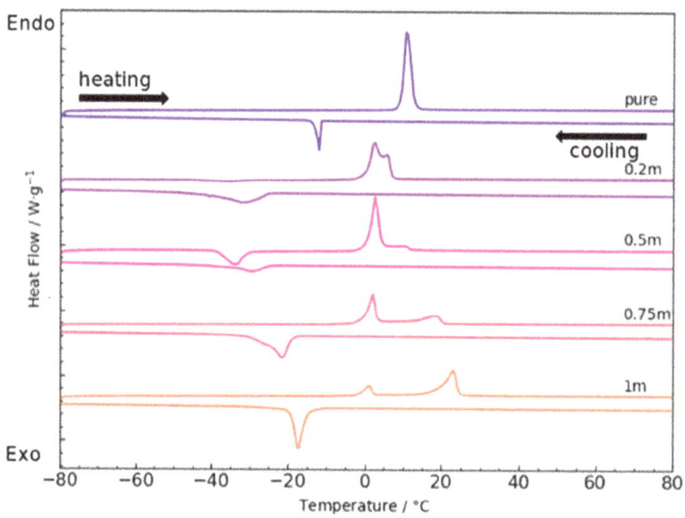

Figure 1. DSC curves of binary mixtures $[C_3C_1Pyrr][TFSI]$ with [Li][TFSI] at different concentrations (endo up).

Table 3 summarizes the values of the characteristic temperatures and enthalpies, determined as the onset temperature and area of the corresponding peak respectively. Onset temperatures corresponding to the melting and freezing of the pure IL are in good con-cordance with findings reported by other researchers [26–33].

Table 3. Temperature of glass transition (t_g), melting (t_f), freezing (t_c), and cold crystallisation (t_{cc}) and freezing, melting, and cold crystallisation enthalpies obtained from DSC curves.

Sample	t_g/°C	t_f/°C	$t_{m,1}$/°C	$t_{m,2}$/°C	t_{cc}/°C	$\Delta_{cc}H$ /J g^{-1}	Δ_fH /J g^{-1}	Δ_fH /kJ mol^{-1}	Δ_mH /J g^{-1}	Δ_mH /kJ mol^{-1}
[C_3C_1Pyrr][TFSI]	--	−12	9	--	--	--	−33	−13	36	15
0.2 m	--	−25	0.3	6	−46	−2.6	−22	−10	34	15
0.5 m	−74	−25	0.7	10	−39	−13.9	−10	−5	34	16
0.75 m	−69	−19	−0.1	19	--	--	−24	−12	27	13
1.0 m	--	−15	−1.4	19	--	--	−24	−13	23	12

The thermal behaviour of IL + salt mixtures reflects important differences compared to the pure IL, and they are highly dependent on both the molar fraction of the [Li][TFSI] salt and the scanning rate, as shown in Figures 2 and 3. Thus, DSC curves upon heating corresponding to the 10 °C min^{-1} (blue) and 5 °C min^{-1} (red), presented in Figure 2, show that the scanning rate does not influence the thermal behaviour of pure IL, as the onset and peak area of the melting peak are similar for both experiments, although this is not fulfilled for the mixtures with the lithium salt. As observed, the cycle at 10 °C min^{-1} presents an exothermic cold crystallization peak upon heating for mixtures of 0.2, 0.5 and 0.75 m of [Li][TFSI] samples; but these peaks became smaller or even disappeared after recooling and reheating the sample at 5 °C min^{-1}. In all cases, the onset of cold crystallization as well as the area of the corresponding peak change with the scanning rate. This is due to the fact that this event is a kinetically controlled process, and crystallization is challenging at high scanning rates, even though it is the most stable form; which subsequently causes additional crystallization of the sample upon heating. However, at lower rates, the sample completely crystallizes during the cooling ramp, and therefore the exothermic peak upon heating, the cold crystallization process, disappears. This can be seen in Figure 3, in which the cooling-heating cycles of the mixture of IL + [Li][TFSI] 1 m at 10 °C min^{-1} and 5 °C min^{-1} are presented, and the reduction of the cold crystallization peak with decreasing scanning rate is clearly visible. It is important to highlight that the melting process is a thermodynamically controlled transition, and the onset of the corresponding peak does not depend on the scanning rate.

Another notable observation is the splitting of the initial melting peak into two distinct peaks of different sizes as salt is introduced into the mixture, even though the molar heats associated with the overall process remains similar to those corresponding to the pure IL, as shown in Table 3. With increasing molar fraction of lithium salt, the resolution of both peaks increases, and the second peak even becomes the most intense at the highest salt concentration. This suggests the coexistence of two different polymorphic states, which is in good concordance with the findings of Hendersson and Passerini [29] for these mixtures and recent results from our group with different mixtures of IL + inorganic salt [34,35]. By increasing the salt concentration in the mixture, the crystallinity tends to decrease, that means an increase of amorphous behaviour, reflected by the disappearance and reduction of crystallization and fusion peaks and the appearance of glass transitions at higher temperatures. For pure IL and lower molar fractions, no glass transitions were detected within the experimental temperature range (T > −80 °C). Although, in our case, crystallization and melting peaks appear for all the mixtures, a clear loss of peak definition with higher salt concentration is observed, connected to the increase of amorphous behaviour with salt addition.

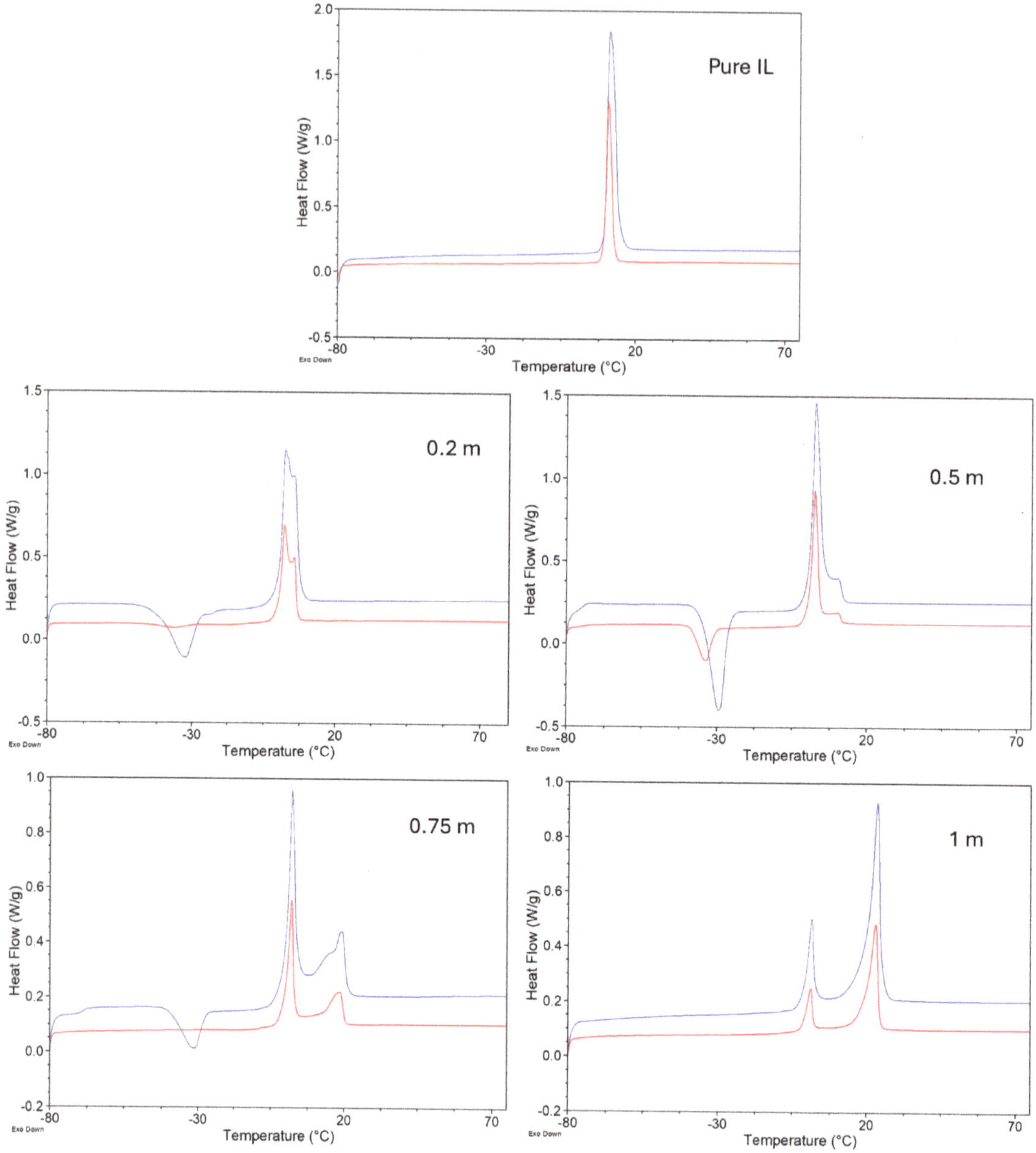

Figure 2. First (blue) and second (red) heating curves corresponding to 10 °C min^{-1} and 5 °C min^{-1} scan rates, respectively, for the pure and mixtures of IL + salt samples (endo up).

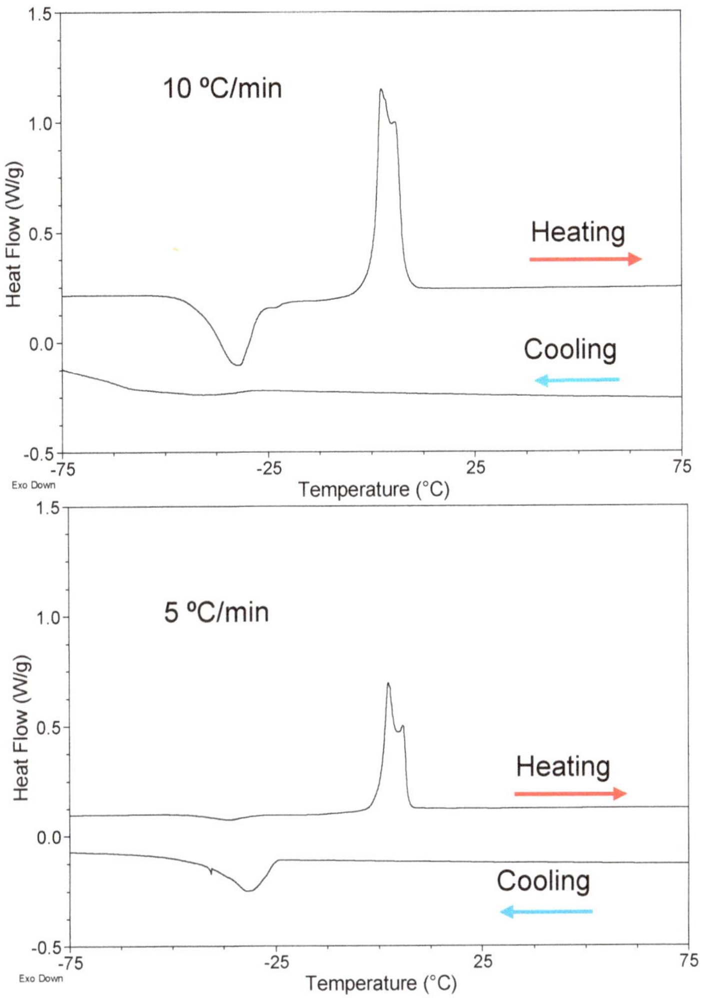

Figure 3. DSC cooling-heating cycles at 10 °C min^{-1} and 5 °C min^{-1} of the mixture IL + [Li][TFSI] 0.2 m (endo up).

3.2. Ionic Conductivity Results

Dielectric measurements were carried out from high to low temperatures to use the supercooling effect of these samples, as previously pointed out. The ionic conductivity of binary mixtures of [C_3C_1Pyrr][TFSI] and [Li][TFSI] are shown in Figure 4a against molality of [Li][TFSI] and Figure 4b against temperature in Arrhenius representation. The ionic conductivity for pure [C_3C_1Pyrr][TFSI] agrees with previously published results [15–17,26]. Regarding [C_3C_1Pyrr][TFSI] + [Li][TFSI] mixtures, their ionic conductivity follows the expected pattern, a decrease with higher molalities of salt [36], reaching 0.1180 S/m at 298 K at the highest molal concentration. This decrease is related with the increment of the viscosity. However, other authors pointed that the formation of diffusing $[Li(TFSI)_n]^{-(n-1)}$ ionic species in the solutions might lead to a decrease in conductivity [17,28]. Table 4 shows the ionic conductivities of all mixtures.

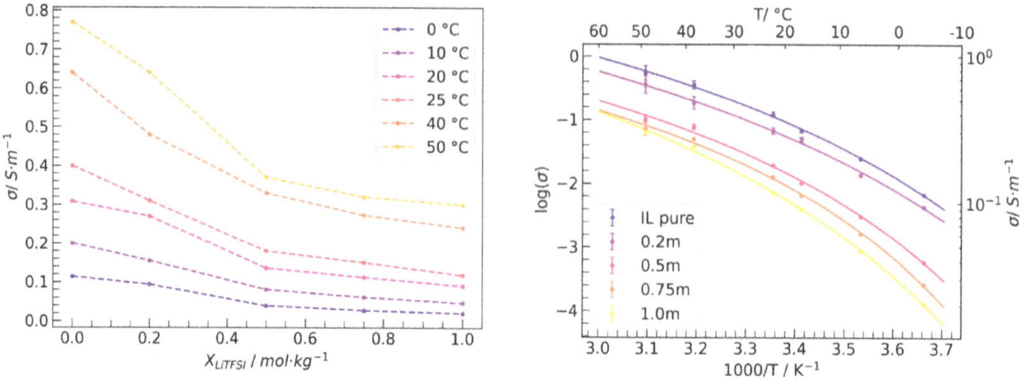

(a) (b)

Figure 4. (**a**) Ionic conductivity of binary mixtures [C_3C_1Pyrr][TFSI] and [Li][TFSI] against the [Li][TFSI] molar fraction at different temperatures. (**b**) Arrhenius plot showing the ionic conductivity vs. temperature for mixtures with different [Li][TFSI] content and their fitting to the Arrhenius equation.

Table 4. Ionic conductivities (in S.m^{-1}) of pure IL and IL + [Li][TFSI] salt for different concentrations and temperatures.

Temperature/°C	IL Pure	0.2 m	0.5 m	0.75 m	1.0 m
0	0.1138(37)	0.0936(71)	0.0394(15)	0.02764(17)	0.020539(95)
10	0.200(11)	0.155(20)	0.0810(50)	0.06148(62)	0.04694(32)
20	0.308(21)	0.267(41)	0.1364(99)	0.1125(19)	0.0905(10)
25	0.397(38)	0.305(48)	0.180(15)	0.1504(30)	0.1180(16)
40	0.645(59)	0.48(10)	0.325(40)	0.2726(89)	0.2412(57)
50	0.77(11)	0.64(14)	0.373(49)	0.32(11)	0.2998(79)

The ionic conductivity behaviour against temperature (shown in Figure 4b) is widely described in the literature by the Vogel-Fulcher-Tammann (VFT) equation (Equation (1)). At low temperatures the relaxation and transport properties become slower due to the glass transitions that these ILs typically present [37] giving a non-monotonic behaviour against temperature, described by this VFT-equation:

$$\log \sigma = \log \sigma_\infty - \frac{B}{(T - T_0)} \tag{1}$$

In the VFT equation, σ_∞ represents the ionic conductivity at infinite temperature, B is the fragility index or strength index, which inversely correlates with liquid´s kinetic fragility, and T_0, also known as Vogel temperature, is the temperature at which mobility vanishes. The VFT parameters obtained from the individual fitting of the samples are compiled in Table 5. The conductivity at infinite temperature shows small variations with salt concentration, as previously reported [35]. A similar behaviour is observed for B and T_0.

Although VFT model fits to experimental data, the resulting dependence on salt concentration shows behaviours that are challenging to explain physically. None of the fitted parameters represent a clear relation with salt concentration, as also observed by Tsamopoulos & Wang [38]. In their work, they demonstrated that the fitted values of T_0 fails to follow the variations in glass transition temperature with salt concentration, and likewise, unphysical fluctuations in σ_∞ and B are found.

Table 5. Fitting parameters of VFT and Arrhenius equations for all samples.

Sample	VFT-Equation			Arrhenius Equation		
	Log σ_∞	B/K^{-1}	T_0/K	log (σ_∞)	E_a (10^{-1})/eV	R^2
IL pure	2.49(54)	326(98)	203(13)	10.36(71)	2.93(18)	0.98
0.2 m	2.38(72)	351(95)	199(21)	10.01(69)	2.89(17)	0.98
0.5 m	1.66(69)	274(87)	217(11)	11.6(11)	3.46(28)	0.97
0.75 m	1.40(47)	246(64)	223.3(86)	12.5(12)	3.75(33)	0.96
1.0 m	2.19(54)	372(89)	211.8(94)	13.7(11)	4.10(29)	0.98

In the case where $T >> T_0$, a linear behaviour in the Arrhenius plot can be obtained, and the approximation of the VFT equation by the Arrhenius equation is valid.

$$\log \sigma = \log \sigma_\infty - \frac{E_a}{k_B T} \quad (2)$$

where E_a represents the activation energy per molecule and k_B the Boltzmann constant.

The fittings for all the samples to the Arrhenius equation were calculated and presented in Table 5 and represented in Figure 5a.

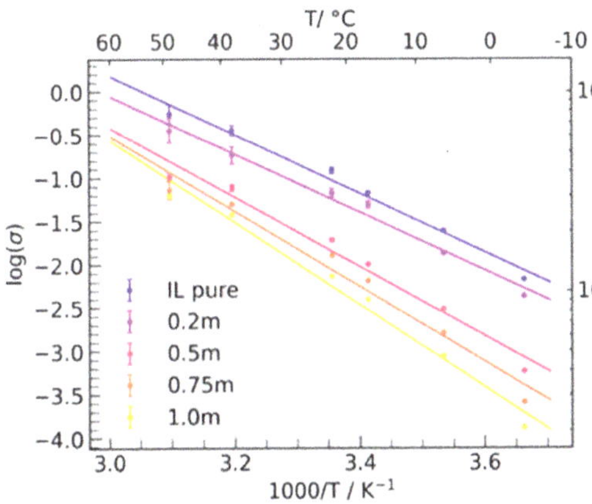

(**a**)

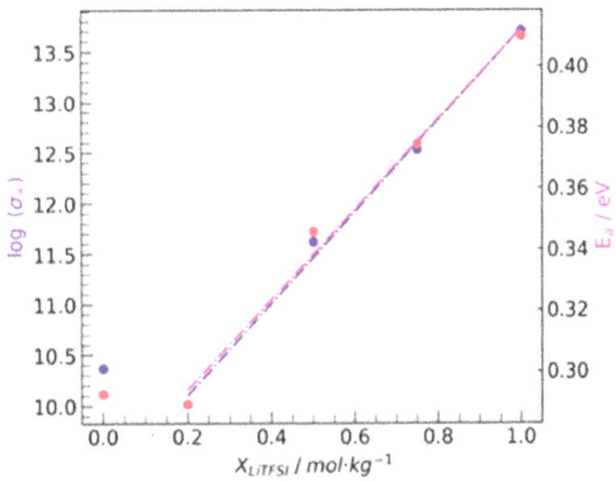

(**b**)

Figure 5. (**a**) Arrhenius plot of the ionic conductivity vs. temperature for mixtures with different [Li][TFSI] contents and their fitting to the VFT equation, and (**b**) activation energy and log (sigma) dependence on the molal concentration of lithium salt.

Regarding the physical meaning of the fitted parameters, it is clear that both activation energy and $\log \sigma_\infty$ follow an increasing monotonic behaviour with salt concentration for all IL and salt mixtures, as shown in Figure 5b. The monotonic increase with salt concentration of the exponential factor in the Arrhenius equation is related to the increase in the dielectric constant [39]. The linear increase in activation energies with salt concentration indicates that ion conduction in the electrolyte requires more energy to take place, which is consistent with the decrease in ionic conductivity.

3.3. Toxicity Evaluation

Figure 6 shows the bioluminescence inhibition of bioluminescence in *A. fischeri* after 30 min of exposure to varying concentrations of the pure IL and its mixture with lithium 0.5 m. This concentration was selected due to its melting point being slightly lower than that of the other concentrations. Despite a 50% reduction in conductivity at 298 K compared to the pure IL, it still maintains a high conductivity value.

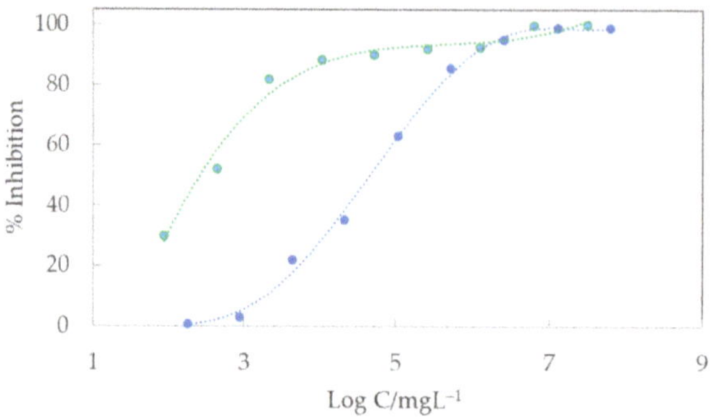

Figure 6. Inhibition of bioluminescence for 30 min of exposure against the logarithm of the concentration of (•) $[C_3C_1Pyrr][TFSI]$ and (•) $[C_3C_1Pyrr][TFSI]$ + [Li][TFSI] 0.5 m.

The inhibition responses for the analysed samples were fitted to a logistic equation. EC_{50}, EC_{20}, and EC_{10} values after 5, 15, and 30 minutes of exposure are calculated from these fitting equations and presented in Tables 6 and 7. As noted earlier, EC_{10} and EC_{20} provide initial benchmarks on the estimation of the lowest observed effect concentration, with EC_{10} being a particularly useful for assessing minimal environmental risks independently of concentration. Bacterial bioluminescence serves as a key indicator of cellular metabolism, where a decrease in luminescence reflects reduced cellular respiration [40,41]. Remarkably, the observed trend in toxic effects remains consistent across the different exposure times, suggesting a uniform mechanism of action on bacteria throughout the entire exposure period.

Table 6. Effective concentration values in mg/L for $[C_3C_1Pyrr][TFSI]$ and the corresponding 95% confidence intervals, determined after of the three selected exposure times to the marine bacteria *A. fischeri*.

	$[C_3C_1Pyrr][TFSI]$		
Exposure Time/min	**EC_{50}/mg L^{-1}**	**EC_{20}/mg L^{-1}**	**EC_{10}/mg L^{-1}**
5	925.88 (741.93; 1109.83)	400.03 (258.09; 541.96)	244.70 (123.27; 366.13)
15	629.53 (423.12; 835.94)	291.98 (123.92; 460.04)	186.18 (39.69; 332.68)
30	516.32 (299.31; 733.32)	246.77 (63.47; 430.07)	160.13 (15.24; 312.40)

Table 7. Effective concentration values in mg/L for $[C_3C_1Pyrr][TFSI]$ + [Li][TFSI] 0.5 m and the corresponding 95% confidence intervals, determined after of the three selected exposure times of the marine bacteria *A. fischeri*.

	$[C_3C_1Pyrr][TFSI]$ + [Li][TFSI] 0.5 m		
Exposure Time/min	**EC_{50}/mg L^{-1}**	**EC_{20}/mg L^{-1}**	**EC_{10}/mg L^{-1}**
5	178.72 (152.97; 204.48)	79.19 (59.39; 98.98)	49.16 (32.62; 65.70)
15	129.34 (107.09; 151.59)	56.92 (39.86; 73.97)	35.19 (21.04; 49.33)
30	109.88 (85.34; 134.42)	47.19 (28.78; 65.61)	28.77 (13.75; 43.79)

Although no literature results were found for this compound or its mixture with lithium for *A. fischeri*, the EC_{50} values at 30 min are slightly lower than $[C_4C_1Pyrr][TFSI]$. This observation correlates well with the trend that the longer the alkyl chain, the higher the toxicity [24,41–43].

As highlighted in a previous study of our research group [23], the EC values decrease approximately 4 times after the addition of [Li][TFSI] for all exposure times. This effect was also observed for [C_4C_1Pyrr][TFSI] and [$C_4C_1C_1$Im][TFSI] [23]. Even though the toxicity of the salt itself is lower than that of the pure IL [43], this reduction can be attributed to the increased concentration of [TFSI]$^-$ anions resulting from the salt addition. The highest [TFSI]$^-$ concentration enhances the hydrophobicity of the sample, which is directly related to the bioluminescence inhibition in A. Fischeri. Despite this, according to the classification by Passino and Smith [24], both compounds can be classified within the same toxicity level: practically harmless.

4. Conclusions

In this work, the thermal and electric properties of pure [C_3C_1Pyrr][TFSI] IL and its binary mixtures with [Li][TFSI] at four different concentrations were studied, as well as the toxicity of these compounds, which was also analysed in the study. The main conclusions of the work are as follows:

- The increase in concentration of [Li][TFSI] makes the crystallization of the samples more challenging, particularly at highest scanning rate. This results in increased amorphous behaviour of the mixtures, evident from the broadening of the peaks and the appearance of the glass transition that takes place at higher temperatures when salt concentration is increased. Despite this, peaks related to crystalline behaviour (freezing and melting) remains observable for all the salt concentrations.
- The ionic conductivity of [C_3C_1Pyrr][TFSI] + [Li][TFSI] mixtures, whose behaviour against temperature is well described by the Vogel-Fulcher-Tammann (VFT) equation, decreases with [Li][TFSI] concentration. A linear increase in activation energies with salt concentration was found when fitting to Arrhenius equation, indicating that ion conduction in the electrolyte requires more energy to take place.
- The addition of salt significantly increases the toxicity of the mixture, as indicated by EC_{50} values. However, this increase does not change the toxicity classification according to the criteria used in this study.

Author Contributions: Conceptualization, A.S.-A., J.S. and J.J.P.; methodology, M.V., J.M.S.-P., A.S.-A., R.S.E. and J.J.P.; validation, A.S.-A., J.J.P. and J.S.; writing—original draft preparation, J.M.S.-P., J.J.P., A.S.-A. and J.S.; writing—review and editing, J.J.P., M.V. and J.S.; funding acquisition, J.S. and J.J.P. All authors have read and agreed to the published version of the manuscript.

Funding: This research was founded by Xunta de Galicia through GRC ED431C 2020/10 and ED481D 2023/014 projects.

Data Availability Statement: The original contributions presented in the study are included in the article; further inquiries can be directed to the corresponding author.

Acknowledgments: The authors acknowledge M. Gómez (RIAIDT-USC) for the technical support in DSC measurements, and also C. Gracia (TA-Instruments) is deeply appreciated by the authors for his invaluable assistance in interpreting the complex DSC curves. This work was supported by Xunta de Galicia through the GRC ED431C 2020/10 project. A. Santiago-Alonso thanks funding to the Doutoramento Industrial program from GAIN-Xunta de Galicia, and J. J. Parajó (ED481D 2023/014) thanks the I2C postdoctoral program of the Xunta de Galicia for their support in funding the study.

Conflicts of Interest: The authors declare no conflicts of interest.

References

1. Chen, Y.; Kang, Y.; Zhao, Y.; Wang, L.; Liu, J.; Li, Y.; Liang, Z.; He, X.; Li, X.; Tavajohi, N.; et al. A Review of Lithium-Ion Battery Safety Concerns: The Issues, Strategies, and Testing Standards. *J. Energy Chem.* **2021**, *59*, 83–99. [CrossRef]
2. Xu, K. Nonaqueous Liquid Electrolytes for Lithium-Based Rechargeable Batteries. *Chem. Rev.* **2004**, *104*, 4303–4417. [CrossRef] [PubMed]
3. Lee, J.; Jeon, A.R.; Lee, H.J.; Shin, U.; Yoo, Y.; Lim, H.D.; Han, C.; Lee, H.; Kim, Y.J.; Baek, J.; et al. Molecularly Engineered Linear Organic Carbonates as Practically Viable Nonflammable Electrolytes for Safe Li-Ion Batteries. *Energy Environ. Sci.* **2023**, *16*, 2924–2933. [CrossRef]

4. Meng, T.; Young, K.H.; Wong, D.F.; Nei, J. Ionic Liquid-Based Non-Aqueous Electrolytes for Nickel/Metal Hydride Batteries. *Batteries* **2017**, *3*, 4. [CrossRef]
5. Balducci, A. Ionic Liquids in Lithium-Ion Batteries. *Top. Curr. Chem.* **2017**, *375*, 20. [CrossRef] [PubMed]
6. Parajó, J.J.; Villanueva, M.; Salgado, J. 1.Thermal Stability of Ionic Liquids. In *Synthesis, Properties, Technologies and Applications*; Fehrmann, R., Santini, C., Eds.; De Gruyter: Berlín, Germany, 2019; pp. 1–16, ISBN 9783110583632.
7. Parajó, J.J.; Otero-Mato, J.M.; Lobo Ferreira, A.I.M.C.; Varela, L.M.; Santos, L.M.N.B.F. Enthalpy of Solvation of Alkali Metal Salts in a Protic Ionic Liquid: Effect of Cation Charge and Size. *J. Mol. Liq.* **2022**, *360*, 119228. [CrossRef]
8. Kim, H.T.; Kang, J.; Mun, J.; Oh, S.M.; Yim, T.; Kim, Y.G. Pyrrolinium-Based Ionic Liquid as a Flame Retardant for Binary Electrolytes of Lithium Ion Batteries. *ACS Sustain. Chem. Eng.* **2016**, *4*, 497–505. [CrossRef]
9. Dupré, N.; Moreau, P.; De Vito, E.; Quazuguel, L.; Boniface, M.; Kren, H.; Bayle-Guillemaud, P.; Guyomard, D. Carbonate and Ionic Liquid Mixes as Electrolytes to Modify Interphases and Improve Cell Safety in Silicon-Based Li-Ion Batteries. *Chem. Mater.* **2017**, *29*, 8132–8146. [CrossRef]
10. Salvador, M.A.; Maji, R.; Rossella, F.; Degoli, E.; Ruini, A.; Magri, R. Structural and Dynamic Characterization of Li–Ionic Liquid Electrolyte Solutions for Application in Li-Ion Batteries: A Molecular Dynamics Approach. *Batteries* **2023**, *9*, 234. [CrossRef]
11. Kerner, M.; Johansson, P. Pyrrolidinium Fsi and TFSI-Based Polymerized Ionic Liquids as Electrolytes for High-Temperature Lithium-Ion Batteries. *Batteries* **2018**, *4*, 10. [CrossRef]
12. McGrath, L.M.; Rohan, J.F. Pyrrolidinium Containing Ionic Liquid Electrolytes for Li-Based Batteries. *Molecules* **2020**, *25*, 6002. [CrossRef] [PubMed]
13. Ravi, M.; Kim, S.; Ran, F.; Kim, D.S.; Lee, Y.M.; Ryou, M.H. Hybrid Gel Polymer Electrolyte Based on 1-Methyl-1-Propylpyrrolidinium Bis(Trifluoromethanesulfonyl) Imide for Flexible and Shape-Variant Lithium Secondary Batteries. *J. Memb. Sci.* **2021**, *621*, 119018. [CrossRef]
14. Yang, B.; Li, C.; Zhou, J.; Liu, J.; Zhang, Q. Pyrrolidinium-Based Ionic Liquid Electrolyte with Organic Additive and LiTFSI for High-Safety Lithium-Ion Batteries. *Electrochim. Acta* **2014**, *148*, 39–45. [CrossRef]
15. Makino, T.; Kanakubo, M.; Umecky, T.; Suzuki, A.; Nishida, T.; Takano, J. Pressure–Volume–Temperature–Composition Relations for Carbon Dioxide+pyrrolidinium-Based Ionic Liquid Binary Systems. *Fluid. Phase Equilib.* **2013**, *360*, 253–259. [CrossRef]
16. Liu, Q.; Zhao, L.; Ma, L.; Chu, J.; Wang, J.; Zang, Y. Hydroxyethyl Group Effect on Properties of Bis[(Trifluoromethyl)Sulfonyl]Imide-Type Ionic Liquids. *J. Chem. Eng. Data* **2020**, *65*, 4780–4789. [CrossRef]
17. Huang, Q.; Lourenço, T.C.; Costa, L.T.; Zhang, Y.; Maginn, E.J.; Gurkan, B. Solvation Structure and Dynamics of Li+ in Ternary Ionic Liquid–Lithium Salt Electrolytes. *J. Phys. Chem. B* **2019**, *123*, 516–527. [CrossRef]
18. Villanueva, M.; Parajó, J.J.; Sánchez, P.B.; García, J.; Salgado, J. Liquid Range Temperature of Ionic Liquids as Potential Working Fluids for Absorption Heat Pumps. *J. Chem. Thermodyn.* **2015**, *91*, 127–135. [CrossRef]
19. Parajó, J.J.; Villanueva, M.; Sánchez, P.B.; Salgado, J. Liquid Window of Some Biologically-Active Ionic Liquids. *J. Chem. Thermodyn.* **2018**, *126*, 1–10. [CrossRef]
20. Salgado, J.; Parajó, J.J.; Villanueva, M.; Rodríguez, J.R.; Cabeza, O.; Varela, L.M. Liquid Range of Ionic Liquid–Metal Salt Mixtures for Electrochemical Applications. *J. Chem. Thermodyn.* **2019**, *134*, 164–174. [CrossRef]
21. Leys, J.; Wübbenhorst, M.; Preethy Menon, C.; Rajesh, R.; Thoen, J.; Glorieux, C.; Nockemann, P.; Thijs, B.; Binnemans, K.; Longuemart, S. Temperature Dependence of the Electrical Conductivity of Imidazolium Ionic Liquids. *J. Chem. Phys.* **2008**, *128*, 064509. [CrossRef]
22. Parajó, J.J.; Santiago-Alonso, A.; Vallet, P.; Teijeira, T.; Emeterio, R.S.; Villanueva, M.; Salgado, J. Comprehensive Analysis of the Acute Toxicity of Ionic Liquids Using Microtox® Bioassays. *Appl. Sci.* **2024**, *14*, 2480. [CrossRef]
23. Parajó, J.J.; Vallet, P.; Varela, L.M.; Villanueva, M.; Salgado, J. Ecotoxicity of Binary Mixtures of ILs and Inorganic Salts of Electrochemical Interest. *Environ. Sci. Pollut. Res.* **2022**, *29*, 24983–24994. [CrossRef] [PubMed]
24. Passino, D.R.M.; Smith, S.B. Acute Bioassays and Hazard Evaluation of Representative Contaminants Detected in Great Lakes Fish. *Environ. Toxicol. Chem.* **1987**, *6*, 901–907. [CrossRef]
25. Shi, B.; Wang, Z.; Wen, H. Research on the Strengths of Electrostatic and van Der Waals Interactions in Ionic Liquids. *J. Mol. Liq.* **2017**, *241*, 486–488. [CrossRef]
26. Salminen, J.; Papaiconomou, N.; Kumar, R.A.; Lee, J.-M.; Kerr, J.; Newman, J.; Prausnitz, J.M. Physicochemical Properties and Toxicities of Hydrophobic Piperidinium and Pyrrolidinium Ionic Liquids. *Fluid. Phase Equilib.* **2007**, *261*, 421–426. [CrossRef]
27. MacFarlane, D.R.; Meakin, P.; Sun, J.; Amini, N.; Forsyth, M. Pyrrolidinium Imides: A New Family of Molten Salts and Conductive Plastic Crystal Phases. *J. Phys. Chem. B* **1999**, *103*, 4164–4170. [CrossRef]
28. Huang, Q.; Lee, Y.-Y.; Gurkan, B. Pyrrolidinium Ionic Liquid Electrolyte with Bis(Trifluoromethylsulfonyl)Imide and Bis(Fluorosulfonyl)Imide Anions: Lithium Solvation and Mobility, and Performance in Lithium Metal–Lithium Iron Phosphate Batteries. *Ind. Eng. Chem. Res.* **2019**, *58*, 22587–22597. [CrossRef]
29. Henderson, W.A.; Passerini, S. Phase Behavior of Ionic Liquid−LiX Mixtures: Pyrrolidinium Cations and TFSI- Anions. *Chem. Mater.* **2004**, *16*, 2881–2885. [CrossRef]
30. Appetecchi, G.B.; Montanino, M.; Carewska, M.; Moreno, M.; Alessandrini, F.; Passerini, S. Chemical–Physical Properties of Bis(Perfluoroalkylsulfonyl)Imide-Based Ionic Liquids. *Electrochim. Acta* **2011**, *56*, 1300–1307. [CrossRef]
31. Gómez, E.; Calvar, N.; Domínguez, Á.; Macedo, E.A. Thermal Behavior and Heat Capacities of Pyrrolidinium-Based Ionic Liquids by DSC. *Fluid. Phase Equilib.* **2018**, *470*, 51–59. [CrossRef]

32. Annat, G.; Forsyth, M.; MacFarlane, D.R. Ionic Liquid Mixtures—Variations in Physical Properties and Their Origins in Molecular Structure. *J. Phys. Chem. B* **2012**, *116*, 8251–8258. [CrossRef] [PubMed]
33. González, B.; González, E.J. Physical Properties of the Pure 1-Methyl-1-Propylpyrrolidinium Bis(Trifluoromethylsulfonyl)Imide Ionic Liquid and Its Binary Mixtures with Alcohols. *J. Chem. Thermodyn.* **2014**, *68*, 109–116. [CrossRef]
34. Vallet, P.; Parajó, J.J.; Santiago-Alonso, A.; Villanueva, M.; Varela, L.M.; Salgado, J. Thermal Characterization of [C2Im][NO_3] and Multivalent Nitrate Salts Mixtures. *Crystals* **2024**, *14*, 502. [CrossRef]
35. Parajó, J.J.; Vallet, P.; Villanueva, M.; Cabeza, O.; Fernández-Carretero, F.; García Luis, A.; Di Pietro, M.E.; Mele, A.; Castiglione, F.; Salgado, J.; et al. Ionogels Based on Protic Ionic Liquid-Lithium Salt Mixtures. *J. Mol. Liq.* **2024**, *397*, 124093. [CrossRef]
36. Martinelli, A.; Matic, A.; Jacobsson, P.; Börjesson, L.; Fernicola, A.; Scrosati, B. Phase Behavior and Ionic Conductivity in Lithium Bis(Trifluoromethanesulfonyl)Imide-Doped Ionic Liquids of the Pyrrolidinium Cation and Bis(Trifluoromethanesulfonyl)Imide Anion. *J. Phys. Chem. B* **2009**, *113*, 11247–11251. [CrossRef]
37. Vallet, P.; Bouzón-Capelo, S.; Méndez-Morales, T.; Gómez-González, V.; Arosa, Y.; de la Fuente, R.; López-Lago, E.; Rodríguez, J.R.; Gallego, L.J.; Parajó, J.J.; et al. On the Physical Properties of Mixtures of Nitrate Salts and Protic Ionic Liquids. *J. Mol. Liq.* **2022**, *350*, 118483. [CrossRef]
38. Tsamopoulos, A.J.; Wang, Z.-G. Ion Conductivity in Salt-Doped Polymers: Combined Effects of Temperature and Salt Concentration. *ACS Macro Lett.* **2024**, *13*, 322–327. [CrossRef]
39. Petrowsky, M.; Frech, R. Salt Concentration Dependence of the Compensated Arrhenius Equation for Alcohol-Based Electrolytes. *Electrochim. Acta* **2010**, *55*, 1285–1288. [CrossRef]
40. Abbas, M.; Adil, M.; Ehtisham-ul-Haque, S.; Munir, B.; Yameen, M.; Ghaffar, A.; Shar, G.A.; Asif Tahir, M.; Iqbal, M. Vibrio Fischeri Bioluminescence Inhibition Assay for Ecotoxicity Assessment: A Review. *Sci. Total Environ.* **2018**, *626*, 1295–1309. [CrossRef]
41. Silva, F.A.; Coutinho, J.A.P.; Ventura, S.P.M. *Aquatic Toxicology of Ionic Liquids (ILs) in Encyclopedia of Ionic Liquids*, 1st ed.; Springer Nature: Berlín, Germany, 2019.
42. Villanueva, M.; Vallet, P.; Teijeira, T.; Santiago-Alonso, A.; Amigo, A.; Tojo, E.; Varela, L.M.; Parajó, J.J.; Salgado, J. Effect of Alkyl Chain Length on the Thermal Properties and Toxicity of N-Alkyl-Ammonium Nitrate Ionic Liquids (n = 2, 3, 4, 5, 6, 8) for Energy Applications. *J. Therm. Anal. Calorim.* **2024**. [CrossRef]
43. Viboud, S.; Papaiconomou, N.; Cortesi, A.; Chatel, G.; Draye, M.; Fontvieille, D. Correlating the Structure and Composition of Ionic Liquids with Their Toxicity on Vibrio Fischeri: A Systematic Study. *J. Hazard. Mater.* **2012**, *215–216*, 40–48. [CrossRef] [PubMed]

batteries

MDPI

Article

The Multi-Parameter Fusion Early Warning Method for Lithium Battery Thermal Runaway Based on Cloud Model and Dempster–Shafer Evidence Theory

Ziyi Xie [1,†], Ying Zhang [1,†], Hong Wang [1], Pan Li [2], Jingyi Shi [3], Xiankai Zhang [3,*] and Siyang Li [1]

1 School of Safety Science and Emergency Management, Wuhan University of Technology, Wuhan 430070, China; 321705@whut.edu.cn (Z.X.); yzhang@whut.edu.cn (Y.Z.); whwang@whut.edu.cn (H.W.); siyangli2020@whut.edu.cn (S.L.)
2 Wuhan Second Institute of Ship Design and Research, Wuhan 430205, China; lipan@mail.ustc.edu.cn
3 EVE Power Co., Ltd., Jingmen 516006, China; 074852@evebattery.com
* Correspondence: 070863@evebattery.com
† These authors contributed equally to this work.

Abstract: As the preferred technology in the current energy storage field, lithium-ion batteries cannot completely eliminate the occurrence of thermal runaway (TR) accidents. It is of significant importance to employ real-time monitoring and warning methods to perceive the battery's safety status promptly and address potential safety hazards. Currently, the monitoring and warning of lithium-ion battery TR heavily rely on the judgment of single parameters, leading to a high false alarm rate. The application of multi-parameter early warning methods based on data fusion remains underutilized. To address this issue, the evaluation of lithium-ion battery safety status was conducted using the cloud model to characterize fuzziness and Dempster–Shafer (DS) evidence theory for evidence fusion, comprehensively assessing the TR risk level. The research determined warning threshold ranges and risk levels by monitoring voltage, temperature, and gas indicators during lithium-ion battery overcharge TR experiments. Subsequently, a multi-parameter fusion approach combining cloud model and DS evidence theory was utilized to confirm the risk status of the battery at any given moment. This method takes into account the fuzziness and uncertainty among multiple parameters, enabling an objective assessment of the TR risk level of lithium-ion batteries.

Keywords: lithium-ion battery; thermal runaway; early warning; multi-source data fusion

1. Introduction

The lithium-ion battery (LIB), prized for its high energy density, lightweight design, and eco-friendly attributes, stands as the preferred technology in the realm of energy storage. Its applications span across various domains, including mobile electronic devices, medical equipment, electric vehicles, and energy storage systems. Yet, amidst this broad range of utilization, safety concerns must not be underestimated. Between 2017 and 2022, over 14 incidents of fire and explosion accidents occurred in energy storage stations. Present warning methodologies for LIBs hinge on distinct threshold models for diverse parameters like voltage, temperature, and gas production. Based on the evolution of internal reaction, thermal runaway (TR) can be categorized into different stages. Randolph et al. [1] segmented the TR of LIB into four stages based on internal reactions. At present, there are several methods for monitoring and warning based on multiple parameters, with the conventional single-parameter threshold determination being a prevalent approach.

1.1. Early Warning Technology Based on External Parameters

Sensors enable monitoring of parameters such as voltage and temperature during battery use, which allows for thermal runaway warning based on anomalies detected in these monitored parameters. Xia et al. [2] utilized multi-voltage sensors and algorithms to monitor the operational status of each battery in a battery pack. Jia et al. [3] employed K-type thermocouples and fiber Bragg grating (FBG) sensors to monitor temperature variations at three different positions of LIB, enabling temperature warnings for TR. Nascimento et al. [4,5] proposed a network comprising 37 FBG sensors to prevent severe consequences of TR incidents and enhance battery safety. However, accurately assessing internal chain reactions remains difficult due to the challenge of measuring the difference between internal and surface temperatures. Current monitoring of battery voltage and surface temperature is insufficient to effectively prevent battery failure [6].

1.2. Early Warning Technology Based on Surface Strain of the Battery

Peng et al. [7] developed a high-precision strain sensor with 11.5 times greater sensitivity compared to FBG sensors. Chen et al. [8] used strain gauges to detect surface strain changes in batteries under thermal abuse, finding earlier detection compared to conventional electrical signals. However, due to the closed nature of lithium-ion batteries, external parameters cannot accurately reflect internal electrochemical reactions or assess thermal runaway risks precisely.

1.3. Early Warning Technology Based on Internal Temperature, Pressure, and Impedance

Raghavan et al. [9] showed that embedded FBG temperature sensors can accurately measure internal battery temperatures. Lyu et al. [10] identified a shift in impedance spectrum from negative to positive within the 30 Hz to 90 Hz frequency range and used this change to develop an early warning system for thermal runaway based on impedance frequency slope variations. Mei et al. [11] developed a fiber optic sensor tailored for insertion into commercial 18650 batteries, which enables continuous monitoring of internal temperature and pressure during TR, issuing warnings before safety valves are triggered.

1.4. Early Warning Technology Based on Characteristic Gases

Cai et al. [12] suggested CO_2 as a characteristic gas in TR scenarios, verified and established thresholds with sensors. Jin et al. [13] demonstrated that H_2 can serve as an early safety warning indicator through overcharging experiments with LIB packs. Threshold determination methods involve setting upper and lower limits for various battery parameters. When parameters collected by on-site sensors exceed the established range, an alert is triggered.

1.5. Early Warning Technology Based on Neural Networks

Conversely, AI-based monitoring and warning predict potential battery anomalies by establishing various types of network models. Jiang et al. [14] employed LSTM-RNN and GRU-RNN to achieve accurate real-time battery temperature estimation, with a maximum absolute error of about 0.75 °C. Zhang et al. [15] established a data-driven MMTPFNN model for forecasting TR propagation when battery temperature exceeds 60 °C, while employing an LSTM model for regular temperature prediction. This method enables more precise prediction of battery temperature, but still relies on a single parameter for early warning, which may not avoid errors such as false negatives.

In the realm of monitoring and early warning technologies for lithium-ion battery thermal runaway, parameters such as voltage, temperature, gas, impedance, and pressure are typically monitored to facilitate early warning. However, these parameters are often treated as independent, and effective integration methods for comprehensive analysis remain lacking. When employing a solitary parameter threshold model for monitoring and warning of battery TR, there is a lack of adequate description regarding the interplay of parameters' fuzziness. Dealing with this fuzziness issue involves subjectivity, posing challenges in acquiring a thorough and objective understanding of TR. And this undermines the precision and dependability of TR warning system.

The cloud model is an evaluation method that transforms qualitative concepts into quantitative descriptions, addressing the fuzziness and uncertainty between qualitative concepts and quantitative descriptions. The Dempster–Shafer (DS) evidence theory is a typical method used to analyze and fuse uncertain information. Xiang et al. [16] proposed a multi-sensor data fusion method based on the cloud model and improved evidence theory. By integrating evidence similarity and credibility, this method enhances fusion accuracy. Validation through early indoor fire detection demonstrates that the approach achieves higher detection precision and reduced false alarm rates.

To address the issues of ambiguity and randomness in assessment, a novel multi-parameter warning method based on cloud model and DS evidence theory was proposed to evaluate the TR risk of LIB in this paper. Based on the empirical model derived from previous TR experiments, three-level warning thresholds are established to categorize thermal runaway risk levels. By applying cloud models to illustrate the degree of involvement within each of the three risk levels and employing Dempster's combination rule for the fusion of data from various sensors, uncertainty was converted into basic probability assignment (BPA). This procedure produces the mass function, belief function, and plausibility function for the TR risk level of LIBs. As a result, the TR risk level of LIBs can be evaluated at any specific moment based on these findings.

Based on the voltage, temperature, gas multi-parameter variations, a multi-level warning strategy was developed to map thermal runaway risks through different warning stages. By integrating the cloud model with DS evidence theory, this approach combines multiple parameters to assess risk levels, reducing subjective influences and enhancing early warning accuracy. This method offers new insights and techniques for thermal runaway monitoring and early warning.

2. Thermal Runaway Experiment for Lithium-Ion Batteries

2.1. TR Test Apparatus Setup

In the previous study, a thermal runaway experiment was conducted on LIBs using 20 Ah LFP batteries (with a cutoff current of 2 A and a cutoff voltage of 2.5 V). Figure 1 shows the setup of the overcharge test platform. The entire overcharge test process was carried out in an explosion-proof box, and a camera positioned in front of the explosion-proof glass was used to record the experimental phenomena. A battery testing system (BTS, Neware BTS-50 V/20 A, Shenzhen, China) was employed for the charging and discharging cycles of the battery, as well as for recording the voltage data during the overcharge test. During the test, 3 K-type thermocouples (1 mm in diameter) were used to measure the surface temperature of the battery, with measurement points indicated as Tc1-Tc3 in Figure 1. A gas analyzer (Shenzhen Qi'an Technology Co., Ltd., Shenzhen, China) was used, with a gas pipe inserted above the battery's safety valve. An air pump was used to draw the gas produced by the battery into H_2 sensor (0–2000 ppm, resolution: 1 ppm), VOC sensor (0–100 ppm, accuracy: 0.1%), and CO sensor (0–2000 ppm, resolution: 1 ppm) of the instrument to monitor real-time changes in gas concentration.

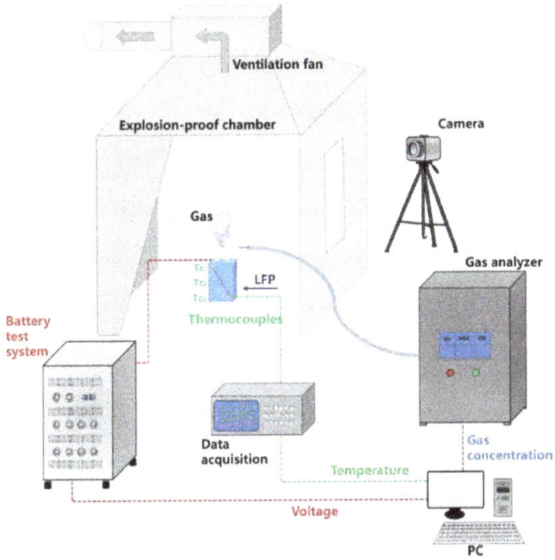

Figure 1. Setup of the overcharge test platform.

2.2. Experimental Method for TR Test

Before initiating the overcharge tests, the LFP batteries were charged to 100% State of Charge (SOC) using a constant current-constant voltage (CC-CV) protocol with the BTS. Following a resting period, the batteries were placed within an explosion-proof chamber. The BTS was programmed to apply various charge rates for overcharging experiments, including rates of 0.5 C, 0.75 C, and 1 C, as specified in Table 1. The overcharging was continued until the battery voltage experienced a sudden drop to 0, at which point the BTS terminated the overcharging process. During the experiment, the BTS recorded the battery voltage, data acquisition collected the surface temperature data of battery, and the gas analyzer monitored changes in gas concentration. After the cessation of gas emission and a subsequent 4 h stabilization period, data on voltage, temperature, and gas concentrations were gathered to analyze the variations in characteristic parameters throughout the thermal runaway process.

Table 1. Experimental conditions of the overcharge thermal runaway tests.

NO.	Overcharge C-Rate	Repetition Time
1	0.5	3
2	0.75	3
3	1	3

2.3. Experimental Results and Discussion

In the TR experiment concerning the overcharge of LIBs, real-time monitoring of key parameters in-cluding voltage, temperature, and gas production was conducted [17]. As shown in Figure 2, the voltage, temperature, and gas concentration trends under three different overcharge rates (0.5 C, 0.75 C, and 1 C) exhibit similar patterns and the entire process of the LIBs overcharge experiment was segmented into four stages, delineated by specific events such as the emergence of the voltage turning point (V_{TP}), the activation of the safety valve, and the voltage decline.

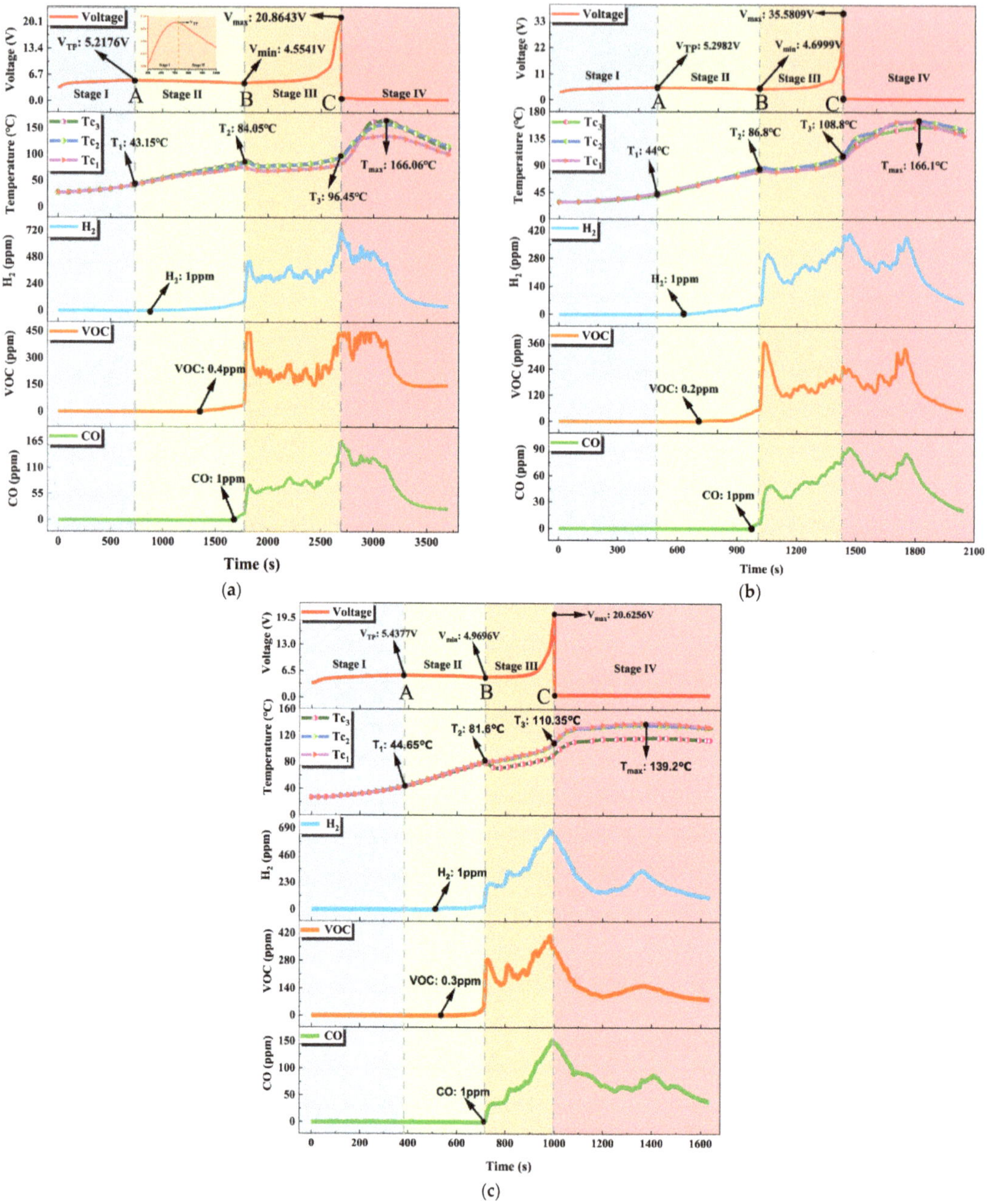

Figure 2. The segment of the four stages of overcharge TR and the changes in voltage, temperature, and gas concentration in various overcharge C-rate: (**a**) 0.5 C, (**b**) 0.75 C, (**c**) 1 C. [17].

During stage 1, the lithium-ion battery begins overcharging, leading to a gradual voltage increase until reaching a voltage turning point (V_{TP}). The surface temperature rises slowly to 50 °C, while the voltage curve shows a brief plateau, indicating active material loss.

As overcharging continues in stage 2, the voltage gradually decreases to the minimum voltage (V_{min}). Heat generation accelerates, raising the surface temperature above 80 °C and reaching the first temperature peak (T_2). Gas analyzers detect H_2, VOC, and CO due to gas generation and battery swelling.

In stage 3, the safety valve opens due to excessive electrolyte decomposition, releasing gas and heat and causing a temporary drop in surface temperature. As internal reactions intensify, the temperature rises, and the voltage increases to the maximum voltage (V_{max}) associated with thermal runaway. Temperature differences at measurement points indicate internal instability.

During stage 4, separator damage causes lithium dendrite short-circuiting, leading to rapid heat release and intensified side reactions. The battery experiences a voltage drop and thermal runaway, releasing all gases and reaching a second temperature peak (T_{max}). The voltage drop precedes the temperature rise, providing a potential early warning indicator.

3. Warning Method Based on Cloud Model and DS Evidence Theory

3.1. Selection of Evaluation Indicators for TR Risk State

Taking the experimental results under a 0.5 C overcharge rate as an example, Table 2 depicts the voltage, temperature, and gas concentration variations throughout the four stages of TR. Meanwhile, Table 3 presents the gas concentration values at critical nodes.

Table 2. Changes in voltage, temperature, and gas concentration at each stage.

Stage	Voltage	Temperature	Gas Concentration
I	Slow rise to V_{TP} V_{TP} = 5.22 V	Slow rise to T_1 T_1 = 43.15 °C	No significant change
End of Stage I		V_{TP} appearance	
II	Slow decrease to V_{min} V_{min} = 4.55 V	Rapid rise to T_2 T_2 = 84.05 °C	H_2, VOC, CO gradually increase
End of Stage II		safety valve opening	
III	Rapid rise to V_{max} V_{max} = 20.86 V	Brief decrease and rise to T_3 T_3 = 96.45 °C	continue to increase
End of Stage III		V_{max} appearance	
IV	Rapid drop to 0	Rapid rise to T_{max} T_{max} = 166.06 °C	continue to increase

Table 3. Gas concentration at nodes where an upward trend occurs and end of stage III.

State	H_2 (ppm)	VOC (ppm)	CO (ppm)
Upward trend	1	0.4	1
End of Stage III	74	42	12

Based on the experimental results under different operating conditions, thermal runaway can be divided into four stages, leading to the proposal of a three-level early warning strategy for thermal runaway:

1. The first-level warning selects the V_{TP} at the transition between the first and second stages as the characteristic parameter, with a value range of (5.2 V, 5.4 V). Considering a certain margin of error, the voltage threshold is set within the range of (5.1 V, 5.5 V).

2. The second-level warning is determined by the maximum safe temperature of 50 °C for the battery pack. With an error margin, the temperature range is set between (48 °C, 55 °C). Based on this temperature range and the time-varying curves of temperature and hydrogen concentration, the hydrogen concentration range is determined to be (2 ppm, 5 ppm).

3. The third-level warning is based on the critical temperature of 80 °C before the battery enters the third stage of thermal runaway after the safety valve opens. Considering an error margin, the temperature range is set between (78 °C, 85 °C). The carbon monoxide concentration range is then determined to be (2 ppm, 10 ppm) based on this temperature range and the time-varying curves of temperature and carbon monoxide concentration.

In this study, overcharge TR is defined as an overcharge TR incident. Three key parameters—voltage, temperature, and gases (H_2, CO)—are selected to form the evaluation indicator system, represented by voltage, temperature, H_2 concentration, and CO concentration. Based on the three-level warning strategy for TR of LIBs, the risk level is categorized into three levels: L_1 for low risk, L_2 for medium risk, and L_3 for high risk. Figure 3 illustrates the three-level warning strategy for monitoring and warning overcharge TR of LIBs.

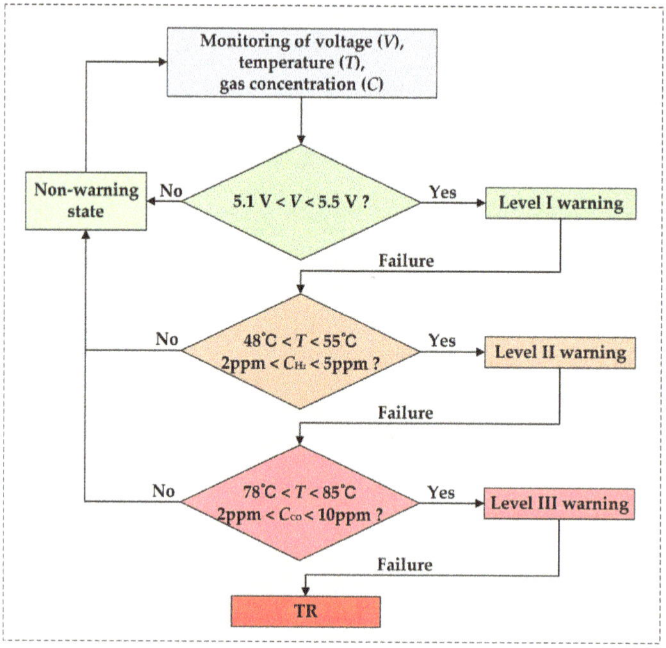

Figure 3. The three-level warning strategy for TR of LIBs.

3.2. Construction of Multi-Parameter Fusion Model Based on Cloud Model and DS Evidence Theory

In this study, the multi-parameter fusion model was constructed using cloud model and DS evidence theory. Utilizing cloud model facilitates the transformation of qualitative concepts into quantitative descriptions, effectively handling the inherent fuzziness and uncertainty between qualitative and quantitative information. The forward cloud model adeptly converts qualitative concepts into precise quantitative values, while the reverse cloud model transforms quantitative values back into qualitative concepts. The forward cloud model is utilized for safety status of LIB, which is characterized by three characteristic parameters: E_x, E_n, and H_e.

$$E_x = (b_{\max} + b_{\min})/2 \tag{1}$$

$$E_n = (b_{\max} - b_{\min})/6 \tag{2}$$

$$H_e = E_n/10 \tag{3}$$

where $b_{\max}$ and $b_{\min}$ denote the upper and lower critical values of the threshold interval $[b_{\min}, b_{\max}]$ for LIB under different warning levels.

The membership degree μ of each sensor measurement value (X_1, X_2, ..., X_n) in the cloud model can be calculated using Equation (4).

$$\mu = e^{-\frac{(x-E_x)^2}{2{E_n}^2}} \tag{4}$$

The DS evidence theory is a typical approach for fusing and analyzing uncertain information in a same identification framework, without relying on conditional probabilities. $\Theta = \{\theta_1, \theta_1, \ldots, \theta_i, \ldots, \theta_n\}$ represent the decision identification framework, consisting of a finite set of mutually exclusive elements. The basic probabilities $m : 2^{\Theta} \to [0,1], \forall A \subseteq \Theta$, according to Equation (5):

$$\begin{cases} m(\phi) = 0 \\ \sum_{A \subseteq \Theta} m(A) = 1 \end{cases} \tag{5}$$

The variable m represents the basic probability assignment (BPA) function for the recognition framework Θ, also known as the mass function. The mass function is used to calculate the basic probability for each element in the hypothesis space, as shown in Equation (6):

$$m_i(A_k) = \frac{\mu_{ik}}{\sum_{k=1}^{m} (\mu_{ik} | i)} \tag{6}$$

where k denotes the risk level of the evaluated LIB; A is the set of risk levels for the evaluated object; i denotes the sensor index.

According to the BPA, the belief function $\mathrm{Bel}(A)$ and plausibility function $\mathrm{Pl}(A)$ form a trust interval $[\mathrm{Bel}(A), \mathrm{Pl}(A)]$, representing the degree of confirmation for a specific hypothesis. On the recognition framework Θ, the belief function indicates the possibility of proposition A being true, calculated as the sum of the basic probabilities of all subsets of hypothesis A by Equation (7). The plausibility function on the recognition framework Θ represents the possibility that proposition A is not false, calculated as the sum of the basic probabilities of sets with a non-empty intersection with hypothesis A by Equation (8).

$$\mathrm{Bel}(A) = \sum_{B \subseteq A} m(B) \tag{7}$$

$$\mathrm{Pl}(A) = \sum_{B \cap A \neq \phi} m(B) \tag{8}$$

The relationship diagram between the Bel function, Pl function, and the trust interval is shown in Figure 4, where the degree of belief represents the lower bound estimate of the hypothesis credibility, and the degree of plausibility represents the upper bound estimate of the hypothesis credibility.

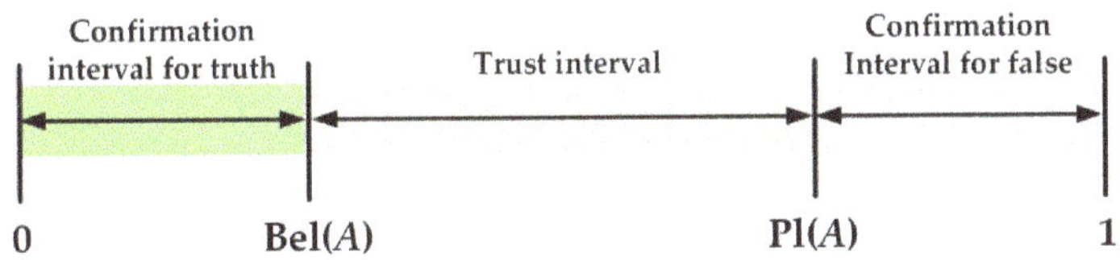

Figure 4. The distribution of belief function, plausibility function, and the belief interval.

For n pieces of evidence with n BPAs, the basic probability distribution (BPD) is calculated using Dempster's rule of combination, as shown in Equation (9). The normalization constant K is calculated by Equation (10).

$$m(A) = (m_1 \oplus m_2 \oplus \ldots \oplus m_n)(A) = \frac{1}{K} \sum_{A_1 \cap A_2 \cap \ldots \cap A_n = A} m_1(A_1) \cdot m_2(A_2) \cdot \ldots \cdot m_n(A_n) \tag{9}$$

$$K = \sum_{A_1 \cap A_2 \cap \ldots \cap A_n \neq \phi} m_1(A_1) \cdot m_2(A_2) \cdot \ldots \cdot m_n(A_n) = 1 - \sum_{A_1 \cap A_2 \cap \ldots \cap A_n = \phi} m_1(A_1) \cdot m_2(A_2) \cdot \ldots \cdot m_n(A_n) \quad (10)$$

For the evaluation of TR risk levels of LIBs, three risk levels L_1, L_2 and L_3 are defined, and four evaluation factors, voltage, temperature, H_2, and CO concentration, are set for assessment. The distribution set of mass functions for each evaluation factor is given by Equation (11).

$$M = (\{\phi\}, \{L_1\}, \{L_2\}, \{L_3\}, \{L_1, L_2\}, \{L_1, L_3\}, \{L_2, L_3\}, \{\Theta\}) \quad (11)$$

where $\{\Theta\}$ is equivalent to the set $\{L_1, L_2, L_3\}$, which represents the discernment framework set.

3.3. Decision Procedure for Three-Level TR Early Warning of LIBs

The process for evaluating the TR risk levels of LIBs based on the cloud model and DS evidence theory is illustrated in Figure 5.

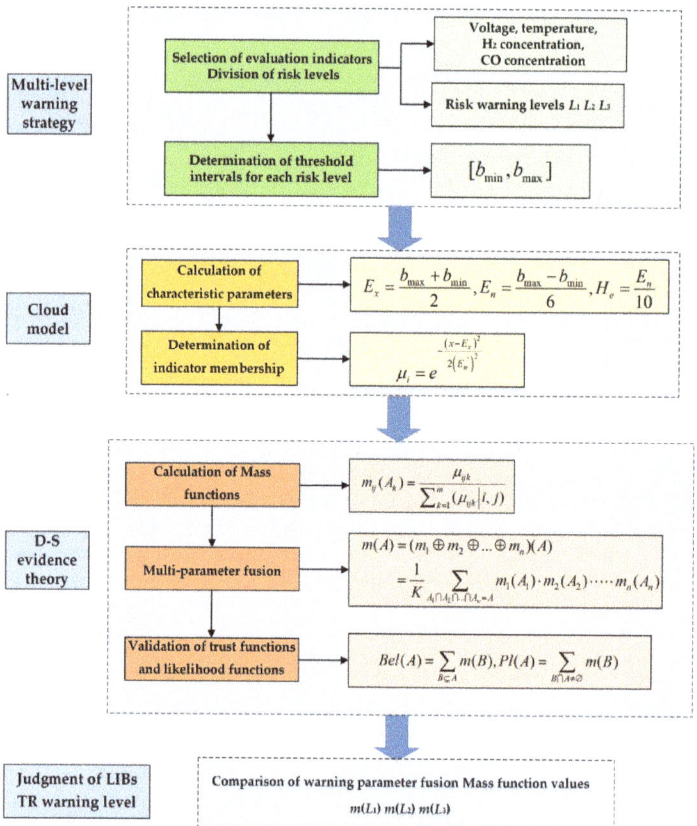

Figure 5. The process for three-level TR early warning of LIBs.

The process for determining TR risk level of LIBs, combining the cloud model and DS evidence theory, follows this concise procedure:

Step 1: Utilizing actual environmental sensor data at any given moment, determine the cloud characteristic parameters (E_x, E_n, H_e) for each evaluation factor under every risk warning level by Equations (1)–(3). Calculate the respective cloud model membership degree based on the measured values and cloud characteristic parameter values for each evaluation factor. This process aims to quantify the degree of attribution of each factor under different conditions, providing foundational data for subsequent risk assessment.

Step 2: Input the membership degrees into DS evidence theory. Construct the BPA function by Equation (6), obtaining the probability distribution of each evaluation factor in the corresponding risk zone. This reflects the possibility of each evaluation factor in different risk areas. Subsequently, use the Dempster combination rule with Equation (9) to obtain the fusion probabilities of each evaluation factor in each risk zone. Combine the information of various evaluation factors, use belief and plausibility functions to validate the credibility of the judgment, improving the accuracy of risk assessment for a more precise determination of the lithium battery's risk level.

Step 3: Based on different risk levels, determine the corresponding warning levels and take appropriate emergency measures. Develop specific response strategies based on risk assessment to ensure the safe operation of the LIBs system.

4. Empirical Analysis of TR Risk Assessment

To achieve comprehensive monitoring and risk assessment of LIBs during operation, the characteristic parameters—voltage, temperature, and gas—exhibiting patterns during the transition from a normal state to TR are defined as evaluation evidence. Four evaluation factors, including voltage, temperature, H_2, and CO concentration, are established. The risk level of thermal runaway is categorized into three levels: L_1, L_2, and L_3.

In Equation (1), the interval represents the evaluation factor threshold range of LIBs under different warning levels, with specific values detailed in Table 4. The calculation of E_x, E_n, and H_e for each evaluation factor under each risk warning level is determined by Equations (1)–(3), as shown in Table 5.

Table 4. The threshold ranges of four evaluation factor at three risk levels.

Risk Evaluation Factors		Risk Level		
		Low Risk L_1	Middle Risk L_2	High Risk L_3
F_1	Voltage/V	(5.1, 5.5)		
F_2	Temperature/°C		(48, 55)	(78, 85)
F_3	H_2 concentration/ppm		(2, 5)	
F_4	CO concentration/ppm			(2, 10)

Table 5. Cloud characteristic parameters of evaluation factors at three risk levels.

Evaluation Factors	Low Risk L_1			Middle Risk L_2			High Risk L_3		
	E_x	E_n	H_e	E_x	E_n	H_e	E_x	E_n	H_e
F_1	5.3	0.07	0.007						
F_2				51.5	1.17	0.117	81.5	1.17	0.117
F_3				3.5	0.5	0.05			
F_4							6	1.33	0.133

The specific values of battery voltage, temperature, and gas parameters at a certain moment during the overcharge-induced TR experiment are shown in Table 6, serving as an example to validate the multi-parameter fusion early warning method for TR of LIB based on cloud model and DS evidence theory. Subsequently, the membership degrees for each evaluation factor are calculated by Equation (4) and converted into BPAs, as shown in Table 7. Figure 6 shows the marking of this moment on the voltage, temperature, and gas concentration over time curves.

Table 6. Values of each evaluation factor at a certain moment.

Evaluation Factors	Values
F_1	5.2
F_2	50
F_3	0
F_4	2

Table 7. The distribution of BPAs for each evaluation factor at the certain moment.

Evaluation Factors	Values	BPAs				Risk Levels
		m_{L_1}	m_{L_2}	m_{L_3}	m_Θ	
F_1	5.2	0.36	0.00	0.00	0.64	unknown
F_2	50	0.00	0.69	0.00	0.31	Middle risk
F_4	2	0.00	0.00	0.76	0.24	High risk

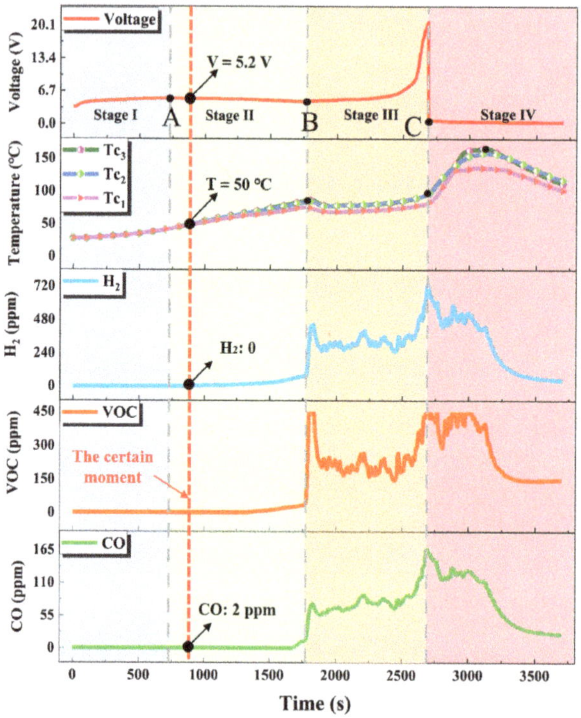

Figure 6. Voltage, temperature, and gas concentration variation curves with a time marker at the specific moment.

Based on the results from Table 7, the temperature and CO gas concentration indicators have the highest credibility in the middle and high-risk levels. During the TR monitoring and warning process of LIB, it is advisable to enhance real-time monitoring and control of these indicators to improve battery safety management and reduce the probability of TR accidents of LIB. Utilizing the multi-source fusion method, the fusion of evaluation factors is performed by Equations (9) and (10). In Equation (11), the assignment results for the set of mass functions distribution for evaluation factors F_1, F_2, and F_4 are shown in Table 8. The values of K and mass functions for the battery at the three risk levels are derived, as shown in Table 9. Finally, the belief functions and plausibility functions for each risk level are computed by Equations (7) and (8), and the results are displayed in Table 10.

Table 8. Mass function distribution of sets of evaluation factors at the certain moment.

Mass Function	Sets							
	$\{\phi\}$	$\{L_1\}$	$\{L_1\}$	$\{L_1\}$	$\{L_1,L_1\}$	$\{L_1,L_1\}$	$\{L_1,L_1\}$	$\{\Theta\}$
M_1	0	0.36	0	0	0	0	0	0.64
M_2	0	0	0.69	0	0	0	0	0.31
M_4	0	0	0	0.76	0	0	0	0.24

Table 9. Values of K and mass function for three risk levels at the certain moment.

Values	BPAs					Risk Levels
	K	$m(L_1)$	$m(L_2)$	$m(L_3)$	$m(\Theta)$	
The certain moment	0.3312	0.081	0.320	0.455	0.144	High risk

Table 10. Values of belief and plausibility function for three risk levels at the certain moment.

Risk Level	Functions	
	Belief Function Bel (L_x)	Plausibility Function Pl (L_x)
L_1	0.081	0.225
L_2	0.320	0.464
L_3	0.455	0.599

Figure 6 shows that at this time point, the battery is in the second stage of thermal runaway. The voltage of 5.0 V falls within the first-level warning threshold (5.1 V to 5.5 V), the temperature of 50 °C is within the second-level threshold (48 °C to 55 °C), and the CO concentration of 2 ppm is outside the third-level threshold (2 ppm to 10 ppm). Individually, these parameters do not clearly indicate the risk level. However, using the multi-parameter fusion model based on the cloud model and DS evidence theory, the calculated risk levels indicate $m(L_3) > m(L_2) > m(\Theta) > m(L_1)$.

Constructing the BPA distribution of different evaluation factors based on actual measurement values, integrating data from multiple sensors, calculating the mass function, and validating by belief functions and plausibility functions, the results indicate the following: at the choosing moment, the LIB is at high risk. Therefore, level III warning is required, and corresponding measures such as cooling the battery, releasing gases, and inspecting and replacing the LIB should be implemented.

5. Conclusions

Based on experiments of overcharge-induced TR of LIB, three warning levels for TR have been established:

1. Level I warning: Voltage threshold (5.1 V, 5.5 V);
2. Level II warning: Temperature threshold (48 °C, 55 °C) and H_2 concentration threshold (2 ppm, 5 ppm);
3. Level III warning: Temperature threshold (78 °C, 85 °C) and CO concentration threshold (2 ppm, 10 ppm).

Associating these with low risk (L_1), medium risk (L_2), and high risk (L_3), and using Dempster's rule of combination along with the voltage, temperature, and gas evaluation factors' threshold intervals, evidence fusion was executed. Considering the uncertainty of each evaluation factor, actual measured values, and cloud model feature parameters were used to determine the membership degree of each evaluation factor at different risk levels, constructing BPA function for each evaluation factor at various risk levels. This allowed for a comprehensive evaluation of the TR risk level of LIBs.

Using experimental data at a specific moment, including voltage, temperature, and CO gas concentration, we validated the multi-source data fusion method and determined the membership degrees for risk levels L_1, L_2, and L_3 for each evaluation factor. Applying Dempster's integration rule, we obtained mass function values, with the highest value ($m(L_3) = 0.455$) indicating the high risk of LIB at the moment.

The proposed method, combining cloud model and the DS evidence theory, enhances the accuracy of TR warnings in LIBs. It objectively handles uncertainties in complex environments and sensor measurements, overcoming the limitations of traditional single-parameter threshold models in accounting for the fuzziness of multiple parameters. This approach improves risk evaluation precision and contributes reliable theoretical support for LIB safety evaluation and management.

Author Contributions: Conceptualization, Z.X. and S.L.; methodology, Z.X. and H.W.; validation, H.W.; formal analysis, Z.X.; investigation, H.W. and P.L.; data curation, H.W.; writing—original draft preparation, H.W.; writing—review and editing, P.L. and J.S.; supervision, Y.Z.; funding acquisition, J.S. and X.Z. All authors have read and agreed to the published version of the manuscript.

Funding: This research was supported by the National Key Research and Development Program of China, grant number 2022YFB3803501; the National innovation and entrepreneurship training program for college students, grant number 20230497002; and the Opening Fund of State Key Laboratory of Fire Science (SKLFS), grant number HZ2022-KF08.

Data Availability Statement: The data presented in this study are available on request from the corresponding author.

Acknowledgments: Thanks to the EVE Power Co., Ltd. for their support.

Conflicts of Interest: Authors J.S. and X.Z. were employed by the company EVE Power Co., Ltd. The remaining authors declare that the research was conducted in the absence of any commercial or financial relationships that could be construed as a potential conflict of interest.

References

1. Leising, R.A.; Palazzo, M.J.; Takeuchi, E.S.; Takeuchi, K.J. A study of the overcharge reaction of lithium-ion batteries. *J. Power Sources* **2001**, *97–98*, 681–683. [CrossRef]
2. Xia, B.; Mi, C. A fault-tolerant voltage measurement method for series connected battery packs. *J. Power Sources* **2016**, *308*, 83–96. [CrossRef]
3. Jia, T.; Zhang, Y.; Ma, C.; Li, S.; Yu, H.; Liu, G. The early warning for overcharge thermal runaway of lithium-ion batteries based on a composite parameter. *J. Power Sources* **2023**, *555*, 232393. [CrossRef]
4. Nascimento, M.; Ferreira, M.S.; Pinto, J.L. Temperature fiber sensing of Li-ion batteries under different environmental and operating conditions. *Appl. Therm. Eng.* **2019**, *149*, 1236–1243. [CrossRef]
5. Nascimento, M.; Paixão, T.; Ferreira, M.S.; Pinto, J.L. Thermal Mapping of a Lithium Polymer Batteries Pack with FBGs Network. *Batteries* **2018**, *4*, 67. [CrossRef]
6. Hong, J.; Wang, Z.; Yao, Y. Fault prognosis of battery system based on accurate voltage abnormity prognosis using long short-term memory neural networks. *Appl. Energy* **2019**, *251*, 113381. [CrossRef]
7. Peng, J.; Zhou, X.; Jia, S.; Jin, Y.; Xu, S.; Chen, J. High precision strain monitoring for lithium ion batteries based on fiber Bragg grating sensors. *J. Power Sources* **2019**, *433*, 226692. [CrossRef]
8. Chen, S.; Wei, X.; Zhang, G.; Wang, X.; Feng, X.; Dai, H.; Ouyang, M. Mechanical strain signal based early warning for failure of different prismatic lithium-ion batteries. *J. Power Sources* **2023**, *580*, 233397. [CrossRef]
9. Raghavan, A.; Kiesel, P.; Sommer, L.W.; Schwartz, J.; Lochbaum, A.; Hegyi, A.; Schuh, A.; Arakaki, K.; Saha, B.; Ganguli, A.; et al. Embedded fiber-optic sensing for accurate internal monitoring of cell state in advanced battery management systems part 1: Cell embedding method and performance. *J. Power Sources* **2017**, *341*, 466–473. [CrossRef]
10. Lyu, N.; Jin, Y.; Xiong, R.; Miao, S.; Gao, J. Real-Time Overcharge Warning and Early Thermal Runaway Prediction of Li-on Battery by Online Impedance Measurement. *IEEE Trans. Ind. Electron.* **2022**, *69*, 1929–1936. [CrossRef]
11. Mei, W.; Liu, Z.; Wang, C.; Wu, C.; Liu, Y.; Liu, P.; Xia, X.; Xue, X.; Han, X.; Sun, J.; et al. Operando monitoring of thermal runaway in commercial lithium-ion cells via advanced lab-on-fiber technologies. *Nat. Commun.* **2023**, *14*, 5251. [CrossRef] [PubMed]
12. Cai, T.; Valecha, P.; Tran, V.; Engle, B.; Stefanopoulou, A.; Siegel, J. Detection of Li-ion battery failure and venting with Carbon Dioxide sensors. *eTransportation* **2021**, *7*, 100100. [CrossRef]

13. Jin, Y.; Zheng, Z.; Wei, D.; Jiang, X.; Lu, H.; Sun, L.; Tao, F.; Guo, D.; Liu, Y.; Gao, J.; et al. Detection of Micro-Scale Li Dendrite via H_2 Gas Capture for Early Safety Warning. *Joule* **2020**, *4*, 1714–1729. [CrossRef]
14. Jiang, Y.; Yu, Y.; Huang, J.; Cai, W.; Marco, J. Li-ion battery temperature estimation based on recurrent neural networks. *Sci. China Technol. Sc.* **2021**, *64*, 1335–1344. [CrossRef]
15. Zhang, W.; Ouyang, N.; Yin, X.; Li, X.; Wu, W.; Huang, L. Data-driven early warning strategy for thermal runaway propagation in Lithium-ion battery modules with variable state of charge. *Appl. Energy* **2022**, *323*, 119614. [CrossRef]
16. Xiang, X.; Li, K.; Huang, B.; Cao, Y. A multi-sensor data-fusion method Based on cloud model and improved evidence theory. *Sensors* **2022**, *22*, 5902. [CrossRef] [PubMed]
17. Zhang, Y.; Li, S.; Mao, B.; Shi, J.; Li, X.; Zhou, L. A multi-level early warning strategy for the $LiFePO_4$ battery thermal runaway induced by overcharge. *App. Energy* **2023**, *347*, 121375. [CrossRef]

batteries

MDPI

Article

Thermal Runaway Warning of Lithium Battery Based on Electronic Nose and Machine Learning Algorithms

Zilong Pu [1,2], Miaomiao Yang [1,2], Mingzhi Jiao [1,2,3,*], Duan Zhao [1,2], Yu Huo [1,2,*] and Zhi Wang [3,4,5,*]

1 School of Information and Control Engineering, China University of Mining and Technology, Xuzhou 221116, China; ts22060164p31@cumt.edu.cn (Z.P.); ts23060020a31tm@cumt.edu.cn (M.Y.); duan.zhao@cumt.edu.cn (D.Z.)
2 The State and Local Joint Engineering Laboratory of Perception Mine, China University of Mining and Technology, Xuzhou 221116, China
3 School of Safety Engineering, China University of Mining and Technology, Xuzhou 221116, China
4 Civil Aircraft Fire Science and Safety Engineering Key Laboratory of Sichuan Province, Guanghan 618300, China
5 China Academy of Safety Science and Technology, Beijing 100012, China
* Correspondence: mingzhi.jiao@cumt.edu.cn (M.J.); yuhuo@cumt.edu.cn (Y.H.); zhiwang@cumt.edu.cn (Z.W.)

Abstract: Characteristic gas detection can be an efficient way to predict the degree of thermal runaway of a lithium battery. In this work, a sensor array consisting of three commercial MOS sensors was employed to discriminate between three target gases, CO, H_2 and a mixture of the two, which are characteristic gases released during the thermal runaway of lithium batteries. In this work, an integrated model that makes the classification stage results one of the feature inputs for the concentration regression stage was employed, successfully reducing the RMSE of the concentration regression results. In addition, we also explored the influence of the selection of the response time length on the classification and regression tasks, achieving the best results in a short time through the optimum algorithm. To assess the impact of time duration sensor data on the results, we selected four time windows of different length and extracted the corresponding sensor response data for subsequent processing. Initially, principal component analysis (PCA) was used to visualise the clustering of the three target gas samples at room temperature, providing a preliminary data analysis. For the classification phase, we chose three classification algorithms—MLP (Multilayer Perceptron), ELM (Extreme Learning Machine), and SVM (Support Vector Machine)—and performed a comprehensive comparison of their classification and generalisation abilities using grid search for hyperparameter optimisation and five-fold cross-validation. The results demonstrated that MLP achieved 99.23% classification accuracy during the 20 s response period. In the concentration regression phase, we combined the classification results with the raw features to create a new feature set, which was then input into a multi-output MLP regression model. The root mean square error (RMSE) employing the new feature set was used to measure the prediction error. Ultimately, the findings showed that the input of combined features significantly reduced the regression error for the mixed gas.

Keywords: lithium battery; thermal runaway warning; MOS sensor; electronic nose; gas classification; concentration regression

1. Introduction

With the continuous progress of technology and the growing demand for environmentally friendly energy, lithium batteries have become one of the indispensable energy mediums in modern society. From smartphones to electric vehicles, portable electronic devices to energy storage systems, the application of lithium batteries has permeated every aspect of our lives [1,2]. However, despite their numerous advantages in energy density, charging speed, and lifespan, the potential dangers associated with thermal runaway in lithium batteries have become increasingly prominent.

Thermal runaway in lithium batteries can lead to fires, explosions, and even personal injury. Lithium-ion batteries may fail to adhere to their intrinsic electrical properties under certain abnormal conditions—such as overcharging or over-discharging, physical damage, internal short circuits, or high-temperature environments [3–6]. This can cause thermal runaway of the electrolyte, triggering a chain reaction that releases a significant amount of heat and harmful gases [3,4], potentially leading to catastrophic outcomes. Despite a series of safety measures implemented during the design and manufacturing of lithium batteries—such as the selection of more reliable anode materials, thermal management systems, and the addition of flame retardants [5–7]—the complexity of their internal chemical characteristics and the unpredictability of external environments mean that thermal runaway remains a significant issue.

Therefore, to promote the continuous development of lithium battery technology and ensure its safety, more in-depth research and exploration are needed to prevent thermal runaway. From a regulatory perspective, the United Nations' proposed Global Technical Regulation No. 20 for Electric Vehicle Safety (EVS) requires that, in the event of thermal runaway, vehicles must provide a warning at least five minutes in advance for the evacuation of passengers from hazardous conditions.

Currently, safety monitoring technologies for commercial lithium-ion batteries primarily focus on detecting temperature rise, pressure distribution changes and impedance variations, etc. However, there is relatively little research on the analysis and early warning detection of released mixed gases. Koch et al. [8] assembled a sensor array consisting of voltage, temperature, pressure, smoke, creeping distance, expansion stress, and gas sensors to conduct thermal runaway tests on lithium battery packs of different sizes, energy densities, and configurations. The results showed that, compared to other sensors, gas sensors show the fastest and clearest response. Pressure and force sensors also responded quickly but with weaker signal amplitudes, while voltage, smoke, and creeping distance sensors required the longest response times. This further demonstrates the higher reliability and accuracy of gas sensors in the detection and early warning of battery thermal runaway. Evaluating the sensitivity and accuracy of monitoring and detecting thermal runaway using temperature, voltage, and gas sensors. It was proved that gas sensors could detect fault signals approximately 10 min before the actual occurrence of thermal runaway, which is about 2 min earlier than voltage sensors and about 7 min earlier than temperature sensors. Koch et al. [9] evaluated the gases released during thermal runaway in 51 different types of batteries, finding that seven components—CO_2, CO, H_2, C_2H_4, CH_4, C_2H_6, and C_3H_6—accounted for more than 99% of the total emissions. Essl et al. [10] developed a custom battery thermal runaway reaction test bench and a gas sensitivity platform to analyse the thermal runaway process, primarily detecting CO_2, CO, H_2, C_2H_4, CH_4, along with small amounts of electrolyte vapour and water. Golubkov et al. [11] analysed the gas composition during thermal runaway of 18,650 commercial batteries using three different cathode materials. Although the volume fractions of gases released varied with different materials, CO_2, CO, H_2, C_2H_4, and CH_4 were consistently detected.

We can utilise an electronic nose to monitor the mixed gases released during the thermal runaway of lithium batteries and achieve early warning. The electronic nose is an artificial sensor system that simulates the biological nose to detect and identify gas components. It simulates the basic principles of the human olfactory system by using an array of gas sensors to capture odours and convert them into quantifiable signals [12]. These signals can then be analysed and processed to identify the presence and concentration of specific gas components. In practical applications, electronic noses are widely used in food processing, medical diagnosis, environmental monitoring, and other fields [13–15]. Based on the research and analysis of previous related work, compared to other models, MLP, ELM, and SVM are commonly used in the classification and regression of gas sensor array data and have achieved desirable results. MLP is an artificial neural network that uses the backpropagation method to iteratively calculate the optimal parameters for network nodes. It is widely used in the field of electronic noses. Zhang et al. [16] achieved simultaneous interference detection of CO and CH_4 semi-quantitatively under interfering environments using an MLP-based electronic nose model. Chu et al. [17] used a genetic algorithm-optimised MLP model to classify and identify mixtures of 11 different concentrations of nitrogen dioxide (NO_2) and carbon monoxide (CO). Li et al. [18] constructed a sensor array and utilised a PSO-MLP algorithm to qualitatively identify multiple mixed toxic gases (formaldehyde, n-butanol, and isopropanol). ELM, proposed by Guang-Bin and Qin-Yu [19], is designed to train single hidden layer feedforward networks (SLFN). ELM is widely applied in electronic noses for tasks such as classification, regression, clustering, and feature mapping. Wang et al. [20] used an ELM-ELM ensemble structure to identify and predict the concentration of six VOC gases, achieving a classification accuracy of 99% and a regression R^2 score of 0.94. Ma et al. [21] presented a novel hierarchical classifier (HC) with a tree structure. This presented ELM-HC has demonstrated higher classification accuracy and better generalisation performance for the quantitative analysis of gases, including six toxic gases (hydrogen sulphide, carbon monoxide, ammonia, toluene, formaldehyde, acetone) and three binary gas mixtures. Ouyang et al. [22] utilised the ELM algorithm to process features from supervised learning, implementing a deep learning network for NO_x concentration regression. SVM was widely employed for gas classification. Zhao et al. [23] applied SVM, BP, and ELM to identify formaldehyde in three mixed gases, with SVM achieving the best classification accuracy. Khan et al. [24] employed four supervised machine learning algorithms: Decision Tree, SVM, Naive Bayes (kernel), and k-Nearest Neighbour (k-NN) for the classification of gas type. Results indicated that the optimised SVM and NB classifier models exhibited the highest classification accuracy on the test dataset.

In this paper, we conducted data processing and analysis on the time series data collected by the sensor array in the experiment and analysed qualitatively and quantitatively the three different gases that can characterise the stage of thermal runaway. Combined with the five-fold cross-validation, we compared the classification accuracy of the three algorithms, MLP, ELM, and SVM, and combined the classification results with the raw features to obtain new combined eigenvalues. Regression accuracy is improved, and the influence of selection of different length time window data is discussed. The specific process is shown in Figure 1.

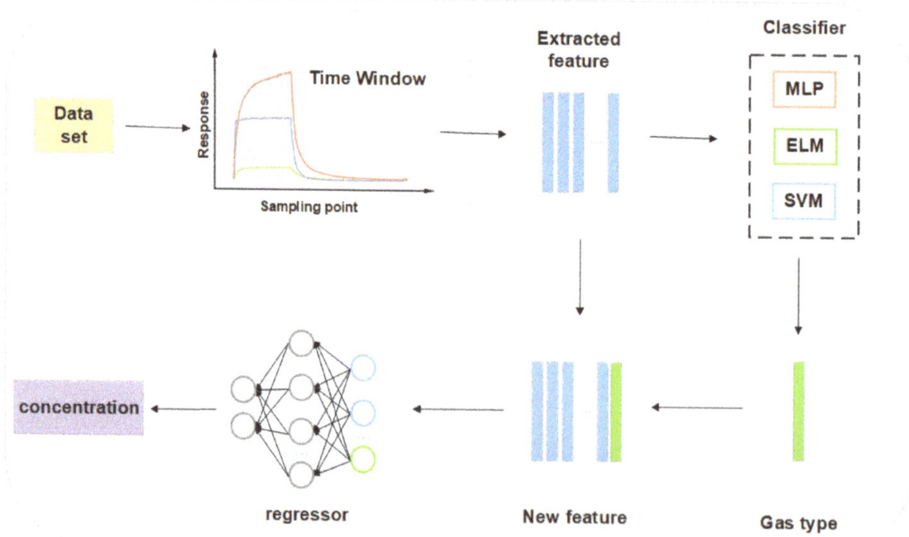

Figure 1. Gas classification and regression model diagram.

2. Experiment

2.1. Sensor Array

Due to the cross-sensitivity phenomenon in MOS sensors, where a single sensor may exhibit similar response patterns to different types and concentrations of gases [25], it is challenging for a single sensor to achieve precise pattern recognition. Therefore, selecting sensors with good selectivity, constructing sensor arrays, and extracting appropriate feature values are important methods to enhance selectivity.

Based on the sensor performance, cost, and power consumption, three types of commercial MOS sensors were selected to establish the sensor array. The HGS1000, HGS1001, and HGS1002 sensors used in this study were all purchased from Micro-Nano Sensing Co. LTD (Hefei, China), which are named S1, S2 and S3 below, respectively, shown in Figure 2. For MOS sensors, the physicochemical reactions between the target gases and different sensitive materials result in changes in the electrical conductivity of the sensitive materials, leading to different response patterns for different gases. Table 1 presents the target detection gases applicable to the three sensors used in this experiment and their detection ranges for the gases. All sensors are cross-sensitive to target gases, so we will train the data processing model to perform classification identification and concentration regression for different target gases.

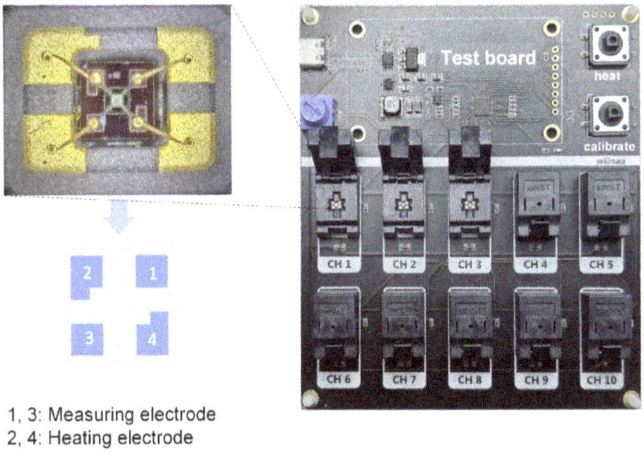

Figure 2. The square array plate of the detection module and MOS sensor.

Table 1. The selected sensor parameters.

Title 1	Sensor Type	Target Gas	Measurement Range
HGS1000	MOS	VOC, smog, etc.	1~500 ppm
HGS1001	MOS	ethanol, acetone, hydrogen, hydrogen sulfide, etc.	0.1~500 ppm
HGS1002	MOS	Ethanol, acetone, hydrogen, hydrogen sulfide, etc.	0.05~300 ppm

2.2. Preparation of the Samples

The experiment in this paper records the response time sequence signals of the constructed sensor array to two single gases, hydrogen and carbon monoxide, as well as their mixed gases. As demonstrated in Table 2, for the target gases in the experiment, the concentration levels for hydrogen are set to 20, 50, 100, 150, and 200 ppm. The concentration levels for carbon monoxide are set to 2, 5, 10, 15, and 20 ppm. The concentration levels for the mixed gas (CO, H_2) are set to (18 ppm, 200 ppm), (10 ppm, 1000 ppm), and (2 ppm, 1800 ppm).

Table 2. Concentration settings of the Characteristic gases.

Setting No.	H_2 (ppm)	CO (ppm)	Total Gas (ppm)
1	20	0	20
2	50	0	50
3	100	0	100
4	150	0	150
5	200	0	200
6	0	2	2
7	0	5	5
8	0	10	10
9	0	15	15
10	0	20	20
11	200	18	218
12	1000	10	1010
13	1800	2	1802

All gas sensitivity tests throughout the work were conducted under atmospheric pressure, relative humidity of (60 ± 10)%, and room temperature of (25 ± 2) °C. The response of the sensor is defined as follows:

$$S = \frac{R_a}{R_g} \tag{1}$$

where R_a and R_g are the resistance of the sensor in air and target gas, respectively.

The gas sensing testing system mainly comprises three parts: the gas distribution module, the detection module, and the signal processing and pattern recognition module. The square array plate of the detection module, purchased from Micro-Nano Sensing Co. LTD, is used to collect the response signals from the sensor array at a sampling frequency of 10 Hz (ten times per second). During the experiment, the sensor array plate is placed in a test chamber with a volume of 1 L and interacts with the target gas to generate a response time sequence signal, which is transmitted to a computer for further processing.

Before testing, the working voltage of the sensor array is set to 1.8 V, and a preheating period of 24 h is conducted to ensure the stable operation of the sensor array. During the sample testing phase, standard air is first introduced into the test chamber for 30 min to clean the gas chamber, remove interferences, and ensure the stability of the sensor response baseline. Since metal oxide semiconductor sensors cannot operate without oxygen, the three target gases are mixed with standard air to different target concentrations and then introduced into the test chamber at a constant flow rate for 2.5 min to ensure complete interaction with the sensors, thereby obtaining response phase data.

After the response phase, standard air is introduced at a constant flow rate to clean the sensor array during the response recovery phase, which lasts for 5 min, resulting in a complete experimental sample. The above process is repeated ten times for each concentration to study the reproducibility of the sensors and improve the accuracy of the results. Finally, a dataset containing 130 gas samples was established.

2.3. Sensor Response

Figure 3a–c illustrates the response and recovery processes of sensors S1, S2, and S3, respectively, to hydrogen, carbon monoxide, and the mixed gases. As the concentration of the target gases increases during the experiment, the response values of all sensors also increase accordingly. For each target gas, sensor S3 shows the highest response. Especially at higher concentrations of hydrogen, S3's response far exceeds the other two sensors. When the hydrogen concentration increases from 20 ppm to 200 ppm, its response increases from 2.85 to 64.13. This indicates that S3 is highly sensitive to hydrogen and can be used when detecting or measuring high hydrogen concentrations is required. For hydrogen, the response of S2 is higher than that of S1. However, for carbon monoxide, the responses of all three sensors are closer, especially at lower concentrations. When dealing with mixed gases, the response values of all sensors significantly increased, with S3 showing the highest response. In summary, the sensor array demonstrates good detection performance and cross-sensitivity characteristics for the three experimental gases. These sensors exhibit different responses to different target gases, meeting the basic requirements for sensor array selection and making them suitable for constructing a sensor array.

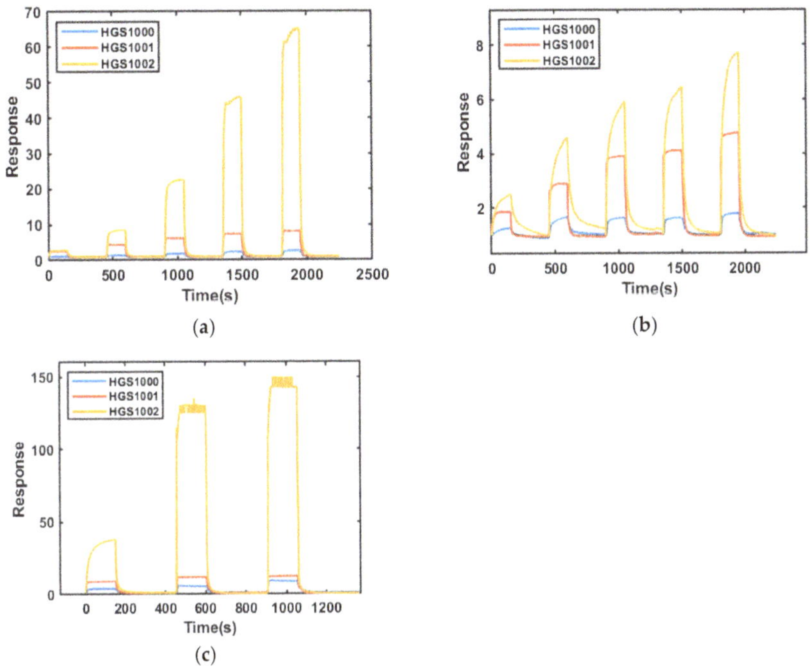

Figure 3. Response of sensor array to three gases: (**a**) H_2; (**b**) CO; (**c**) the gas mixture.

3. Data Processing

3.1. Feature Extraction

In the data processing phase, feature extraction is crucial for improving the performance of subsequent pattern recognition. Feature extraction involves processing the sensor responses to extract and select useful information. The goal is to obtain robust information with minimal redundancy from the sensor responses to ensure the effectiveness of subsequent pattern recognition algorithms. Extracting single features from the sensor array is likely to overlook the selectivity and specificity of the sensors. In this study, the integral value, first derivative, and mean value of specific response time are extracted from the original response curve of the sensor array. The first derivative represents the reaction rate of the sensor array to the gases, while the mean of the sensor response data represents the sensor's sensitivity.

These manually extracted features are then normalised, with each feature value scaled between 0 and 1. This normalisation process eliminates the impact of dimensions, reduces calculation errors, and prevents large numerical differences from affecting the accuracy of subsequent pattern recognition.

$$F' = \frac{F - F_{min}}{F_{max} - F_{min}} \tag{2}$$

3.2. Data Division

During the model training process, a five-fold cross-validation method is employed. This involves randomly dividing the sample data into five mutually exclusive subsets of equal size. Each time, four subsets are used as the training set, and the remaining subset is used as the test set for model training and validation. The cross-validation process is repeated five times, with each subset used as the test set exactly once. In total, this produces five test errors, which are averaged to generate a single estimate. This estimate is used to evaluate the model's performance under the current parameters.

3.3. Time Window Selection

In real-time applications, we need to perform high-precision pattern recognition of the target gas as soon as possible after the response phase begins. To evaluate the impact of time duration on the final results, feature extraction and normalisation were carried out on the data within different lengths of time windows starting from the time the gas is injected. We then use PCA (Principal Component Analysis) for visualisation and observe the clustering of samples, which aids in selecting the optimal time window length.

We classified the samples into three groups: H_2, CO, and gas mixture, then used PCA to visualise the clustering of the sample data. The figure below shows 2D scatter plots of the first two principal components (PC1 and PC2) for data within the 150 s, 40 s, 20 s, and 10 s time windows after performing PCA. Since the sum of the first two principal components for each sample group explains at least 94.84% of the variance in the original data, the visualisation results of the principal component analysis are acceptable, which means clustering separation in two-dimensional space often reflects, to some extent, the clustering separation trends in high-dimensional space. In this work, we initially selected 150 s (whole response phase), 40 s, 20 s, and 10 s after the target gas injection as the time windows.

As shown in Figure 4a–c, the PCA results for the data within the 150 s, 40 s, and 20 s time windows demonstrate that while shorter time windows lead to slightly more dispersed clustering, the three target gases can still be linearly separated. However, as shown in Figure 4d, only extracting the first 10 s of response data results in noticeable overlap between samples, reducing overall accuracy. This may be because during the first 10 s of the response process, the sensitive material of the MOS sensor has not fully reacted with the target gases, resulting in a lower amount of useful information. Consequently, the sensor array's ability to distinguish between different categories and concentrations of gases is less sensitive, increasing the difficulty of subsequent pattern recognition.

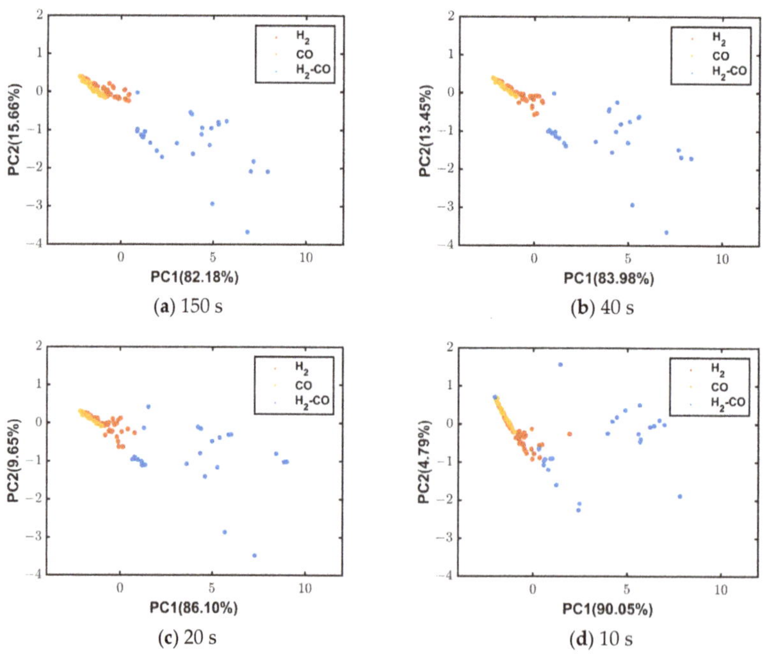

(**a**) 150 s (**b**) 40 s (**c**) 20 s (**d**) 10 s

Figure 4. PCA result plots of data in different time windows.

4. Algorithm Selection

4.1. SVM

Support Vector Machine (SVM) is a classic supervised learning model used for classification and regression analysis. Its principle is to find a hyperplane that can be used to separate data points from different categories, the illustration of hyperplanes is shown in Figure 5. The optimal hyperplane separating the data can be obtained by solving the following optimisation problem:

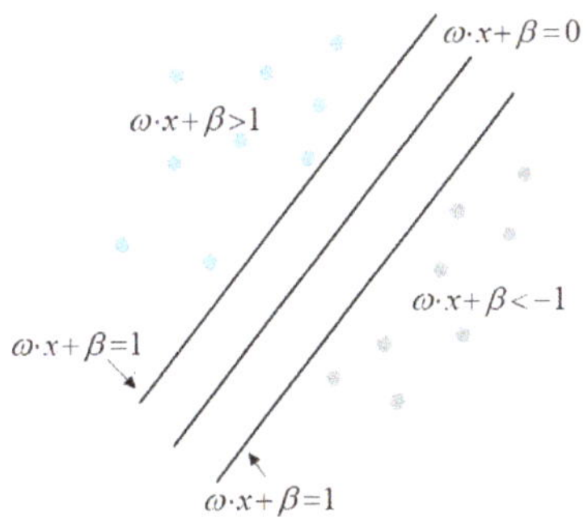

Figure 5. Illustration of hyperplanes.

Minimise

$$\tau(\omega) = \min_{\omega,\beta}\left\{\frac{1}{2}\|\omega\|^2\right\} \tag{3}$$

subject to

$$y_i((\omega \cdot x) + \beta) - 1 \geq 0 \tag{4}$$

where the vector ω defines the boundary and β is a scalar threshold. In case there is an overlap between the two classes, a slack variable ($\xi_i, i = 1, 2, \ldots, k$) can be introduced. The optimisation problem changes as follows:

Minimise

$$\tau(\omega) = \min_{\omega,\beta}\left\{\frac{1}{2}\|\omega\|^2 + \gamma\sum_{i=1}^{k}\xi_i\right\} \tag{5}$$

subject to

$$y_i((\omega \cdot x) + \beta) - 1 + \xi_i \geq 0 \tag{6}$$

where γ is a penalty coefficient that determines the trade-off between minimising the training error and minimising model complexity.

The four most renowned kernels are linear, polynomial, radial basis function (RBF), and sigmoid. In this work, we utilised the RBF kernel function, a commonly used high-performance function. We performed a grid search to find the optimal parameters for the regularisation parameter C and kernel coefficient γ. A larger penalty factor C implies a higher penalty for misclassified samples, resulting in higher accuracy on the training samples but reduced generalisation ability, meaning lower classification accuracy on the test data. Conversely, reducing C allows for some misclassification errors in the training samples, enhancing the generalisation ability.

A lower γ value indicates a larger radius of similarity, leading to more points being grouped together. For higher γ values, points need to be very close to be considered in the same group (or class). In this study, the grid search is employed to find the best parameters. Grid search involves setting a range of values for different hyperparameters and exhaustively exploring all possible combinations within that range. Cross-validation is used to evaluate the model's performance for each combination. After completing the training and test for all parameter combinations, the best combination is selected based on performance metrics (such as accuracy or F1 score), and the model is retrained using the optimal parameters. Consequently, models with very high γ values tend to overfit. Therefore, we optimised C in the range of 0.01 to 100 and γ in the range of 0.01 to 10, with the step size set to 0.1.

4.2. ELM

Extreme Learning Machine (ELM) is a training algorithm for single hidden layer feedforward neural networks (SLFN) proposed by Guang-Bin and Qin-Yu [19]. It is widely used for classification, clustering, and regression in many real-time learning tasks. ELM consists of a three-layer structure: input layer, hidden layer, and output layer, as shown in Figure 6.

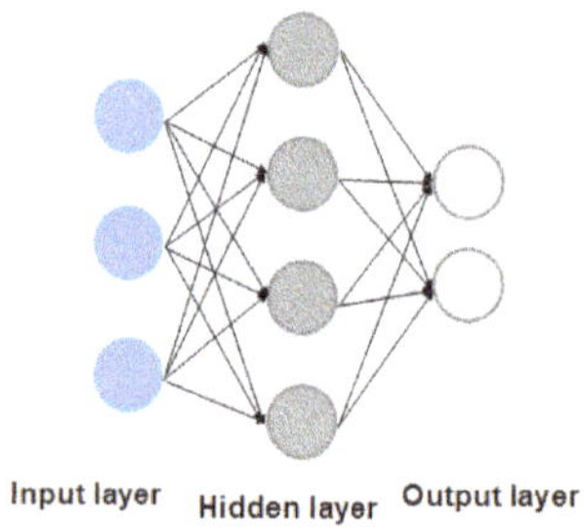

Figure 6. Structure of ELM.

Unlike gradient-based methods, ELM assigns random values ω_i and b_i to the weights and biases between the input and hidden layers. The input vectors are mapped to a random feature space with randomly set and nonlinear activation functions. The nonlinear activation function in the hidden layer provides the system's nonlinearity, thus it can be considered a linear system.

The parameters between the input and hidden layers remain fixed, and the only parameters that need to be learnt are the weights between the hidden and output layers. The mathematical model of ELM is:

$$\sum_{i=1}^{\tilde{N}} \beta_i g_i(x_j) = \sum_{i=1}^{\tilde{N}} \beta_i g(\omega_i \cdot x_j + b_i) = t_j, j = 1, \ldots, N \tag{7}$$

$g()$ is the activation function, $\omega_i = \omega_{i1}, \omega_{i2}, \ldots, \omega_{inT}$ is the weight vector connecting the ith hidden node and the input nodes, $\beta_i = \beta_{i1}, \beta_{i2}, \ldots, \beta_{inT}$ is the weight vector connecting the ith hidden node and the output nodes, and b_i is the bias of the ith hidden node. Given N arbitrary samples (X_i, t_i), where $X_i = X_{i1}, X_{i2}, \ldots, X_{inT} \in R_n$, $t_i = t_{i1}, t_{i2}, \ldots, t_{imT} \in R_m$.

The equation can be succinctly written as:

$$H\beta = T \tag{8}$$

The value of β is obtained by solving Equation (9).

$$\beta = H^{-1}T \tag{9}$$

During the testing process, the H and β values obtained during training are used to calculate the approximations for the test samples. We selected the sigmoid function as the activation function for the ELM and performed a grid search over the range of 0 to 500 for the number of hidden layer nodes to achieve the optimal accuracy.

4.3. MLP

MLP (Multilayer Perceptron) is a classical artificial neural network consisting of an input layer, an output layer, and one or more hidden layers in a feedforward structure. The hidden layer consists of S neurons, each with R weights. These weights can be represented in the $S \times R$ input weight matrix I, as shown in Equation (10). The input vector P has R elements $[p_1, p_2, \ldots, p_R]^T$, which are multiplied by I, and the resulting matrix is added to the bias vector b to form the vector n, as shown in Equation (11). The output of the hidden layer a is the result of applying the activation function $f(\cdot)$ to n.

$$I = \begin{bmatrix} w_{1,1} & w_{1,2} & \cdots & w_{1,R} \\ w_{2,1} & w_{2,2} & \cdots & w_{2,R} \\ & \cdots\cdots & & \\ w_{S,1} & w_{S,2} & \cdots & w_{S,R} \end{bmatrix} \tag{10}$$

$$n = I \cdot p + b \tag{11}$$

$$a = f(n) \tag{12}$$

The same operations applied in the hidden layer are used in the output layer, which consists of K neurons, with a serving as the input vector. The hyperparameters of the MLP include learning rate, number of iterations, number of layers, and hidden layer units. The activation function for the hidden layer is the Sigmoid function. For concentration regression, the activation function for the output layer is Softmax, which provides quantitative predictions for sample gases. We train the neural network using the backpropagation algorithm, allowing the model parameters to be automatically obtained without prior knowledge, regardless of the sensors or target gases.

We set the number of iterations to 1000. Additionally, since the dataset used in this study is relatively small, we set the target training error to 1×10^{-6} and automatically monitor the training error to prevent overfitting. If the set value is reached, the training process is terminated early. Moreover, we set the learning rate to 0.01 and two hidden layers, as illustrated in Figure 7, with a maximum of 30 neurons per layer. Each neural network is automatically trained to search for the optimal structure and parameters [26].

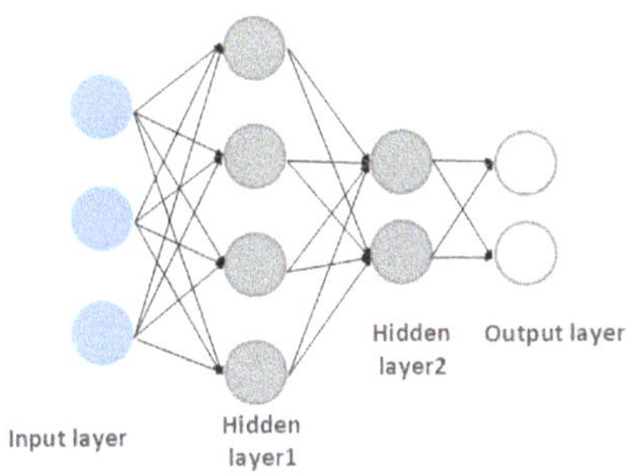

Figure 7. Structure of double hidden layer MLP.

5. Results and Discussion

5.1. Classification Recognition

In this study, we selected three classification models—MLP, ELM, and SVM—to train the data within different time windows. To verify the performance of the classification system, all samples were randomly shuffled, and 5-fold cross-validation and grid search hyperparameter optimisation were performed on the extracted feature data for each classification model. The optimal classification accuracy results and parameters of each algorithm for different time Windows are recorded and compared in Table 3. The confusion matrix for the classification results of data from different time windows is shown in Figure 8.

Considering the classification accuracy of the data in the 150 s time window, MLP performed the best, achieving a classification accuracy of 99.23%, as shown in Table 3. This indicates that the MLP model has the strongest fitting and generalisation abilities among the three models in this context. ELM followed with a classification accuracy of 94.62%. While

ELM has certain advantages in fast learning and processing large-scale data, and it has the fastest training and testing speed among the models, its classification accuracy is slightly lower than that of MLP. SVM had the lowest classification accuracy at 76.92%, particularly struggling with recognising mixed gas and hydrogen, with accuracies of only 73.3% and 60%, respectively. Although SVM performs well in specific problems such as small sample sizes, nonlinearity, and high-dimensional pattern recognition, its classification accuracy in this comparison was significantly lower than the other two models. The detailed confusion matrix is shown in Figure 8a–c.

The data in Table 3 also show that when the time window is shortened from 150 s to 40 s, the MLP maintains a high recognition accuracy of 99.23%, indicating that the MLP model can still provide a very high classification accuracy when processing the characteristic data in the first 40 s response stage. The classification accuracy of ELM decreased from 94.62% to 92.31%. The classification accuracy of SVM decreased slightly, from 76.92% to 75.38%. The detailed confusion matrix can be found in Figure 8d–f.

When the time window is shortened to 20 s, the classification accuracy of ELM decreases again. The classification accuracy of the MLP remained unchanged at 99.23%. Still, the number of hidden layer nodes deployed was slightly higher than that in the first two cases, and the training time increased. The classification accuracy of ELM was reduced from 92.31% to 91.54%. SVM classification accuracy barely changed, increasing slightly from 75.38% to 76.92%. This may be because SVM has reached the limit of its learning effectiveness for experimental sample data. The detailed confusion matrix can be found in Figure 8g–i.

After extracting the characteristic data of the first 10 s response stage and training the model, the best classification accuracy of MLP, ELM, and SVM is 94.61%, 88.46%, and 76.92%, respectively. Although the classification accuracy of the MLP model remained at a high level, reaching 94.61%, it decreased by 4.62% compared with the feature data extracted in the early 20 s and 40 s. This suggests that the predictive power of the MLP model may decline for features extracted within a shorter time window in the response phase. Similarly, the classification accuracy of the ELM model also decreases, reaching 88.46%, but the decline is smaller than that of MLP. The classification accuracy of the SVM model remained at 76.92%, indicating that the classification accuracy of the SVM model was not greatly affected by the data features with shorter response times. The detailed confusion matrix can be found in Figure 8j–l.

Table 3. Comparison of performance of the three algorithms using data within different time windows.

Time Window	Algorithm	Accuracy Rate	Hyperparameter		Time	
					Training Time (s)	Test Time (s)
150 s	MLP	99.23%	Layer1*: 4	Layer2*: 19	7340×10^{-5}	320×10^{-5}
	ELM	94.62%	Layer1*: 20	-	3.518×10^{-5}	0.82×10^{-5}
	SVM	76.92%	C*: 62.01	γ*: 37.01	19.03×10^{-5}	6.394×10^{-5}
40 s	MLP	99.23%	Layer1: 5	Layer2: 17	6550×10^{-5}	345×10^{-5}
	ELM	92.31%	Layer1: 22	-	3.588×10^{-5}	1.136×10^{-5}
	SVM	75.38%	C: 6.01	γ: 3.51	18.8×10^{-5}	6.36×10^{-5}
20 s	MLP	99.23%	Layer1: 25	Layer2: 11	$26{,}922 \times 10^{-5}$	347.61×10^{-5}
	ELM	91.54%	Layer1: 30	-	4.218×10^{-5}	1.02×10^{-5}
	SVM	76.92%	C: 52.51	γ: 36.51	38.73×10^{-5}	16.676×10^{-5}
10 s	MLP	94.61%	Layer1: 20	Layer2: 8	$23{,}300 \times 10^{-5}$	338×10^{-5}
	ELM	88.46%	Layer1: 14	-	4.24×10^{-5}	0.656×10^{-5}
	SVM	76.92%	C: 8.51	γ: 50.51	20.5×10^{-5}	7.14×10^{-5}

* The number behind the layer is the number of neurons in that layer, C refers to regularisation parameter, γ refers to kernel coefficient.

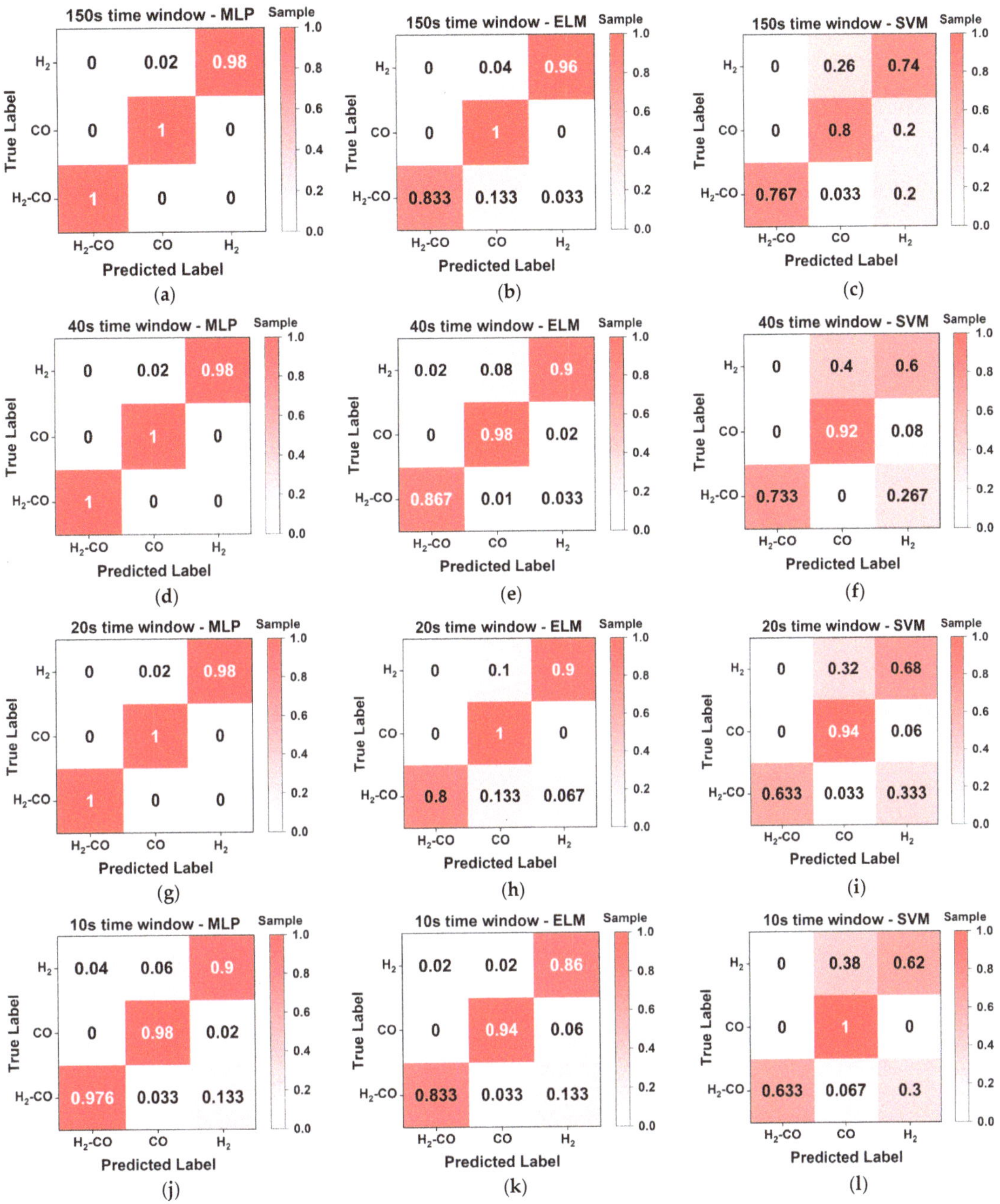

Figure 8. Comparison of confusion matrix of the three algorithms using 150 s data (**a**) MLP; (**b**) ELM; (**c**) SVM; 40 s data (**d**) MLP; (**e**) ELM; (**f**) SVM; 20 s data (**g**) MLP; (**h**) ELM; (**i**) SVM; 10 s data (**j**) MLP; (**k**) ELM; (**l**) SVM; where column instances are prediction class, row instances are real class.

On the whole, the classification accuracy of all models eventually decreases when the time window shortens. According to the experimental data in this paper, when the time window length is gradually shortened, MLP still maintains the best classification accuracy, indicating its high robustness to the time window length, and the classification accuracy reaches 99.23% under the 150 s, 40 s, and 20 s time windows. The classification accuracy of ELM is always slightly lower than that of MLP and is greatly affected by the length of the time window, from 94.62% to 88.46%. The classification accuracy of SVM is the lowest, only about 76%, especially for the identification of hydrogen and mixed gas. For multiclass problems, SVM needs to decompose multiclass problems into multiple binary classification problems for multiple training and prediction, which may not be as effective as other algorithms that can deal with multiclass problems. In addition, the SVM algorithm may have classification bias for datasets with unbalanced data volume.

In summary, for the classification stage, considering the high safety and efficiency requirements of the studied scenario and the fact that the accuracy of the classification stage in the integrated model affects the results of the subsequent concentration regression, we recommend training the MLP model to fit the data within the 20 s time window, which had higher recognition accuracy, as the best outcome.

5.2. Concentrations Regression

Considering the best fit and generalisation ability of the MLP model to the data sample in the classification stage, a multiple output MLP model was constructed to estimate the concentration of the target gas, respectively, after the classification identification stage. Two different feature inputs are selected: one is the raw features extracted in the data processing stage, and the other is the combined features of the raw feature data and the results obtained in the classification stage. The regression is fit considering the RMSE (root mean square error). The specific flowchart is demonstrated below in Figure 9 and the RMSE results with raw features and combined features as input are shown in Tables 4 and 5 respectively.

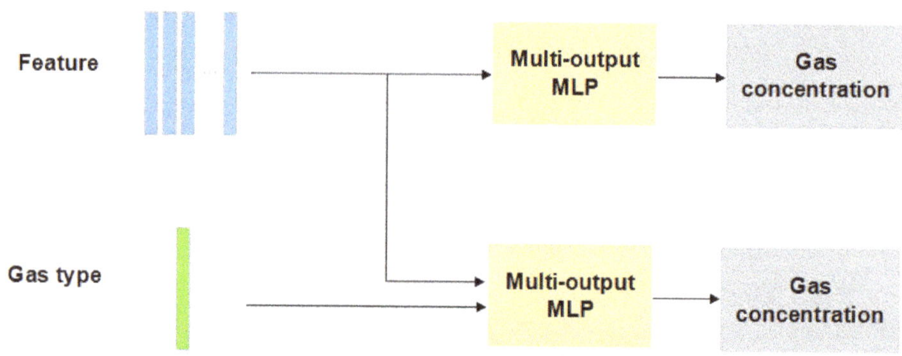

Figure 9. Gas concentration regression model diagram.

Table 4. RMSE results with raw features as input.

RMSE	150 s	40 s	20 s	10 s
H_2	0.0593	0.0840	0.0790	0.1285
CO	0.1289	0.1208	0.2077	0.2916
hybrid gas	0.1208	0.1478	0.1811	0.2424

Table 5. RMSE results with combined features as input.

RMSE	150 s	40 s	20 s	10 s
H_2	0.0537	0.0607	0.0534	0.0921
CO	0.1056	0.1156	0.1361	0.1620
hybrid gas	0.0956	0.1048	0.1137	0.1433

These data indicate the root mean square error (RMSE) of the MLP fitting results for the three gas concentrations in different time windows. RMSE is a standard index to measure the accuracy of model prediction. The lower the RMSE value, the higher the accuracy of model prediction. The RMSE trends of different gases are shown in Figure 10a–c.

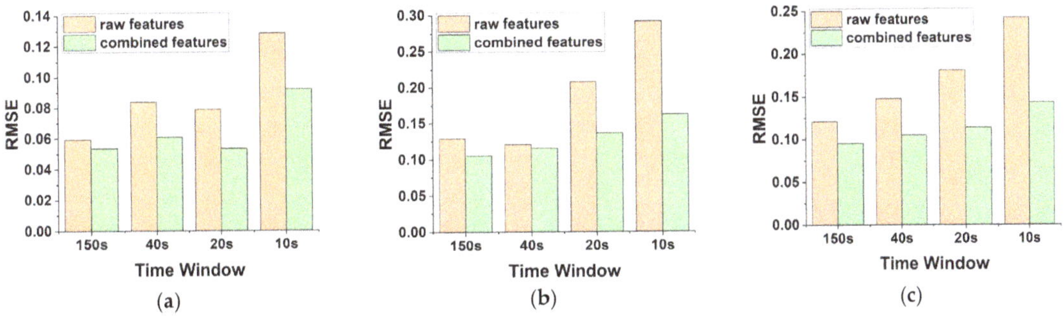

Figure 10. Comparison of regression results of (**a**) H_2, (**b**) CO, (**c**) mixed gas concentration with different inputs.

From the result data, we can make the following observation: For the results of the two input features, the fitting accuracy of all gases deteriorates with the shortening of the time window (from 150 s to 10 s), and the RMSE increases, which is in line with expectations. The data information in a relatively short period may not be enough for the MLP model to accurately capture the change in gas concentration.

In the same time window, the RMSE of the combined feature input on the regression results of all gas categories decreased, indicating that the combined feature may provide better prediction information and is more suitable for the MLP model to perform the concentration regression prediction of mixed gas. Similarly, although the RMSE of the combined feature input results increases gradually with the shortening of the time range, it increases more slightly than that of the original feature. With the shortening of the time window from 150 s to 10 s, the RMSE of the mixed gas regression result of the combined feature input increased from 0.0956 to 0.1433, while that of the original feature regression results risen from 0.1208 to 0.2424.

Overall, MLP models with combined features outperformed MLP models with original features over all time ranges, especially over shorter time ranges. This may be because the combined features contain more information, which helps the model capture more complex data patterns. In addition, predictions for hydrogen are generally more accurate than those for carbon monoxide and mixed gases, which may be related to the physical properties of hydrogen or the characteristics of the dataset.

In real lithium battery thermal runaway scenarios, various background interference gases, such as CO_2 and C_2H_4, are present. The impact of these gases on the performance of our specific sensors remains to be studied. Additionally, the effects of environmental fluctuations causing noise or partial sensor failures on algorithm performance also warrant further investigation.

6. Conclusions

In this paper, we collected data on the responses of a sensor array to three different gases at various concentrations in the laboratory, and these three gases were identified as characteristic gases for monitoring thermal runaway in lithium-ion batteries. We also conducted classification and concentration regression of the gases. Three methods (SVM, ELM, and MLP networks) were employed to classify three types of gases. Due to mutual interference among gases in the mixture, sensor responses are not simply the mathematical sum of responses to individual gases. Among the three methods, ELM exhibited the fastest training time with high accuracy of three gases, while MLP achieved the highest classification accuracy of three gases in 20 s. In the concentration regression stage, we used the multi-output MLP to combine the two input features and compare the root mean square error of the fitting results. The results show that the regression result of combined features is better than the original feature input effect, proving the method's feasibility.

Author Contributions: Conceptualization, M.J. and Z.P.; methodology, Z.P. and M.Y.; software, Z.P. and Z.W.; validation, D.Z. and Y.H.; formal analysis, Z.W.; investigation, M.J.; resources, M.J.; data curation, Z.P.; writing—original draft preparation, Z.P.; writing—review and editing, M.J.; visualisation, M.Y.; supervision, M.J.; project administration, Y.H.; funding acquisition, M.J. and D.Z. All authors have read and agreed to the published version of the manuscript.

Funding: This research was funded by the Department of Science and Technology of Anhui Province (2022CSJGG0703), National Natural Science Foundation of China (No. 52204253), the Opening Foundation of Civil Aircraft Fire Science and Safety Engineering Key Laboratory of Sichuan Province (MZ2023KF06), the Fundamental Research Funds for the Central Universities (2021ZDPY0208). Dr. Zhi Wang is funded by the Jiangsu Provincial Double-Innovation Doctor Program (No. 140923027), and the China Postdoctoral Science Foundation (2023M733766).

Data Availability Statement: The data that supported the findings of this study are available from the corresponding author upon reasonable request.

Conflicts of Interest: The authors declare no conflicts of interest.

References

1. Dunn, B.; Kamath, H.; Tarascon, J.M. Electrical Energy Storage for the Grid: A Battery of Choices. *Science* **2011**, *334*, 928–935. [CrossRef] [PubMed]
2. Wang, Q.; Mao, B.; Stoliarov, S.I.; Sun, J. A review of lithium ion battery failure mechanisms and fire prevention strategies. *Prog. Energy Combust. Sci.* **2019**, *73*, 95–131. [CrossRef]
3. Fernandes, Y.; Bry, A.; de Persis, S. Identification and quantification of gases emitted during abuse tests by overcharge of a commercial Li-ion battery. *J. Power Sources* **2018**, *389*, 106–119. [CrossRef]
4. Yamanaka, T.; Takagishi, Y.; Tozuka, Y.; Yamaue, T. Modeling lithium ion battery nail penetration tests and quantitative evaluation of the degree of combustion risk. *J. Power Sources* **2019**, *416*, 132–140. [CrossRef]
5. Gao, D.; Xu, J.B.; Lin, M.; Xu, Q.; Ma, C.F.; Xiang, H.F. Ethylene ethyl phosphate as a multifunctional electrolyte additive for lithium-ion batteries. *RSC Adv.* **2015**, *5*, 17566–17571. [CrossRef]
6. Zhang, H.Y.; Pinjala, D.; Teo, P.S. Thermal management of high power dissipation electronic packages: From air cooling to liquid cooling. In Proceedings of the 5th Electronics Packaging Technology Conference (EPTC 2003), Singapore, 10–12 December 2003; pp. 620–625.
7. Zhang, S.S. A review on electrolyte additives for lithium-ion batteries. *J. Power Sources* **2006**, *162*, 1379–1394. [CrossRef]
8. Sascha, K.; Kai, B.; Robert, K. Fast Thermal Runaway Detection for Lithium-Ion Cells in Large Scale Traction Batteries. *Batteries* **2018**, *4*, 16. [CrossRef]
9. Koch, S.; Fill, A.; Birke, K.P. Comprehensive gas analysis on large scale automotive lithium-ion cells in thermal runaway. *J. Power Sources* **2018**, *398*, 106–112. [CrossRef]
10. Essl, C.; Seifert, L.; Rabe, M.; Fuchs, A. Early Detection of Failing Automotive Batteries Using Gas Sensors. *Batteries* **2021**, *7*, 25. [CrossRef]
11. Golubkov, A.W.; Fuchs, D.; Wagner, J.; Wiltsche, H.; Stangl, C.; Fauler, G.; Voitic, G.; Thaler, A.; Hacker, V. Thermal-runaway experiments on consumer Li-ion batteries with metal-oxide and olivin-type cathodes. *RSC Adv.* **2014**, *4*, 3633–3642. [CrossRef]
12. Fan, S.R.; Li, Z.; Xia, K.W.; Hao, D.X. Quantitative and Qualitative Analysis of Multicomponent Gas Using Sensor Array. *Sensors* **2019**, *19*, 3917. [CrossRef] [PubMed]
13. Chapman, E.A.; Thomas, P.S.; Stone, E.; Lewis, C.; Yates, D.H. A breath test for malignant mesothelioma using an electronic nose. *Eur. Respir. J.* **2012**, *40*, 448–454. [CrossRef] [PubMed]
14. Chen, L.Y.; Wong, D.M.; Fang, C.Y.; Chiu, C.I.; Chou, T.I.; Wu, C.C.; Chiu, S.W.; Tang, K.T. Development of an Electronic-Nose System for Fruit Maturity and Quality Monitoring. In Proceedings of the 4th IEEE International Conference on Applied System Invention (IEEE ICASI), Tokyo, Japan, 13–17 April 2018; pp. 1129–1130.
15. Men, H.; Yin, C.B.; Shi, Y.; Liu, X.T.; Fang, H.R.; Han, X.J.; Liu, J.J. Quantification of Acrylonitrile Butadiene Styrene Odor Intensity Based on a Novel Odor Assessment System With a Sensor Array. *IEEE Access* **2020**, *8*, 33237–33249. [CrossRef]
16. Zhang, J.Y.; Xue, Y.Y.; Sun, Q.Y.; Zhang, T.; Chen, Y.T.; Yu, W.J.; Xiong, Y.Z.; Wei, X.W.; Yu, G.T.; Wan, H.; et al. A miniaturized electronic nose with artificial neural network for anti-interference detection of mixed indoor hazardous gases. *Sens. Actuators B Chem.* **2021**, *326*, 128822. [CrossRef]
17. Chu, J.F.; Li, W.J.; Yang, X.; Wu, Y.; Wang, D.W.; Yang, A.J.; Yuan, H.; Wang, X.H.; Li, Y.J.; Rong, M.Z. Identification of gas mixtures via sensor array combining with neural networks. *Sens. Actuators B Chem.* **2021**, *329*, 129090. [CrossRef]

18. Li, M.H.; Ge, S.K.; Gu, Y.L.; Zhang, Y.F.; Li, X.; Zhu, H.C.; Wei, G.F. Qualitative Identification of Single/Mixture Gases Based on Fe-ZnO Sensor Array and PSO-BP Neural Network. *IEEE Sens. J.* **2023**, *23*, 20037–20043. [CrossRef]
19. Huang, G.B.; Zhu, Q.Y.; Siew, C.K. Extreme learning machine: A new learning scheme of feedforward neural networks. In Proceedings of the IEEE International Joint Conference on Neural Networks (IJCNN), Budapest, Hungary, 25–29 July 2004; pp. 985–990.
20. Wang, T.; Zhang, H.X.; Wu, Y.; Chen, X.Y.; Chen, X.W.; Zeng, M.; Yang, J.H.; Su, Y.J.; Hu, N.T.; Yang, Z. Classification and Concentration Prediction of VOCs With High Accuracy Based on an Electronic Nose Using an ELM-ELM Integrated Algorithm. *IEEE Sens. J.* **2022**, *22*, 14458–14469. [CrossRef]
21. Ma, H.L.; Wang, T.; Li, B.L.; Cao, W.Y.; Zeng, M.; Yang, J.H.; Su, Y.J.; Hu, N.T.; Zhou, Z.H.; Yang, Z. A low-cost and efficient electronic nose system for quantification of multiple indoor air contaminants utilizing HC and PLSR. *Sens. Actuators B Chem.* **2022**, *350*, 130768. [CrossRef]
22. Ouyang, T.H.; Wang, C.W.; Yu, Z.J.; Stach, R.; Mizaikoff, B.; Huang, G.B.; Wang, Q.J. NOx Measurements in Vehicle Exhaust Using Advanced Deep ELM Networks. *IEEE Trans. Instrum. Meas.* **2021**, *70*, 7000310. [CrossRef]
23. Zhao, L.; Li, X.G.; Wang, J.; Yao, P.J.; Akbar, S.A. Detection of Formaldehyde in Mixed VOCs Gases Using Sensor Array With Neural Networks. *IEEE Sens. J.* **2016**, *16*, 6081–6086. [CrossRef]
24. Khan, M.A.H.; Thomson, B.; Debnath, R.; Motayed, A.; Rao, M.V. Nanowire-Based Sensor Array for Detection of Cross-Sensitive Gases Using PCA and Machine Learning Algorithms. *IEEE Sens. J.* **2020**, *20*, 6020–6028. [CrossRef]
25. Zhou, Y.; Heng, Y.; Zhu, J.; Qian, C.; Wang, T.; Nguyen, D.H.; Jiao, M. Enhanced Gas Recognition of Electronic Nose Using 1-D Convolutional Neural Network With Savitzky–Golay Filter. *IEEE Sens. J.* **2024**, *24*, 10769–10778. [CrossRef]
26. Song, H.; Ma, L.; Pei, S.; Dong, C.; Zhu, E.; Zhang, B. Quantitative detection of formaldehyde and ammonia using a yttrium-doped ZnO sensor array combined with a back-propagation neural network model. *Sens. Actuators A Phys.* **2021**, *331*, 112940. [CrossRef]

batteries

MDPI

Article

Thermal Propagation Test Bench for the Study of the Paschen Curve and Lightning Arcs of Venting Gas

Björn Mulder [1], Kai Peter Birke [1,*], Björn Obry [2], Stefan Wigger [2], Ruslan Kozakov [3], Pavel Smirnov [3] and Jochen Schein [3]

[1] Institute for Photovoltaics, University of Stuttgart, Pfaffenwaldring 47, 70569 Stuttgart, Germany
[2] Mercedes-Benz Group AG, Mercedesstraße 120, 70372 Stuttgart, Germany
[3] Institute for Plasma Technology and Basics of Electrical Engineering, Bundeswehr University Munich, Werner-Heisenberg-Weg 39, 85579 Neubiberg, Germany
* Correspondence: peter.birke@ipv.uni-stuttgart.de

Abstract: Thermal propagation events are characterized by fire and thick black smoke, leading to propagation methods with a focus on preventing heat transfer and optimizing gas flow. Yet little attention is being paid to the electric conductivity of the gas, leading to possibly unexpected battery casing openings due to lightning arcs as well as potentially providing the minimum ignition energy. This gas composition (omitting particles) was used at different temperatures and pressures in a lightning arc test bench, leading to the Paschen curve. Using a mini-module cell setup, filtered venting gas was flowed through another lightning arc test bench, allowing for in situ measurements.

Keywords: thermal runaway; thermal propagation; lightning arcs; Paschen's law; test bench

1. Introduction

With the rise in electric vehicles comes the need for the public to understand and predict possible safety issues compared to well-known internal combustion engine cars. A select number of incidents showing the destructiveness of battery fires have been recorded by Sun et al. [1].

The perceived threat of battery fires, typically due to thermal propagation, has caused governments across the globe to implement legislation in order to reduce the potential overall occurrence of said fires as well as their individual severity, most notably the UN Global Technical Regulation GTR 20 [2]. The current legislation aims to delay propagation to ensure sufficient passenger evacuation time windows. For this, the thermal parameters of the battery such as the cells and parts need to be well tailored.

The venting of the initial cell introduces a new environment into the battery: this venting gas contains flammable components at high temperatures as well as particles of various sizes and compositions. The abrasive nature of these particles in the flow as well as the high temperatures may lead to previously protected components losing their safeguards. As a result, previous air and creepage distances in normal operation may not be sufficient or be rendered useless [3].

To better understand the phenomenon, a single NMC 622 lithium ion pouch cell was placed in an autoclave with atmospheric gas composition; thermal runaway was triggered and the venting gas composition was evaluated. The breakdown voltage of this gas at different temperatures was then analyzed and compared to the Paschen curve of air. A breakdown voltage detector for the use of venting gas from cells undergoing thermal runaway was introduced, and the gas was filtered to remove particles. Experiments by Li et al. [4] used a setup with particles in the venting gas.

2. Materials and Methods

In earlier works, single battery cells were placed in autoclaves, triggered into thermal runaway and the released venting gas amount and composition was evaluated, namely in the work of Koch et al. [5]. The most commonly found components are CO_2, CO and H_2 ranging typically around 10–40 %, followed by C_2H_4 and CH_4 at 5–10% and a multitude of trace gases, most notably C_2H_6 and C_3H_6 at around 1% or below.

For the following experiments, a pouch cell with a $LiNi_6Mn_2Co_2O_2$ (NMC 622)-cathode was used. The relevant cell is seen in Figure 1 and its characteristics before testing are as follows:

- $\rho_{vol} \approx 500\,\text{Wh}\ell^{-1}$
- $E_{capacity} \approx 60\,\text{Ah}$

Figure 1. Pouch cell under investigation with cell tabs on the side.

2.1. Venting Gas Mixture from Autoclave Setup

An autoclave has a reinforced container at its center, allowing to observe chemical reactions at different temperatures and pressures with hermetic sealing and pressure resistance of the reaction chamber. The used autoclave has a volume of 80 ℓ and is initially filled with air; the cell is placed inside the autoclave and brought into thermal runaway (TR) using a heater. Once the TR has stopped and the autoclave has cooled to ambient temperatures, a gas sample is taken. The gas analysis is then performed by a gas chromatograph.
The following results for the venting gas were yielded from averaging multiple experiments with cells tested with a SoC (State of Charge) of 100%:

$c_{N_2} = 35.4$ vol%, $c_{CO_2} = 24.0$ vol%, $c_{CO} = 21.0$ vol%, $c_{H_2} = 13.0$ vol%, $c_{C_2H_4} = 3.6$ vol% and $c_{CH_4} = 3$ vol%.

The artificial venting gas used in the experiments that follow was manufactured accordingly, including nitrogen. During thermal propagation, the gas composition within a battery changes rapidly, as the initial nitrogen gets pushed out due to more cells undergoing thermal runaway, and the initial oxygen is quickly reacted. Lightning arc events can already occur during the venting of the initial cell. Thus, the artificial venting gas should contain nitrogen for a better understanding of the breakdown voltage in venting gas. This paper focuses on the gas component; particles were omitted from the artificial venting gas as a first step in analyzing the arcing phenomenon.

2.2. Paschen's Law

Paschen's law describes an equation of the necessary voltage between two electrodes in a gas for a discharge or electric arc to occur ([6–9]). The influencing parameters include the specific type of gas, pressure, temperature, and distance/gap between electrodes, as well as the shape of the electrodes. Graphs typically show the breakdown voltage as a function $f(d \cdot p)$ of distance d and pressure p with the abscissa typically in a log scale. The curves for various gases as well as air are shown in Figure 2, given by Celani et al. [10]. The parameter of Paschen's law for mixtures of gases cannot be calculated from its individual gas components, so testing is unavoidable. The used gas composition was gathered by the autoclave testing described in Section 2.1.

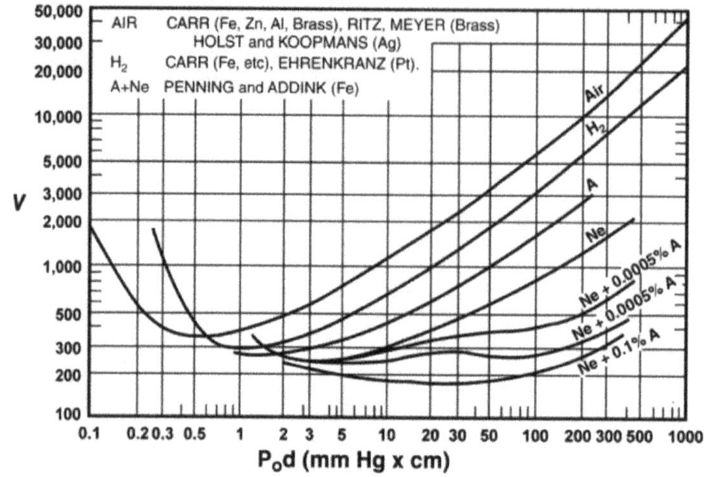

Figure 2. Overview of different Paschen curves, most notably of air; see Lehr and Ron [11]. 1 mm Hg · cm ≈ 13.3 mbar · mm.

According to Lehr and Ron [11], the Paschen curve can be described using the equation

$$V_b = \frac{B \cdot (pd)}{ln(pd \cdot A) - ln(ln(1 + \gamma^{-1}))} \tag{1}$$

with V_b being the breakdown voltage, p the pressure, d the distance between the electrodes, A and B the gas-specific constants, and γ the third Townsend coefficient. For large values of (pd), coefficient γ is negligible, and Equation (2) simplifies to

$$\lim_{(pd) \to \infty} V_b \approx \frac{B \cdot (pd)}{ln(pd \cdot A)} \tag{2}$$

Two points $(pd)_1$ and $(pd)_2$ can be selected with their respective arcing voltages $V_b((pd)_1)$ and $V_b((pd)_2)$, leading to value B being expressed as

$$B \approx \frac{V_b((pd)_1) \cdot ln((pd)_1 \cdot A)}{(pd)_1}; \qquad B \approx \frac{V_b((pd)_2) \cdot ln((pd)_2 \cdot A)}{(pd)_2} \tag{3}$$

These two Equations (3) can now be equated to

$$\frac{V_b((pd)_1)}{(pd)_1} \cdot ln((pd)_1 \cdot A) \approx \frac{V_b((pd)_2)}{(pd)_2} \cdot ln((pd)_2 \cdot A) \tag{4}$$

Rearranging and solving for A allows for the approximate calculations of A and B as

$$A \approx e^{\left[\frac{\frac{V_b((pd)_1)}{(pd)_1} \cdot ln((pd)_1) - \frac{V_b((pd)_2)}{(pd)_2} \cdot ln((pd)_2)}{\left(\frac{V_b((pd)_2)}{(pd)_2} - \frac{V_b((pd)_1)}{(pd)_1}\right)}\right]}; \text{ for large } (pd) \tag{5}$$

$$B \approx \frac{V_b(pd) \cdot ln((pd) \cdot A)}{(pd)}; \text{ for large } (pd) \tag{6}$$

2.3. Breakdown Voltage Detector

The breakdown voltage detector is designed for pressures of p = 10 mbar–1500 mbar and temperatures up to $T_{max} \approx 1000$ °C. The separation between the electrodes is adjusted by a stepper motor to allow distance increments of $d_{step} = 3.125 \cdot 10^{-3}$ mm.

A schematic cross section of the used breakdown voltage detector is shown in Figure 3: A pair of electrodes (1) is placed in a glass cylinder containing a gas specimen. Inner isolators (2) and outer isolators (3) keep the gas within while also allowing the manipulator (4) to change the position of the left electrode; the volume between the left and right electrode is called the test gap. The manipulator is connected to a stepper motor to allow direct discrete distances in multiples of $3.125 \cdot 10^{-3}$ mm. Pressure levels are measured using a Pirani gauge (5); gas inlets (6) allow to attach gas cylinders, and a valve connector (7) connects to a vacuum pump.

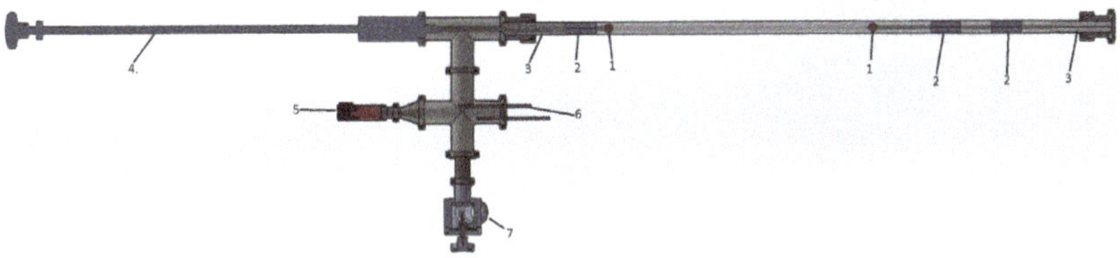

Figure 3. Schematic overview of the breakdown voltage detector. (1) Pair of electrodes, (2) inner isolaters, (3) outer isolators, (4) manipulator, (5) Pirani gauge, (6) gas inlet, and (7) valve connector.

In previous work (see Mulder et al. [12]), it was shown that a module of 12 pouch cells with a SoC of 100% can deliver peak temperatures of around 900 °C. To recreate these temperatures, a small brick box with heating elements acting as a brick oven is used. The test gap of the breakdown voltage detector can be placed within the heated volume, which can deliver temperatures up to the required 900 °C as seen in Figure 4.

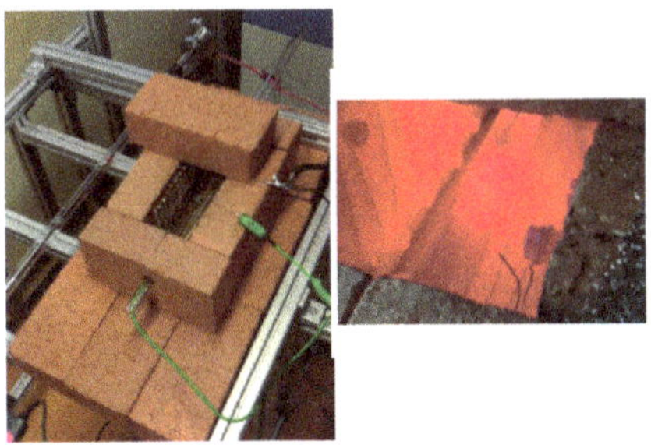

Figure 4. For tests at higher temperatures, the test gap of the breakdown voltage detector can be placed inside a brick oven. (**Left**) Test bench at room temperatures, (**Right**) test bench with activated heating elements.

Before experiment data can be collected, the vacuum pump attached to the outlet valve (7) evacuates the main glass cylinder. Then, the cylinder can be filled using the gas inlets (6) from attached gas bottles; for higher gas purity, the process of gas evacuation and refilling can be repeated. The inner (2) and outer (3) isolators prevent undesired gas from bleeding into the test gap. Initially, a set of glass cylinder and electrodes (set A) was used, which was later replaced with a new glass cylinder, and the electrodes were equipped with additional alignment (set B); both sets can be seen in Figure 5. An overview of the experiments carried out using venting gas with different temperature/pressure points can be found in Table 1.

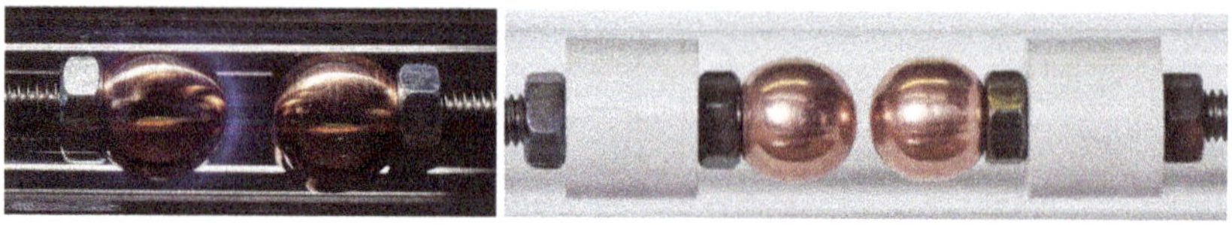

Figure 5. (**Left**) Cylinder and electrode set A, (**Right**) cylinder and electrode set B, with additional supporting alignment. The electrodes are manufactured from copper.

Table 1. Overview of experiments with venting gas: temperature T, pressure p and glass cylinder/electrode set (A or B) combinations.

Gas	$p = 1250$ mbar	$p = 1500$ mbar
Cylinder A		
200 °C	X	X
500 °C	X	X
900 °C	X	X
Cylinder B		
200 °C	X	X
500 °C	X	X
900 °C	X	X

2.4. Venting Gas Breakdown Voltage Detector

The test facility, seen in Figure 6, can be split into three functional groups: the mini-module test bench (red box), the exhaust system with measurement equipment (blue box), and the surrounding environment. All components were selected and designed with robustness in mind to ensure a high level of safety during testing as well as its reusability. A more detailed description of the capabilities and experiment setup can be found in the previous publication by Mulder et al. [12]. A schematic overview of the entire test setup can be seen in Figure 7.

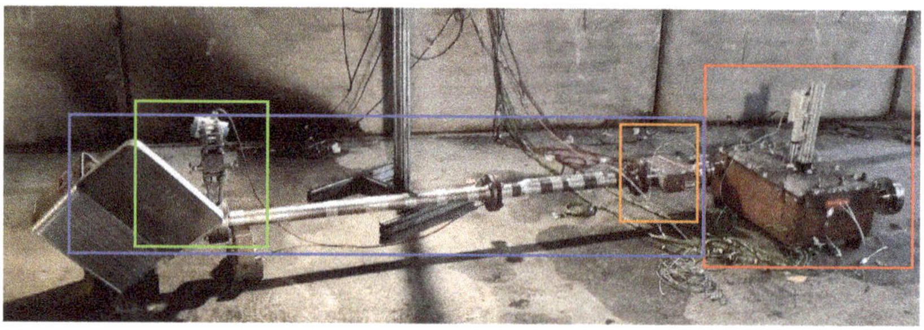

Figure 6. Overview of the test facility: the mini-module test bench (in red box) is connected via flanges to the exhaust system (blue box), which allows the generated venting gas to vent into the surrounding environment—a metal box is placed at the end to allow the capture of particles. Within the exhaust system, the lightning arc box (orange box) is placed, as well as a mass flow meter (green box).

The opened mini-module test bench can be seen in Figure 8, where a custom cell stack of mid-of-life 6 2p3s-connected pouch cells are placed. An edge cell of the pack is triggered, ensuring that cells propagate one after another instead of up to two cells simultaneously. The used trigger method is nail penetration with a wedge-shaped nail parallel to the electrode layers. Other trigger methods can be implemented yet were not chosen for the lightning arc study.

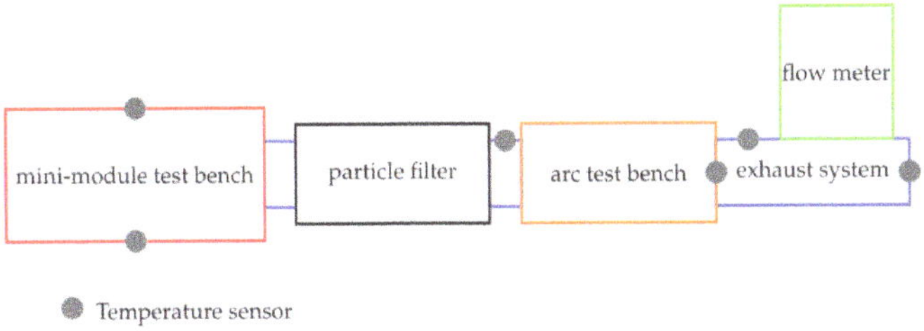

Figure 7. Schematic overview of the test facility: cells can be triggered in the mini-module test bench, and exhaust travels through the particle filter and the arc test bench out of the exhaust pipe into the surrounding environment. Temperature sensors placed outside of the mini-module test bench are shown.

Figure 8. View into the opened test bench: A mini-module of 2p3s-connected pouch cells is placed inside the test mount. The nail gun will trigger the topmost cell marked with a black stripe.

In the exhaust line (see Figure 6), the lightning arc box can be installed using the flange system. The opened box can be seen in Figure 9: two electrodes with a diameter of 25 mm are connected to the power supply and voltage measurement equipment. Using feeler gauges, the appropriate distance can be set, as well as ensuring that the flat surfaces are parallel to each other. The studied gas can flow in a pipe made of two PEEK pieces. Before testing, the other half of the pipe piece is placed and sealed. Loctite SI 5300 is used as a sealant around the electrode rod as well as in between the pipe pieces and the lid of the lightning arc box.

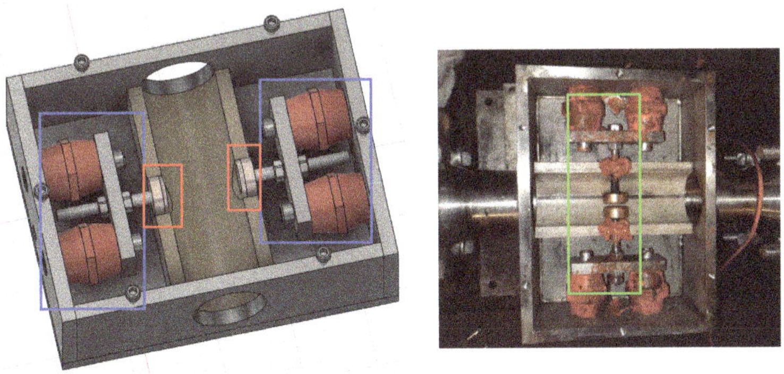

Figure 9. (**Left**) Schematic view of the lightning arc box. (**Right**) View of the lightning arc box before closing. Circular electrodes (red box) are attached via electrode rods (blue box) to the power supply. The bottom half of the PEEK pipe (green box) is already installed.

2.4.1. Testing Procedure Setup

The cell block is charged to 100% SoC and placed in the test bench. The cell mount is adjusted to ensure that the nail gun (later attached to the lid) will only trigger one cell. Before the lid is placed on the test bench, Loctite SI 5300 is applied to create a gastight seal. Screws and a clamping system secures the lid in place. The test bench has two flanges placed diagonally from each other: to one flange of the test bench, an end flange is connected, and to the other flange, the exhaust system is connected. It consists of a pipe system of the lightning arc box and the mass flow meter; a particle filter can be implemented in the exhaust pipe, placed before the arc box. To the lightning arc box, a power supply is wired. Once the mechanical setup is complete, the voltage to the electrodes is initially set at 50 V; after every vented cell, it is to be increased by 50 V. The cell pack is triggered by a nail gun, which is attached to the top side of the lid. Two cameras are set up. The first focuses on the test bench with the attached nail gun; the second camera records the venting gas leaving the end of the exhaust system.

2.4.2. Sensors and Placement

The following measurement data are collected: (1) initial time of nail penetration, (2) temperature within the cell block, test bench, exhaust system and environment, (3) voltage of the cell block and lightning arc box, and (4) massflow in exhaust system.

1. Initial time of nail penetration
   It is recorded when and for how long the nail gun motor is activated—once the nail is fully inserted into the cell, the motor stops.
2. Temperature sensors
   Type K sensors are used; a schematic overview can be seen in Figure 10. Four sensors are placed within the cell mount/cell stack and are numbered, using aluminum plates with sensor cutout to avoid indentation into the cells. Two sensors are placed on the connection between cell mount and test bench, two in the test bench (top and bottom), four along the exhaust system, and one recording the surrounding environment.
3. Voltage metering
   Voltage is measured in the cell pack—every logical cell (2p) as well as the overall stack are measured. The electrodes within the lightning arc box are supplied with the appropriate voltage, yet to record the lightning arc, a sufficiently fast voltmeter needs to be used. To detect the presence of lightning arcs, 100 Hz is found to be sufficient.
4. Massflow
   A SITRANS P320 digital pressure transmitter allows for gas flow (and thus gas creation) measurement of venting gas during the TP process.

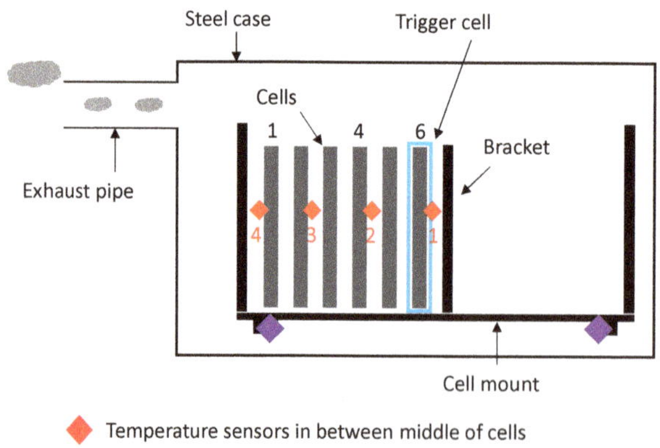

Figure 10. Schematic overview of temperature sensor placement in the test bench: 4 centrally located between cells, respectively between the cells and cell mount, and numbered starting with 1 from the trigger cell towards the exhaust pipe.

3. Results

3.1. Breakdown Voltage Detector Experiments

During one experiment run using the breakdown voltage detector, one combination of gas composition, pressure, and temperature is used (see Table 1). The distance between the electrodes is variable and set by a stepper motor. The procedure of the breakdown curve measurement starts from moving the electrodes together until the short circuit is detected. After that, one electrode is driven away by the stepper motor by one step in 500 ms. The position at which the short circuit disappears is taken as the smallest distance of $3.125 \cdot 10^{-3}$ mm. This procedure makes the accuracy of the electrode separation measurement equal to that of the stepper motor resolution, eliminating the dependence on thermal expansion at different temperatures. For each distance, the voltage is increased until an abrupt drop occurs: a lightning arc with the breakdown voltage is detected and recorded. Then, the stepper motor moves one step further, and the process repeats.

3.1.1. Air

The breakdown voltage detector is filled with air at $T = 100$ °C. In Figure 11, the breakdown voltage for pressures $p = 2.2$ mbar and $p = 5.6$ mbar is recorded. For each pressure, two curves are recorded. It can be seen that the minimum voltage recorded is around 400 V. Lehr and Ron [11] place the lowest breakdown voltage at around 350 V. The range of the breakdown voltage detector presented here therefore has a margin of error of around 10–15 % and is sufficient for the measurement of the breakdown voltage of other gases.

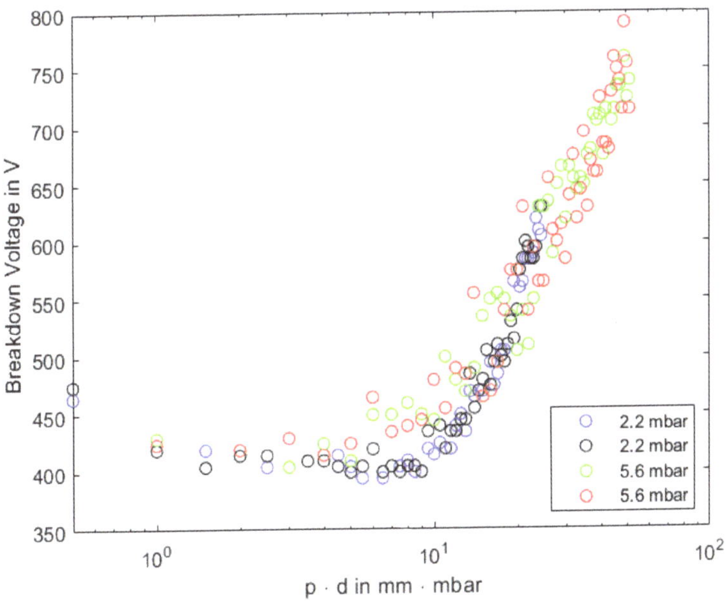

Figure 11. Breakdown voltage detected for Air at $T = 100$ °C for pressures $p = 2.2$ mbar and $p = 5.6$ mbar.

3.1.2. Venting Gas

In Figure 12, the breakdown voltage for the experiments with venting gas (see Table 1) is displayed. It can be seen that higher temperatures result in lower breakdown voltages for higher distances. It needs to be noted that the abscissa is dependent on the pressure and distance ($p \cdot d$) and in log scale. The distances in the experiment are from $d = 3.125 \cdot 10^{-3}$ mm to $d = 1.625 \cdot 10^{-1}$ mm small. The lowest recorded voltage is 330 V, which is lower than the voltage of 400 V recorded for air at 100 °C at $p = 1$ bar. This would suggest that the venting gas mixture could be more susceptible to lightning arcs than air.

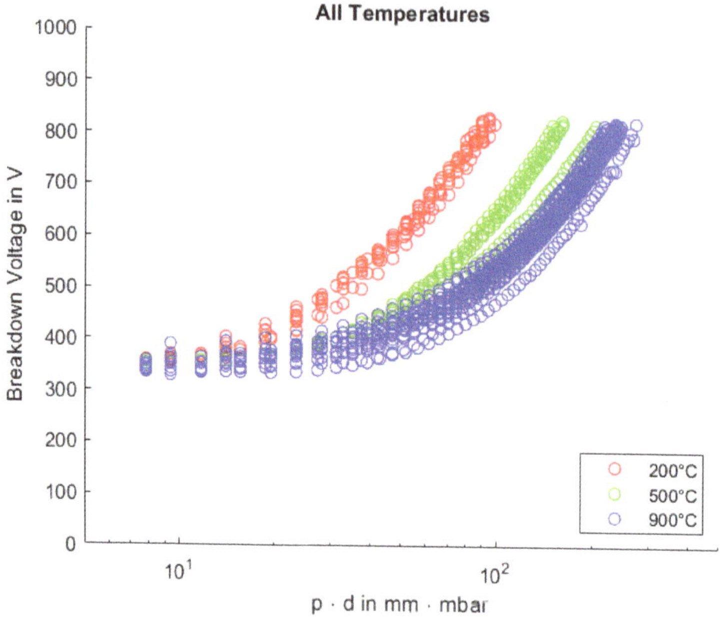

Figure 12. Recorded breakdown voltage for experiments with temperatures $T = 200$ °C (Red), 500 °C (Green) and 900 °C (Blue).

In Figure 13, the averaged Paschen curve for venting gas at 200 °C is visualized. The measurement data are plotted in red, while the blue curve is added using the simplified Paschen's law for large $p \cdot d$ according to Equation (2). The lowest recorded breakdown voltage is 350 V at 4.6875 mm · mbar. Compared to 100 °C air at around 400 V, this is a 12.5% decrease. Lehr et al. have analyzed the dependence of the cathode materials in hydrogen while recording the breakdown voltage. It can be clearly seen that the copper electrodes show a minimum breakdown voltage of 295 V, while commercial aluminum shows a minimum voltage of 255 V—copper has a 15.6% higher breakdown voltage compared to commercial aluminum. As the used electrodes for the breakdown voltage detector are made from copper, the noted breakdown voltage for this experiment at 200 °C should be reduced by that percentage, as the materials within a battery are a mixture of multiple metals with casings mainly made from aluminum [13]. This leaves an adjusted voltage of 295.4 V.

In Figure 14, the same curve for 900 °C is shown. The measurement data are again plotted in red, while the blue curve is added using the simplified Paschen's law for large $p \cdot d$ according to Equation (2). The lowest recorded breakdown voltage is 330 V at 9.375 mm · mbar. To adjust the breakdown voltage according to Figure 15 by 15.6%, an adjusted breakdown voltage of 278.52 V can be noted.

According to DIN EN 60664-1 [14], a suitable distance to prevent breakdown voltages in 400 V environments is 0.8 mm. Assuming no excess pressure during venting gas creation, thus setting the pressure to $p = 1000$ mbar, a 0.8 mm distance would require a voltage of ≈1850 V for a lightning arc to occur. The artificial gas is at pressure levels and near thermal propagation temperatures, yet the necessary voltages for a lightning arc in said artificial gas cannot be achieved by a typical traction battery, even if considering 800 V vehicles. Figure 14 shows low voltages in the 300 V–800 V range, yet these are only recorded at low $p \cdot d \leq 300$ mm · mbar. As the real venting gas also has a variety of particles, and lightning arcs have been observed during TP (Sun et al. [1]), the Paschen's curve analysis indicates that particles need to be part of the observation.

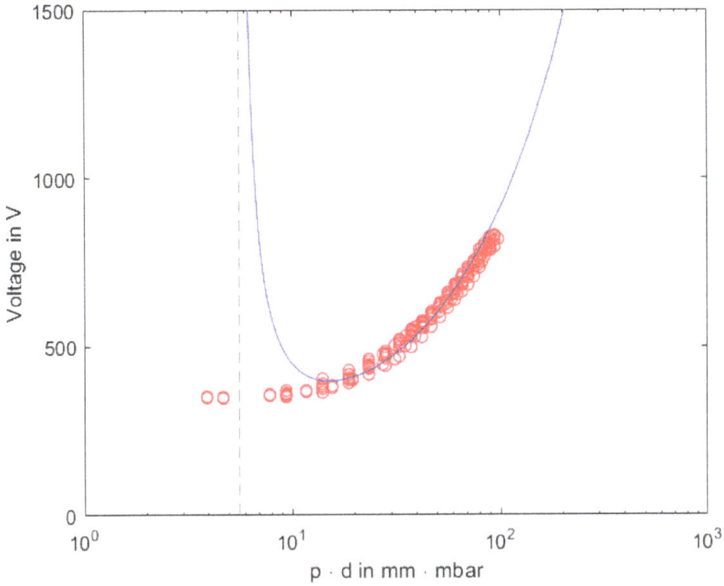

Figure 13. Paschen curve of venting gas at $T = 200\,°C$. The experimental data are shown in red, while the blue curve is generated using the simplified Paschen's law for large $p \cdot d$ (see Equation (2)).

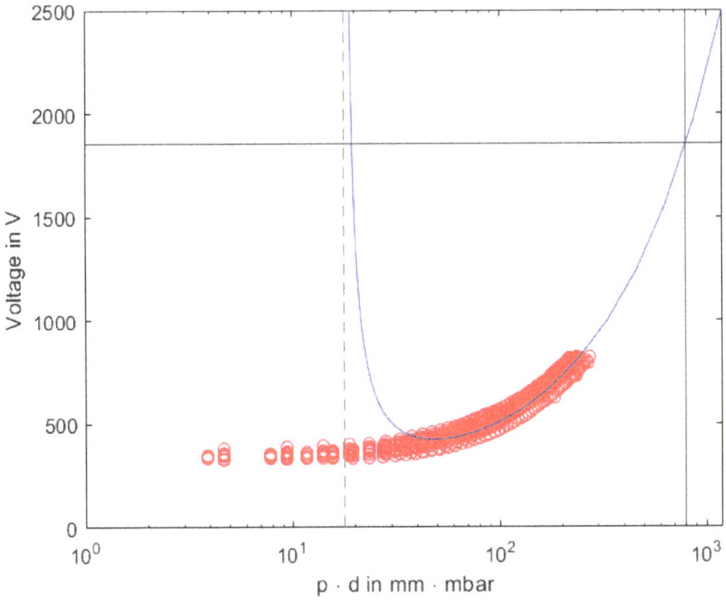

Figure 14. Paschen curve of venting gas at $T = 900\,°C$. The experimental data are shown in red, while the blue curve is generated using the simplified Paschen's law for large $p \cdot d$ (see Equation (2)). For a typical breakdown voltage of 800 V, a distance of 0.8 mm is required for air and creepage distances from the regulatory body [14]—yet the graph shows a required voltage of 1850 V for an arc.

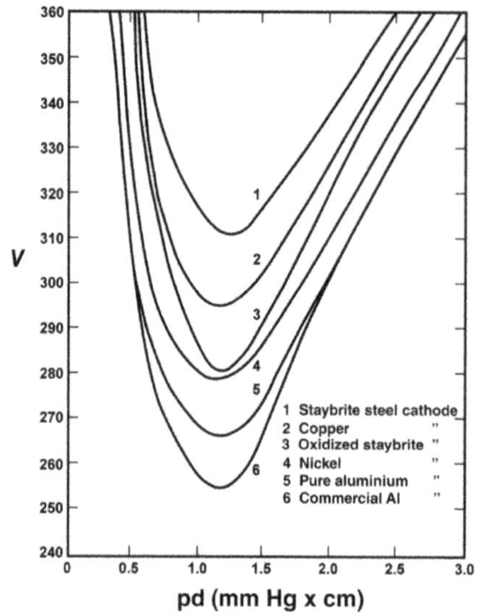

Figure 15. Influence of the cathode materials using Hydrogen when recording the breakdown voltage. Taken from Lehr et al. [11].

3.2. Thermal Propagation Testing

The results of the breakdown voltage detector in Section 3.1 lead to the conclusion that the gaseous part of the venting gas might not be responsible for the lightning arc phenomenon. To verify this, a cell module is triggered to undergo thermal propagation in order to generate real venting gas. In the exhaust pipe, a particle filter is installed, allowing for the study of the venting gas without particles.

A pouch cell mini-module is triggered and undergoes thermal runaway as seen in Figure 16. The electrodes are set to a distance of 2.5 mm ± 0.5 mm. A distance of 4 mm previously saw arcing, yet experiments using 2 mm are discussed in detail by Li et al. [4]. Before triggering the first cell, the voltage of the lightning arc box was raised to 400 V—during the entire experiment, the voltage did not drop—showing that no lightning arc occurred. Experiments short-circuiting the electrodes showed rapid voltage drops and returns to the preset voltage once the short was removed. The temperature sensor 1 next to the triggered cell showed an immediate rise in the temperature. Sensors 2 and 3 later showed the propagation through the module with temperature peaking between 800 °C and 850 °C, which are near the tested temperatures for the Paschen curves Section 3.1. The numbering of these temperature sensors can be seen in Figure 10.

The experiment showed an absence of drops in the applied voltage to the electrode set in the lightning arc box. The distance of the electrodes used is more than sufficient according to DIN EN 60664-1 [14] for clear clean air. Combining the results from Section 3.1 regarding the Paschen curve, where arcing with 400 V requires distances well below 0.8mm, and the results from the lightning arc box testing, it hints at the conclusion that the gaseous part of the venting gas does not contribute to the lightning arcs observed in traction battery testing. Testing regarding venting gas including particles therefore needs to be carried out and is planned for future publication ([15–21]).

Experiments by Li et al. [4] clearly showed significant arcing at a distance of 2 mm using 316 V of DC power. To a lesser degree, it was also shown that a 4 mm distance experienced arcing. The difference between the experiments by Li et al. and the experiment presented here is the absence of particles in the experiment presented here—which aims to show the influence of particle-free venting gas.

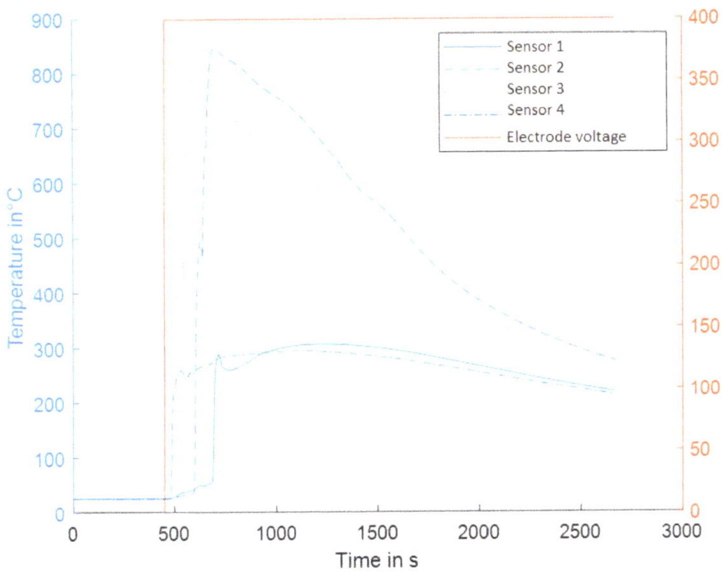

Figure 16. Cell module is triggered and undergoes thermal propagation. The temperature sensors *TB*02 and *TB*03 indicate a reaction, yet the recorded voltage once set to 400 V does not drop. The correlating sensor with its position can be seen in Figure 10.

4. Conclusions

Research regarding thermal propagation is mainly focused on the spread of the event by looking at heat transfer and gas flow. However, researching the electric conductivity of the venting gas can lead to the explanation of battery casing openings due to lightning arcs. Using an autoclave, cells underwent thermal runaway, and the resulting venting gas used was analyzed to contain a mixture of mainly N_2, CO_2, CO and H_2 with a low single-digit percentage of hydrocarbons; artificial venting gas was manufactured accordingly. A breakdown voltage detector was designed with two electrodes in a glass cylinder, allowing for the testing of different gases at pressure levels ranging p = 10–1500 mbar up to gas temperatures of $T \approx 1000$ °C. Paschen curves for $T = 200$ °C and $T = 200$ °C have been determined. For $T = 200$ °C, the lowest recorded breakdown voltage is 350 V at $p \cdot d = 4.6875\,\text{mm} \cdot \text{mbar}$—for $T = 900$ °C, the respective voltage is 330 V at $p \cdot d = 9.375\,\text{mm} \cdot \text{mbar}$.

In order to use real venting gas from a cell pack, a test bench for a pouch cell mini-module with an attached lightning arc box was developed. A particle filter was installed between the test bench and the lightning arc box. The applied voltage to the electrodes of the lightning arc box was set to 400 V, and the distance between the electrodes was set to $d = 2.5\,\text{mm} \pm 0.5$ mm. Using nail penetration, the cell pack was triggered into thermal propagation, and no lightning arcs were detected. This is consistent with the results of the breakdown voltage detector. The literature has shown that a distance of 4 mm with a voltage of 316 V DC in venting gas including particles shows sign of arcing. Both experimental setups confirmed that the required breakdown voltage of venting gas without particles is higher than the typical battery voltage of 400 V. Future experiments should be carried out without the particle filter. The modular setup of the venting gas breakdown voltage detector allows for the easy removal of the particle filter for full venting gas analysis. Additionally, particles will be analyzed for their composition and typical sizes.

batteries

MDPI

Article

Unravelling Lithium Interactions in Non-Flammable Gel Polymer Electrolytes: A Density Functional Theory and Molecular Dynamics Study

Nasser AL-Hamdani [1], Paula V. Saravia [2,3], Javier Luque Di Salvo [2,3], Sergio A. Paz [2,3] and Giorgio De Luca [1,*]

1 Institute on Membrane Technology, ITM-CNR, Ponte P. Bucci, Cubo 17/c, 87036 Rende, Italy; n.alhamdani@itm.cnr.it
2 Departamento de Química Teórica y Computacional, Facultad de Ciencias Químicas, Universidad Nacional de Córdoba, Córdoba X5000HUA, Argentina; pauvsaravia@unc.edu.ar (P.V.S.); javier.luquedisalvo@mi.unc.edu.ar (J.L.D.S.); apaz@unc.edu.ar (S.A.P.)
3 Consejo Nacional de Investigaciones Científicas y Técnicas (CONICET), Instituto de Investigaciones en Fisicoquímica de Córdoba (INFIQC), Córdoba X5000HUA, Argentina
* Correspondence: g.deluca@itm.cnr.it

Abstract: Lithium metal batteries (LiMBs) have emerged as extremely viable options for next-generation energy storage owing to their elevated energy density and improved theoretical specific capacity relative to traditional lithium batteries. However, safety concerns, such as the flammability of organic liquid electrolytes, have limited their extensive application. In the present study, we utilize molecular dynamics and Density Functional Theory based simulations to investigate the Li interactions in gel polymer electrolytes (GPEs), composed of a 3D cross-linked polymer matrix combined with two different non-flammable electrolytes: 1 M lithium hexafluorophosphate ($LiPF_6$) in ethylene carbonate (EC)/dimethyl carbonate (DMC) and 1 M lithium bis(fluorosulfonyl)imide (LiFSI) in trimethyl phosphate (TMP) solvents. The findings derived from radial distribution functions, coordination numbers, and interaction energy calculations indicate that Li^+ exhibits an affinity with solvent molecules and counter-anions over the functional groups on the polymer matrix, highlighting the preeminent influence of electrolyte components in Li^+ solvation and transport. Furthermore, the second electrolyte demonstrated enhanced binding energies, implying greater ionic stability and conductivity relative to the first system. These findings offer insights into the Li^+ transport mechanism at the molecular scale in the GPE by suggesting that lithium-ion transport does not occur by hopping between polymer functional groups but by diffusion into the solvent/counter anion system. The information provided in the work allows for the improvement of the design of electrolytes in LiMBs to augment both safety and efficiency.

Keywords: lithium metal batteries; gel polymer electrolytes; Li^+ interactions; Li^+ solvation; molecular dynamics; density functional theory

1. Introduction and Background

Lithium metal batteries are at the forefront of next-generation energy storage technologies, primarily due to their exceptional energy density and theoretical specific capacity, which significantly exceed those of conventional lithium-ion batteries with graphite anodes. Lithium metal has a theoretical specific capacity of 3860 mAh g^{-1}, compared to graphite's capacity of 372 mAh g^{-1} [1,2]. This makes LiMBs a viable option for high-energy applications such as electric vehicles and large-scale energy storage systems [3,4]. In addition, the lightweight and high electrochemical potential of lithium metal makes it even more appealing for improving the overall energy efficiency of batteries [5].

However, the implementation of LiMBs in real-world scenarios is hindered by significant safety issues. One of the main concerns is the use of flammable organic liquid electrolytes, which carry a substantial risk of flammability, especially when exposed to high temperatures or in case of an internal short circuit [6,7]. Furthermore, the formation of lithium dendrites during the process of repetitive charging and discharging can result in short internal circuits and ultimately cause the battery to fail catastrophically [3–6]. These dendritic structures not only compromise safety but also reduce the lifespan and efficiency of the batteries, presenting a major obstacle to the widespread adoption of LiMBs technology [8,9].

To mitigate these issues, recent advancements have focused on the development of gel polymer electrolytes (GPE), which integrate the advantages of both liquid and solid-state electrolytes. A key innovation in this domain is the use of dendrimers—highly branched, tree-like polymers with well-defined structures. These polymers offer precise control over the electrolyte architecture, enhancing both mechanical stability and ionic conductivity [10,11]. Their non-stackable nature prevents undesirable phase separation and ensures uniform ion distribution, addressing critical challenges such as dendrite formation and electrolyte instability [12–14]. Moreover, dendrimers can be functionalized with various groups to further enhance the performance and safety of GPEs [14–16].

In this framework, the work, recently published by Hu et al. [15], is particularly interesting. In their work, a multifunctional, all-in-one gel polymer electrolyte membrane with a 3D cross-linking network was synthesized with the aim of creating polymer electrolyte membranes with 3D cross-linking networks that improve both performance and safety. The proposed GPE integrates -CN and -$PO(OC_2H_5)_2$ functional groups into a single polymer matrix, formed by acrylonitrile (AN) and diethyl vinyl-phosphonate (DEVP) monomers, cross-linked by poly(ethylene glycol) diacrylate (PEGDA). This integration aims to produce a synergistic effect that enhances ionic conductivity, mechanical stability, and flame resistance. As highlighted by Hu et al. [15], the -CN groups enable the formation of stable Li3N-rich solid electrolyte interphase (SEI), which inhibited the proliferation of lithium dendrites, and high-quality cathode electrolyte interphase (CEI) layers that stabilize the $LiFePO_4$ cathode. Moreover, the addition of -$PO(OC_2H_5)_2$ functional groups improved the flame resistance of the electrolyte, thereby decreasing the likelihood of combustion by capturing reactive radicals. Hence, the combined effect of these functional groups in a 3D cross-linking polymer network showed significant enhancements in both electrochemical performance and safety, showing the need to design at the molecular level more sophisticated, dependable, and safe LiMBs. This GPE is innovative because it combines synergistic advantages not commonly observed in other comparable systems used in Li metal batteries. Therefore, this cross-linked GPE deserves notable attention. To date, no theoretical studies have explored the impact of the various components of this GPE. This underscores the significance of our work, as it provides critical insights that can be used to optimize this novel system.

Despite significant advancements in the development of gel polymer electrolytes, there is still a lack of comprehensive understanding of the molecular-level behavior of the lithium ions in these systems. The transport of Li^+ is affected by various crucial parameters, including the coordination between lithium cations and functional groups present in the polymer matrix, as well as the structural arrangement of the GPE [3–10]. Previous theoretical studies, such as those by Wróbel et al. [17], have investigated the solvation dynamics of lithium ions in various solvents, emphasizing the role of solvent interactions in facilitating Li^+ transport. Similarly, Tasaki [18] explored the decomposition mechanisms and interactions of lithium ions in electrolytic environments through DFT calculations and molecular dynamics simulations, highlighting the significance of these interactions on overall battery performance. Additionally, research by Haskins et al. [19] has provided insights into the electronic structure and transport properties of lithium-doped ionic liquids, further informing the understanding of ion mobility in polymer systems. Nevertheless, the specific molecular interactions and pathways that control ion mobility in GPE are not well comprehended, which makes it difficult to optimize these membranes to meet the demanding performance and safety criteria for future LiMBs [4–18].

This study aims to address the knowledge gap by examining the interactions between Li^+ and the components of the novel 3D cross-linked polymer electrolyte (3D-ADCL-PE) proposed by Hu et al. To achieve this goal, two distinct solvent systems are analyzed: (E_1) 1 M Lithium hexafluorophosphate ($LiPF_6$) in ethylene carbonate (EC)/dimethyl carbonate (DMC) solvent, originally used to test the 3D-ADCL-PE polymer [15], and (E_2) 1 M Lithium bis(fluorosulfonyl)imide (LiFSI) in trimethyl phosphate (TMP) solvent, which is known for its potential in developing high-safety rechargeable Li batteries, including both LiBs and LiMBs [20,21]. These electrolytes were chosen to evaluate the influence of different ionic environments on Li^+ coordination with solvent molecules and counter-anions within the GPE.

Advanced Molecular Dynamics (MD) and Density Functional Theory (DFT) simulations are carried out to assess the Li^+ interactions and how the components affect Li^+ transport [17–23]. MD simulations capture the motions of ions and polymer chains throughout time. From MD trajectories, we can extract frames of structures (snapshots) for further accurate quantum mechanical calculation of the interactions experimented by Li^+, guaranteeing precise outcomes in the subsequent stage of the investigation [17–19]. Building on these MD simulations, DFT calculations provide energies associated with interactions between Li^+ and GPE components. The Li^+ interactions linked to their distribution in the non-stackable gel polymer electrolytes were studied to understand how these molecular interactions govern key electrolyte properties, including conductivity and stability. The combined MD and DFT simulations provide a complete approach to understanding at the molecular scale the GPE behavior. While MD sets up accurate structures as a function of time, DFT delves into quantum aspects, revealing key insights correlated to ion transport in the GPE. This study is useful to receive a more profound comprehension of the transport mechanisms at the molecular level to achieve further optimization of these systems. This research intends to support the design of new materials with optimized functional groups and the development of sustainable and eco-friendly battery technologies by reducing the need for extensive experimental testing [14–19].

2. Computational Approach

2.1. Preparation of GPE Systems

The gel polymer electrolyte proposed by Hu et al. [15] was used as a reference. The polymer model was composed of 820 acrylonitrile (AN) and 270 diethyl vinyl-phosphonate (DEVP) monomers, in addition to 5 poly(ethylene glycol) diacrylate (PEGDA) units used as cross-linkers, shown in Figure 1. The polymer was generated from its monomers and cross-linkers using the HTPolyNet Python package [24–27]. From monomer structures and topologies describing the chemistry of any linking reactions, HTPolyNet builds the appropriate parameter set and initiates a capture-radius-based bond-formation step with relaxation, performing short MD simulations in Gromacs [27] with simultaneous updating of bonding topology.

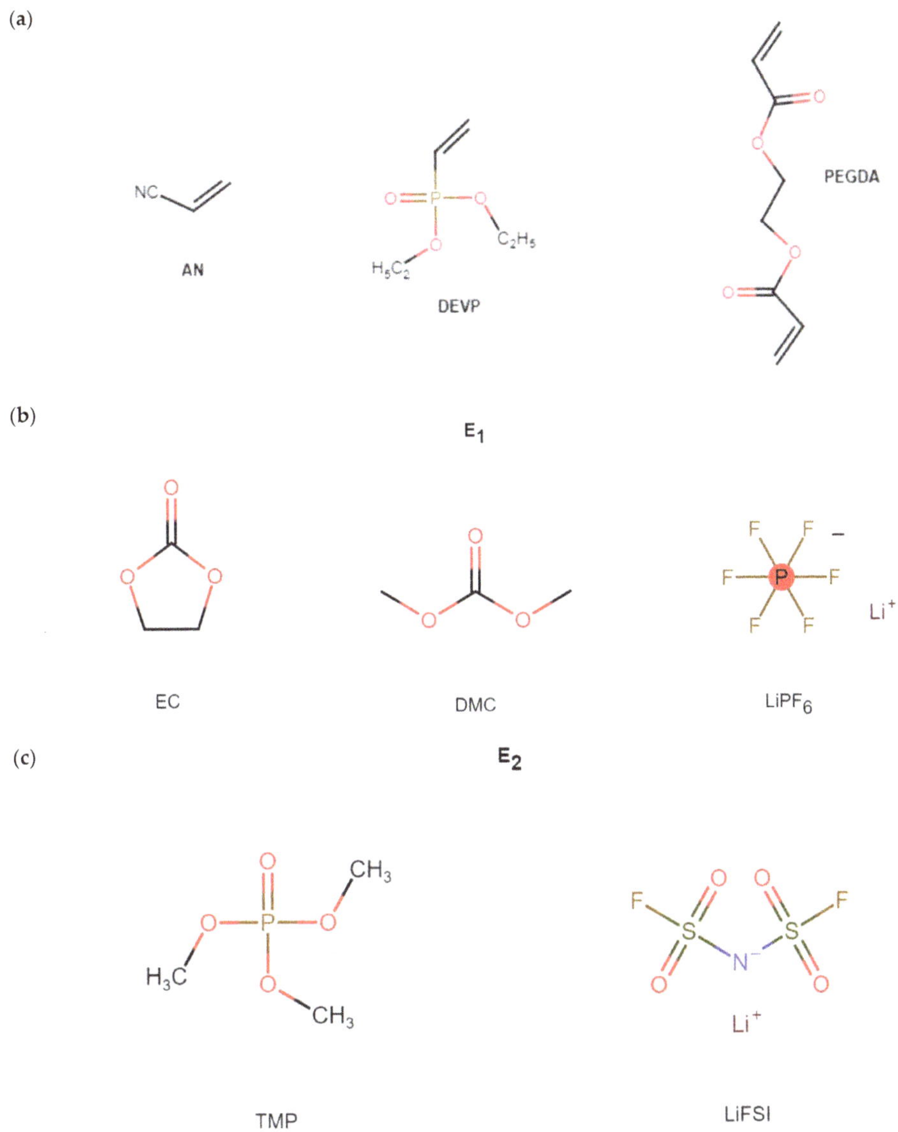

Figure 1. (**a**) Monomer structures of the 3D-ADCL-PE; (**b**,**c**) structures of the solvents and salts used as electrolytes: E_1 and E_2, respectively.

The obtained cross-linked matrix was a dense system of 3D cross-linked polymers with a final density of 1.11 g/cm^3 without electrolytes. Moreover, in order to improve the polymer model, it was subjected to NPT for 10 ns at very low pressure (0.01 bar) to lower the density of the dried polymer and then facilitate the random insertion of electrolytes. Finally, the 3D cross-linked polymer and tested electrolytes were equilibrated for at least 200 ns NPT at 1 bar. The equilibrated GPE box has a 5.92 nm length in each simulation box axis. Full details on the molecular dynamics simulations adopted for the 3D cross-linking step and the liquid electrolyte insertion are given in the Supplementary Materials.

Two different electrolytes, whose structures are shown in Figure 1b,c, were studied in order to assess the effect of solvents and salts. The final systems, E_1 and E_2, consisted of 3D-ADCL-PE, 266 DMC molecules, 340 EC molecules, 90 Li^+ and 90 PF_6^- (E_1) along the polymer matrix, and 427 TMP, 50 FSI, and 50 Li^+ ions (E_2). The number of solvent molecules and salt ions for E_1 corresponds to 1M $LiPF_6$ in DMC:EC 1:1 v/v mixture as adopted in the experimental reference to prepare the GPE [15]. Likewise, the number of solvent molecules and ions for E_2 corresponds to electrolyte solution 1M LiFSI in TMP. Based on the reported electrolyte uptake by the polymer of 298.7 wt.% with a porosity of 65.8%, a polymer/electrolyte weight ratio corresponding to 80 wt.% was adopted in order to take into account without overestimation only the swelled polymer region and exclude the liquid entrapped within the larger pores; in doing this, we assume a homogenous distribution of liquid electrolyte within/near the polymer chains and in the pores.

In order to obtain representative atomic clusters for the subsequent DFT calculations, radial distribution functions (RDF) and the corresponding cumulative coordination numbers (CN) were evaluated by averaging the trajectories over the last 100 ns. Then, the first peak of each RDF was used as a reference for the most probable Li-X distance, with X = N (-CN), O (DEVP), and P (DEVP), referring to atoms of the polymer monomers, respectively, and X = O (DMC), O (EC), F (PF_6^-), and O (TMP), O (FSI), N (FSI), and F (FSI), referring to E_1 and E_2 O, P, and F atoms of the two analyzed electrolyte compositions, respectively. The all-atom General Amber Force Field (GAFF) parameters [28], which have been used in the literature to model similar gel polymer electrolytes [29], were used to model the cross-linked polymer, solvent molecules, as well as the anions and Li^+. This force field was chosen due to its generality in describing interactions between a broad variety of components that make the GPE of our study: polymers with different functional groups, solvent molecules, and ions. This allows us to effectively model the specific system under study considering the two electrolytes chosen in this work, and at the same time, it opens the room for easy transferability to other similar systems.

Radial distribution functions are defined as the probability density of finding a *B* atom at a distance *r* from a reference *A* atom, normalized by the local density of *B* atoms in the system,

$$g_{AB}(r) = \frac{1}{\langle \rho_B \rangle_{local}} \frac{1}{N_A} \sum_{i \in A}^{N_A} \sum_{i \in B}^{N_B} \frac{\delta(r_{ij} - r)}{4\pi r^2} \quad (1)$$

where $g_{AB}(r)$ is the RDF, $\langle \rho_B \rangle_{local}$ is the average local number density of *B* atoms, N_A and N_B are the number of atoms of type A and B, respectively, and $\delta(r_{ij} - r)$ is the Dirac delta function, ensuring contributions only when the distance $r_{ij} = r$.

The cumulative coordination number is obtained by integrating the RDF over the range from 0 to a specific distance *D*,

$$CN\ (D) = 4\pi \langle \rho_B \rangle_{local} \int_0^D r^2 g_{AB}(r) dr \quad (2)$$

While the RDF provides the local probability distribution of an atom type, the *CN* offers a cumulative measure, giving complementary insights into the solvation structure.

It is important to note that the smoothness of the RDF curves in this study arises naturally from averaging over the 100 ns trajectory data. While RDFs are theoretically defined using a Dirac delta function to count pairwise distances exactly at *r*, practical implementations group distances into small bins through histogram binning [30].

When the RDF is calculated for a single snapshot, the limited number of atom pairs contributing to each bin results in noisy fluctuations. However, by averaging over the entire trajectory, the bins are populated with contributions from numerous pairs across time, leading to smoother curves. This standard approach provides reliable statistics and accurately represents the system's equilibrium behavior.

Finally, it is important to emphasize that the deep analysis of the averaged RDFs is important to understand what the chemical environment around the target Li ion in this GPE is, hence, to decide which atomic clusters should be considered in the quantum calculation of the interaction energy.

2.2. Interaction Energy Calculation

The clusters for quantum calculations were extracted based on RDF analysis. As will be demonstrated in Section 3, RDFs revealed that cations predominantly interact with solvent molecules and counter-anions, rather than the polymer matrix. Thus, the clusters were constructed following these observations, ensuring a focus on the solvent and counter-anion interactions surrounding Li^+.

Overall, four clusters were extracted. Clusters C-E_1 and C-E_2 consisted of different Li cations surrounded by cross-linked polymer chains and solvent molecules without PF_6^- or FSI^- counter-anions. Instead, $C_{PF6}{}^-$-E_1 and $C_{FSI}{}^-$-E_2 included lithium ions, polymer chains, and counter-anions. The average total number of atoms included in each cluster is 310. The preparation of the clusters was carried out by identifying Li^+ ions that adhered to the observed RDF behavior, i.e., Li^+ that consistently resided near solvent molecules and counter-ions. It is important to emphasize that the RDF analysis showed the absence of close interactions between Li^+ and the polymer matrix's functional groups (N from AN, O, or P from DEVP), as Li^+ ions were constantly found in proximity to solvent or counter-anion molecules, which act as a shield between Li^+ and polymer chains. Finally, in order to capture the surrounding atomic environment, all atoms within a radius of 1 nm from a target Li^+ were included in the clusters. Remember that the dimension of the complete equilibrated GPE box is 5.92 nm.

After the clusters were extracted, quantum calculations were performed using Density Functional Theory (DFT) with the Northwest Computational Chemistry Package (NWChem 6.6) [31]. Calculations were carried out for both closed-shell and open-shell configurations, with a multiplicity of 2 for open-shell structures. The B3LYP exchange-correlation functional [32] was employed with Grimme's D4 dispersion correction. The self-consistent field (SCF) energy convergence criterion was set to 5×10^{-6} a.u., with a root mean square (RMS) electron density convergence of 1×10^{-5} a.u.

Different basis sets were used for the atoms in each cluster to balance computational efficiency and accuracy. Specifically, a split-valence double ζ basis set with polarization functions (6–31G*) was applied to lithium and atoms directly interacting with Li^+ using the cut-off distance for defining interactions equal to 0.3 nm [33]. While the remaining atoms in the clusters were treated with a standard split-valence double-ζ basis set (4–31G) to optimize computational performance.

Lithium-ion interaction energies were calculated according to the following equation:

$$\Delta E_{int} = Ener_1 - Ener_2 - E_{Li^+}, \tag{3}$$

where $Ener_1$ is the energy of the atomic cluster containing the target Li^+, $Ener_2$ is the energy of the cluster without these cations, while E_{Li^+} is the energy of the target Li^+ alone, placed in the position found in the cluster. It is worth noting that in the considered clusters, more Li ions were found, which determine the total charge of the cluster; however, ΔE_{int} was calculated for the central Li^+ indicated as the target. Furthermore, to enhance the accuracy of the computed energies referred to the interaction between the Li^+ and their surrounding environment, basis set superposition error [34] (BSSE) adjustments were applied utilizing the counterpoise approach. BSSE is important as it facilitates a more precise assessment of the interaction energy by removing irrelevant contributions from basis set deviations [35]. An overview of the simulation protocol of the GPE can be seen in Scheme 1 below.

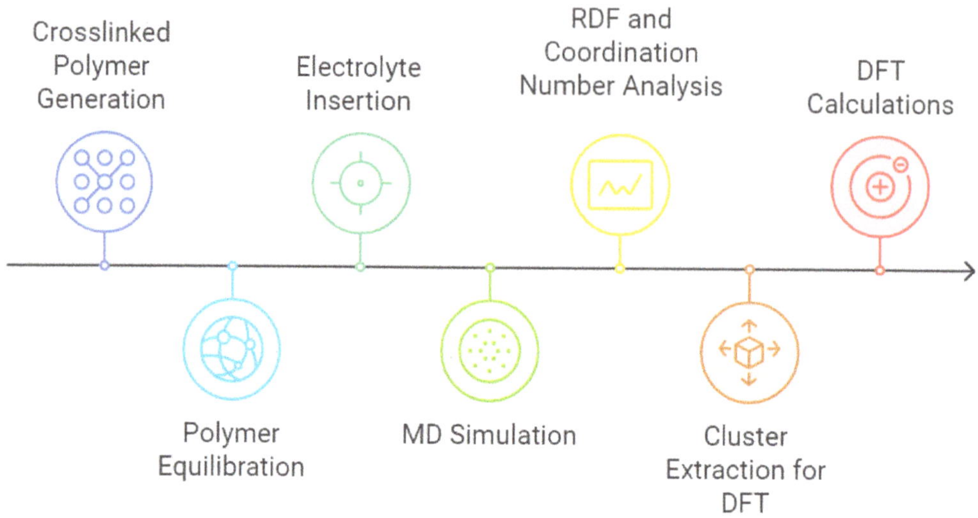

Scheme 1. A diagram illustrating the process flow and stages involved in the simulation.

3. Results and Discussion

3.1. Li^+ Surrounding Environment Through RDF and CN Analyses

This study aimed to investigate the structures and interactions preferred by Li ions in the novel 3D cross-linked polymer electrolyte. MD simulations were first performed to elucidate the Li^+ coordination environments by analyzing the RDFs and CNs obtained from the equilibrated GPE models, shown in Figure 2. The obtained density of the modeled GPEs is 1225 and 1220 kg/m^3 for E_1 and E_2 systems, respectively.

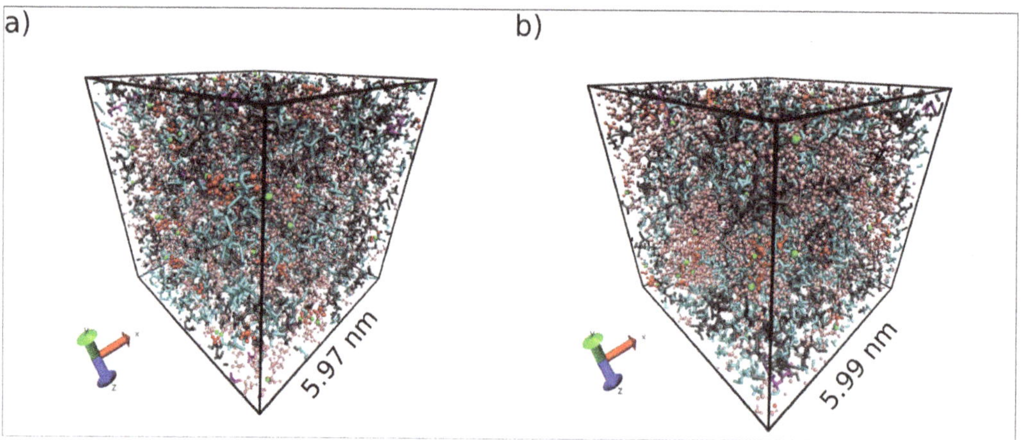

Figure 2. MD equilibrated GPE: (**a**) 3D cross-linked polymer plus 1 M Li PF_6 in EC/DMC; (**b**) 3D cross-linked polymer and 1 M LiFSI in TMP. Boxes are cubic and have the same dimensions in all xyz coordinates.

Figure 3 presents the RDFs for the E_1 system. Figure 3a displays the radial distribution functions for the distances between Li^+ and the key atoms in the electrolyte, while Figure 3b focuses on the RDFs for the polymer matrix. As can be seen in these figures, no peaks were observed below 0.5 nm for the distances between Li^+ ions and N, O, or P atoms of the polymer functional groups. Instead, peaks below this threshold exclusively correspond to interactions between Li^+ and PF_6^-, EC, or DMC. In particular, the RDFs in 3a show strong, well-defined peaks at 0.2 nm associated with the most probable distance between Li^+ and oxygen atoms of the solvent molecules, as well as fluorine atoms from PF_6^-. These results highlight that Li^+ forms stable solvation shells primarily with these mobile components of the GPE. Figure 3b also shows that the RDFs for distances between Li^+ and functional group atoms of the polymer matrix remain zero up to 0.5 nm, only increasing slightly beyond this distance. This is further supported by the broader and less defined curves, indicative of weaker interactions. This means that, during the last 100 ns of the MD simulation, the Li^+ ions are more likely interacting with solvent molecules or PF_6^- than with the functional groups of the 3D cross-linked polymer matrix. This behavior can be attributed to the lower mobility of the 3D cross-linked polymer compared to the solvent molecules, which can dynamically arrange around the Li^+ ions in greater numbers, forming stable solvation shells.

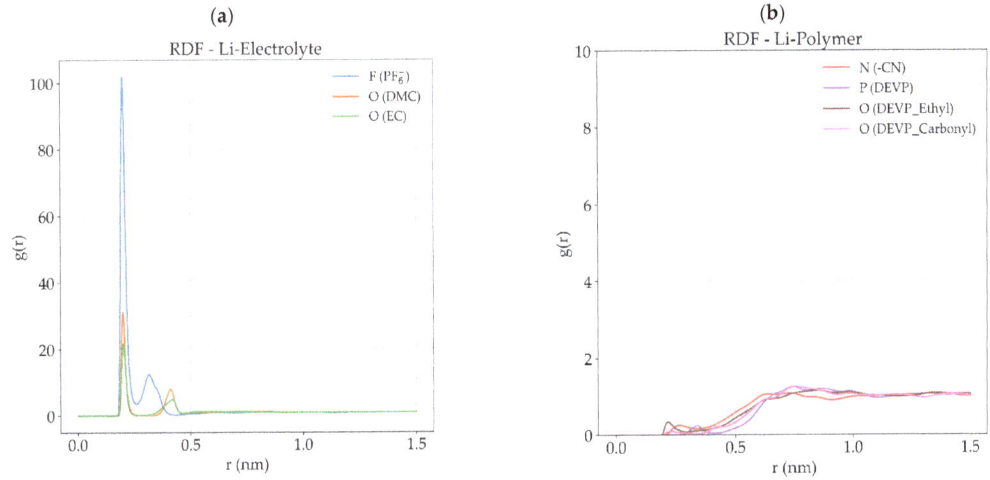

Figure 3. (**a**) Li-X RDFs associated with all interatomic distances among Li ions and atoms of the GPE-E_1 electrolyte components; (**b**) RDF associated with the atoms of the functional groups of the cross-linked polymer.

The CN as a function of the Li-X distance confirms this conclusion. The CN plots allow the direct comparison of how many atoms, on average, are at a certain distance from Li^+ thus giving an idea of its coordination shell. Figure 4a shows that the coordination numbers for the electrolyte components (Li^+-PF_6^-, -EC, and -DMC) remain zero up to 0.2 nm, after which the CN curves increase rapidly. While the CN profiles, related to the atoms of polymer functional groups (Figure 4b), remain at zero up to 0.5 nm and increase afterwards. It should be stressed that g(r) is normalized by the number of atoms and not the number of molecules/ions; thus, at r = 0.5 nm, the CN for Li-F(PF6), Li-O(DMC), and Li-O(EC) are 7.1, 4.7, and 5.1, respectively, indicating the presence of roughly 1 PF6 anion and 3 solvent molecules (the Li-O RDF curves capture the three oxygen atoms per solvent molecule) around Li^+ within 0.5 nm. These trends confirm that DMC, EC, and PF_6^- are the primary contributors to the Li^+ solvation structure. The CN plots make this result apparent, i.e., lower coordination numbers for -CN and -$PO(OC_2H_5)_2$ for short-range distance (at bond length) from Li^+, suggesting that the polymer matrix contributes minimally to Li^+ solvation in this GPE.

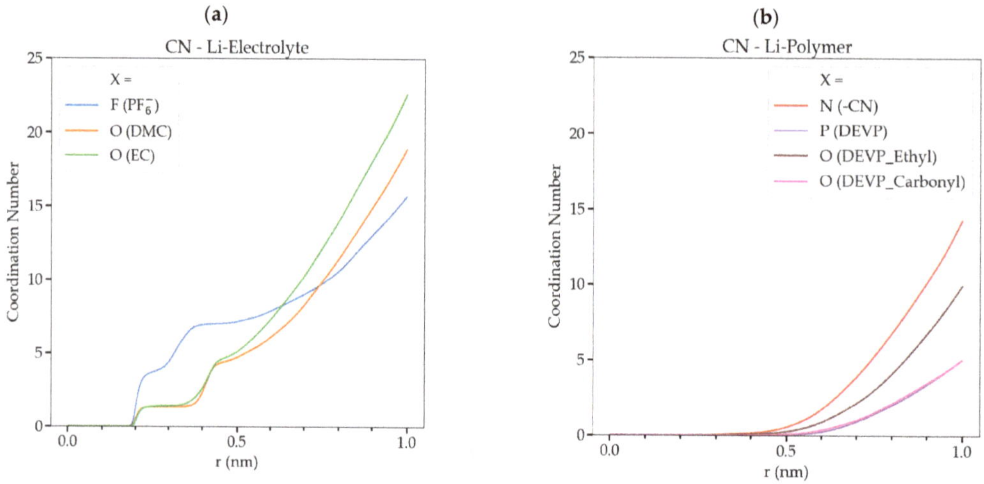

Figure 4. (**a**) CN associated with all atoms interacting with Li^+ in GPE E_1 for the electrolyte atoms; (**b**) CN of the atoms of the functional groups of the cross-linked polymer.

The RDF profiles for the E_2 systems, shown in Figure 5, demonstrate trends that are similar to those found for the E_1 polymer electrolyte in Figure 4. In detail, in Figure 5a (RDF for Li-Electrolyte), sharp peaks in yellow and blue colors at 0.19 nm and 0.21 nm can be noticed, respectively, which reveal strong interactions between Li ions and the oxygen atoms of the FSI counter-anion and solvent molecules. Furthermore, at 0.23 nm and 0.40 nm, two more marked peaks in red and green are observed, which indicates that the other probable atoms closer to the Li cations are N and F of the FSI. As for the previous E_1, no peak corresponding to interatomic distances between Li^+ and functional group atoms of the polymer was found below 0.5 nm (Figure 5b); only a broader line appears above this threshold. The RDFs plots reveal that Li^+ preferentially interacts with electrolyte components, specifically the oxygen atoms of FSI and TMP as well as the fluorine and nitrogen atoms of the counter-anion. Again, the CN profiles reported in Figure 6 confirm this finding. The oxygen TMP-related coordination number increases significantly after 0.2 nm, hence the oxygen-associated coordination number of FSI, underscoring their importance in the formation of the solvation shell around Li^+. In particular, the CN plots indicate that within 0.5 nm, Li^+ is coordinated by roughly 1 anion and 3 solvent molecules. Meanwhile, the CN trends for the functional group atoms exhibit weaker and more diffuse interactions, similar to the E_1 system, further highlighting the minimal role of the polymer in Li^+ solvation. Detailed RDF and CN profiles for each target atom can be found in the Supplementary Materials.

Overall, a coherent picture emerges in which both GPE and Li^+ strongly favor interactions with the electrolyte components—the solvent molecules (DMC, EC, and TMP) and counter-anions (PF_6^- and FSI^-)—over the polymer matrix. The solvation shells formed by these components are critical for efficient Li^+ transport. It should be noted that the RDF and CN analyses do not provide the extent of the interactions between cation and electrolyte components, but they do provide a picture of the environment surrounding the ions in the GPEs. To obtain the strength of the Li^+ interactions, DFT calculations were carried out in the next section using snapshots obtained from the MD simulations based on the results coming from the RDF and CN analysis.

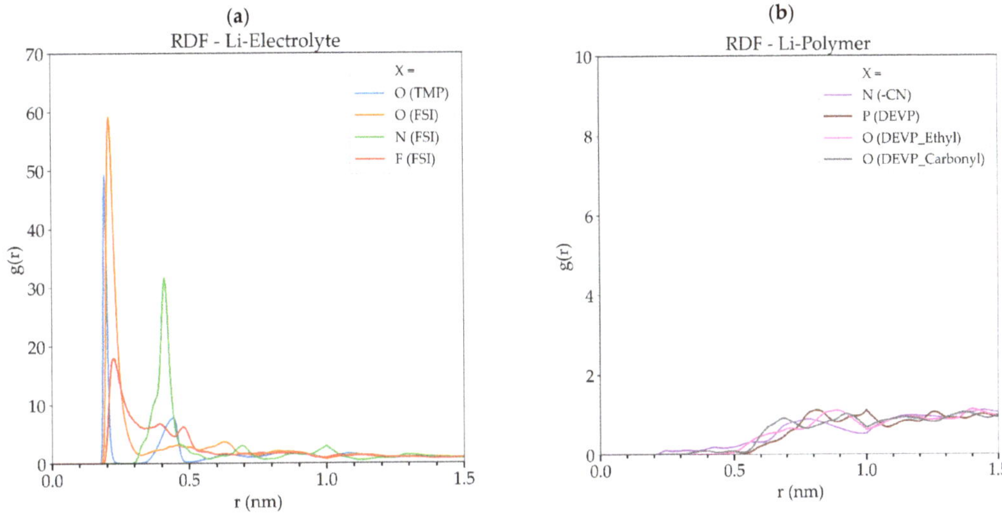

Figure 5. (**a**) Li-X RDFs associated with all inter-atomic distances among Li ions and atoms of the GPE E_2 for electrolyte components; (**b**) RDF associated with the atoms of the functional groups of the cross-linked polymer.

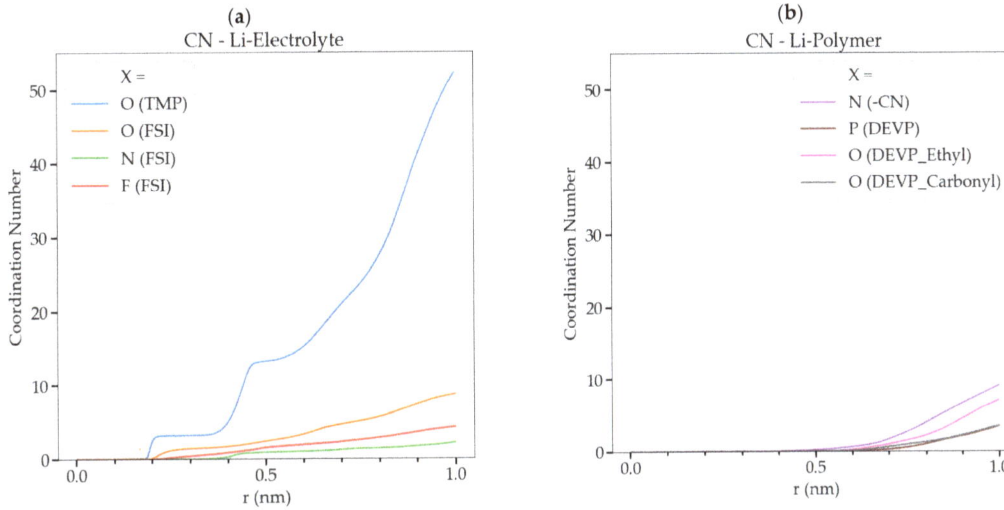

Figure 6. (**a**) CN associated with all the atoms interacting with Li^+ in GPE E_2. for the electrolyte atoms and (**b**) CN of the atoms of the functional groups of the cross-linked polymer.

3.2. Interaction Energies of Li Ions in GPE

MD simulations provided information on the most likely chemical environment experienced by Li ions in the cross-linked GPE. From these simulations, we extracted four representative atomic clusters shown in Figure 7, each depicting distinct solvation environments surrounding the Li^+ and used them to evaluate the interaction energies ΔE_{int}. The first cluster (C-E_1, Figure 7a) comprises Li^+ primarily interacting with adjacent solvent molecules, while in the second one ($C_{PF6}{}^{-}$-E_1, Figure 7b), the short-range interactions with both solvent and counter-anion species were considered. Furthermore, similar clusters associated with the E_2 composition were extracted from the MD trajectories (C-E_2, $C_{FSI}{}^-$-E_2, Figure 7c,d).

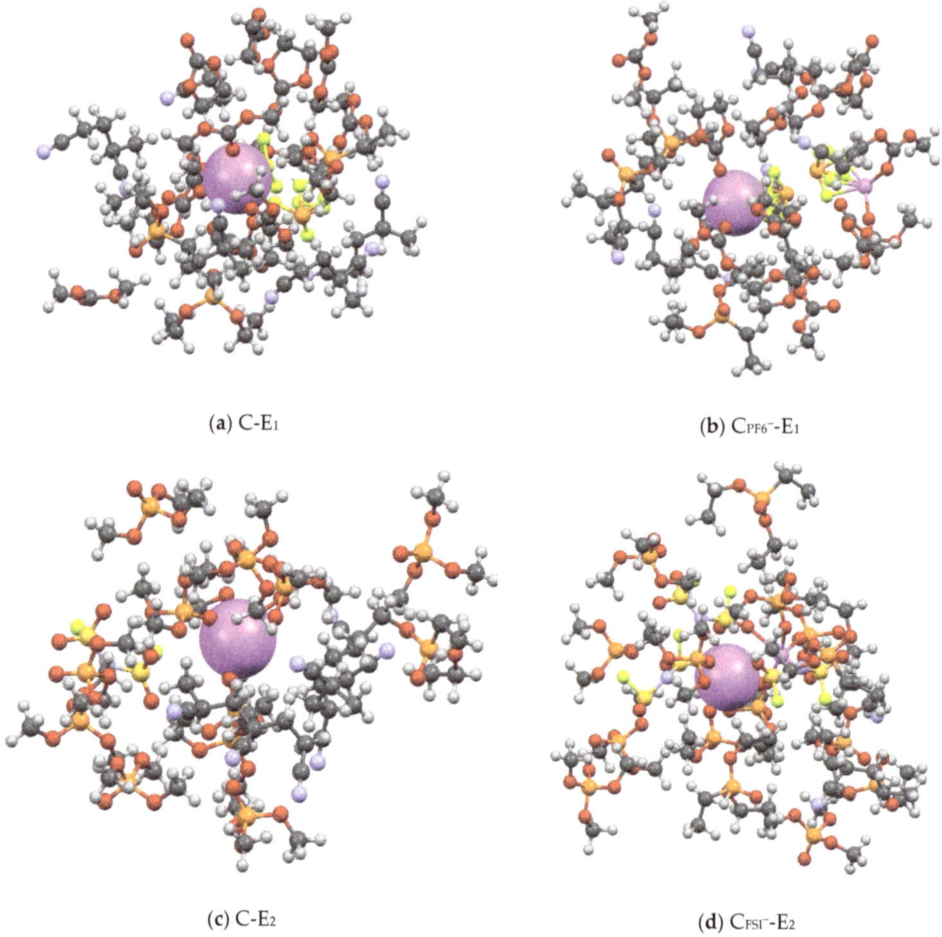

Figure 7. Cluster structures extracted from MD equilibrated systems: (**a**) central Li^+ interacting at short distance with (EC:DMC) solvent, (**b**) central Li^+ interacting at short distance with (DMC:EC) solvent and $PF_6{}^-$ counter-anions, (**c**) Li^+ interacting with TMP solvent, and (**d**) interacting with TMP solvent and FSI^- counter-anions. All 4 clusters had parts of the polymer equilibrated box as well. Colors: Li^+ (violet), C (gray), H (white), O (red), F (yellow), and P (orange). Target Li ion is represented with its van der Waals radii while the rest are visualized by the ball-and-stick representation.

Efforts to identify clusters with locations near the functional group of the polymer matrix (-CN or DEVP) were ineffective, since, as demonstrated from RDFs and CN analysis, Li^+ continually remained near solvent molecules and counter-anions. As previously discussed, Li ions primarily engage with the electrolyte molecules instead of the polymer matrix, highlighting the preeminent influence of solvents and salts in solvation. Thus, these clusters offer insights into the binding characteristics of Li ions in the two electrolyte systems.

Although we were unable to find clusters in which Li^+ interacts at short distances (bond length) with polymer functional groups in agreement with RDFs analysis, the interaction energies evaluated according to Equation (3), and reported in Figure 8, take into consideration not only the molecules nearby to Li^+ but also the overall chemical environment, i.e., polymer fragments with the functional groups. Thus, the indirect effect of the polymer is taken into account in the ΔE_{int} values.

Figure 8. Li^+ interaction energies for E_1 and E_2 electrolyte compositions. The light blue column refers to the C-E_1 and C-E_2 clusters, while the orange column refers to the C_{PF6^-}-E_1 and C_{FSI^-}-E_2 ones.

As shown in Figure 8, the interaction energies referring to the two electrolyte systems show significant differences. In the E_1 system, the ΔE_{int} of Li^+ with the solvent (DMC/EC) was −169.18 kcal/mol, but the interaction with the whole electrolyte, i.e., incorporating the PF_6^- counter anions, yielded a more substantial ΔE_{int} of −214.48 kcal/mol. This demonstrates that PF_6^- plays a crucial role in the solvation shell surrounding Li^+. E_2 exhibited stronger interactions overall compared to E_1. The interaction energy between target Li^+ and solvent molecules (TMP) was −204.82 kcal/mol, but incorporating the effect of FSI^- counter anions further increased, in absolute value, the ΔE_{int} to −252.13 kcal/mol. The increased interaction energies in this system underscore the stabilizing influence of FSI^-, which surpasses that of PF_6^- in the EC/DMC system.

The DFT results clearly show that the LiFSI/TMP system provides a more stable solvation environment than the $LiPF_6$/EC-DMC electrolyte. The stronger Li interaction energies in the FSI^-/TMP system suggest that Li ions form a more stable solvation shell, which could enhance the ion solvation from the SEI or CEI and then ionic conductivity. On the other hand, the weaker interactions in the PF_6^-/EC-DMC system suggest a more dynamic solvation shell, which might support faster ion mobility.

Insights from the RDF and CN analyses, as well as the ΔE_{int} values, offer a deeper understanding of Li^+ transport mechanisms at the molecular level. The absence of RDF peaks for Li-NC and Li-DEVP bond distances below 5 Å indicates that interactions between Li^+ and the functional groups of the cross-linked polymer do not hinder Li transport. Instead, the electrolytes directly interact with Li^+ ions, supporting a solution-diffusion mechanism for transport.

Based on the above consideration, a macroscopic solubility-diffusion mechanism is plausible; thus, Li^+ ion flux is governed by two primary factors: solvation and mobility. Solvation (S) is influenced by the interaction energies between ions and solvents/salts, while the mobility is correlated with the ion diffusion coefficient (D) depending on the activation energy for diffusion, which in turn depends on the overall interactions between the ion and its environment. Stronger solvent-Li interactions are not always advantageous. Depending on the strength of interactions among the solvent, counter ions, and polymer network, a more dynamic solvation sphere may facilitate faster lithium transport.

If ion solvation dominates, systems like FSI^-/TMP, showing higher interaction energies, may improve ionic conductivity. Conversely, ion mobility, correlated to low activation energy for diffusion and high D, takes advantage of a low overall interaction energy with the Li environment, such as, maybe, in the $LiPF_6$/EC-DMC system.

In summary, it is crucial to understand which of these contributions predominates. For instance, high ion solvation can compensate for modest ion mobility in the GPE, ensuring effective diffusion. In the investigated GPE, the solubility of Li^+ is expected to be high, as the GPE incorporates a significant amount of liquid electrolyte, as reported in reference [15], and the interactions between Li^+ and the polymer function groups are not direct. Thus, we anticipate that solvation plays a significant role in the observed Li^+ transport.

However, the -CN and -$PO(OC_2H_5)_2$ functional groups in the GPE have important functions. In fact, the former is necessary to form a stable SEI at the anode interface, while the latter forms a high-quality cathode electrolyte interphase layer as well as an anti-flammability function. It is worth highlighting that the simulations performed in this work refer to the bulk of the cross-linked 3D-ADCL-PE, which is the polymeric part not in contact with electrodes.

The combined DFT and MD protocol used in this study is quite general and adaptable, especially for preparing 3D cross-linked polymers. This flexibility makes them applicable to a variety of systems where cross-linking effects are important [36]. It is worth noting that we applied this protocol to systems with linear, non-cross-linked polymer chains [37,38], showcasing their versatility. Thus, these simulations offer valuable insights into how the cross-linking influences molecular interactions and ion transport. By exploring pore formation, molecular distribution inside the polymer matrices, structural stability, and ion mobility, this approach deepens our understanding of the system under study and opens opportunities for optimizing similar polymer-based electrolyte systems for advanced Li batteries.

4. Conclusions

Using a computational approach combining MD and DFT, our study provides significant insights into the solvation behavior of Li cations in innovative 3D cross-linked gel polymer electrolytes and highlights the influence of the different components of the transport of lithium ions.

The radial distribution functions and coordination numbers obtained from long MD simulations reveal well-defined Li solvation shells predominantly involving solvent molecules and the associated PF_6^- and FSI^-. We found that Li^+ exhibits a pronounced affinity for solvent molecules and counter-anions of the electrolytes ($LiPF_6$/EC/DMC and LiFSI/TMP), while showing negligible interactions with the –CN and -$PO(OC_2H_5)_2$ polymer functional groups of the bulk polymer. This suggests that a solution-diffusion mechanism for lithium-ion transport can be assumed in these systems in agreement with the large values of the computed interaction energies.

The ΔE_{int} further indicates that the LiFSI/TMP shows more robust interactions with Li^+ than the $LiPF_6$/EC/DMC electrolyte. The last finding indicates that LiFSI/TMP more efficiently stabilizes the Li^+ solvation shell and may improve ionic conductivity, assuming a solution-diffusion transport mechanism predominated by the solvation contribution. The $LiPF_6$/EC-DMC electrolytes, although giving lower values of ΔE_{int}, may provide enhanced ion mobility due to their looser solvation structure if the transport is controlled by diffusion contribution, but with associated stability trade-offs. Moreover, the robust interaction of Li^+ with the electrolyte constituents in LiFSI/TMP compared to $LiPF_6$/EC-DMC suggests its potential for enhanced long-term stability and safety while concurrently preserving competitive performance. It is very important to emphasize that the ΔE_{int} was calculated taking into account the effect of the polymer matrix in which the electrolytes are dispersed.

Thus, the findings remark on the significance of choosing suitable solvents and counter-anions to enhance the efficiency and safety of LiMBs using GPE. These observations can guide future research focused on optimizing the electrolyte composition in multifunctional designs and 3D cross-linking networks, specifically for solvent-salt combinations, to attain the ideal equilibrium among ionic conductivity, stability, and safety in lithium metal batteries.

Supplementary Materials: The following supporting information can be downloaded at https://www.mdpi.com/article/10.3390/batteries11010027/s1, Figure S1: (a) RDF plot referring to Li^+-F from PF6- anion, (b) and (c) referring to Li^+-O of EC and DMC solvent molecules, respectively; Figure S2: (a) Coordination Number plot referring to Li^+-F from PF6- anion, (b) and (c) referring to Li^+-O of EC and DMC solvent molecules, respectively. Figure S3: (a), (b) and (c) RDF plots referring to Li^+- F, Li^+- N, Li^+- O of the FSI-, (d) referring to Li^+-O from TMP solvent molecule, respectively. Figure S4: (a), (b) and (c) coordination number as a function of Li^+- F, Li^+- N, Li^+- O of the FSI-, (d) referring to Li^+-O from TMP solvent molecule, respectively. Figure S5. RDF plots referring to Li^+-N of ACN (a), Li^+- O from DEVP carbonyl (b), Li^+- O from DEVP ethyl (c) and Li-P of DEVP (d), respectively. It is very important to note the scale of the g(r) compared with the RDFs referring to E_1 and E_2 components. Figure S6. Coordination number as a function of Li^+-N of ACN (a), Li^+-O from DEVP carbonyl (b), Li^+- O from DEVP ethyl (c) and Li-P of DEVP (d) distances, respectively. Figure S7: RDF plots referring to Li^+-N of ACN (a), Li^+- O from DEVP carbonyl (b), Li^+- O from DEVP ethyl (c) and Li-P of DEVP (d), respectively. It is very important to note the scale of the g(r) compared with the RDFs referring to E1 and E2 components. Figure S8: Coordination number as a function of Li^+-N of ACN (a), Li^+-O from DEVP carbonyl (b), Li^+- O from DEVP ethyl (c) and Li-P of DEVP (d) distances, respectively. the 3D cross-linked polymer is soaked in the 1 M FSI-, TMP electrolyte.

Author Contributions: Conceptualization, G.D.L.; Methodology, G.D.L.; Validation, G.D.L.; Formal analysis, N.A.-H., J.L.D.S., S.A.P. and G.D.L.; Investigation, N.A.-H., P.V.S., J.L.D.S. and G.D.L.; Resources, S.A.P.; Data curation, N.A.-H., P.V.S. and J.L.D.S.; Writing—original draft, N.A.-H. and G.D.L.; Writing—review & editing, N.A.-H., P.V.S., J.L.D.S., S.A.P. and G.D.L.; Visualization, N.A.-H., P.V.S. and J.L.D.S.; Supervision, G.D.L.; Funding acquisition, S.A.P. All authors have read and agreed to the published version of the manuscript.

Funding: This research received fund from CONICET (28720210101190CO, 11220200102306CO), SECyT of the Universidad Nacional de Córdoba, and MinCyT (RESOL-2023-850-APNMCT).

Data Availability Statement: The original contributions presented in the study are included in the article and Supplementary Materials, further inquiries can be directed to the corresponding author.

Acknowledgments: This work used computational resources from CCAD-UNC. We especially acknowledge Cameron Abrams for his useful explanations about the HTPolyNet package. P.V.S. acknowledges support from a Ph.D. fellowship from CONICET. P.V.S. and S.A.P. also acknowledge other financial support from CONICET (28720210101190CO, 11220200102306CO), SECyT of the Universidad Nacional de Córdoba, and MinCyT (RESOL-2023-850-APNMCT).

Conflicts of Interest: The authors declare no conflict of interest.

References

1. Hu, Z.; Zhang, Y.; Zhang, Y.; Luo, J.; Chen, W.; Fan, W.; Huo, S.; Jing, X.; Bao, W.; Long, X.; et al. In-situ construction of high-temperature-resistant 3D composite polymer electrolyte membranes towards high-performance all-solid-state lithium metal batteries. *J. Power Sources* **2022**, *548*, 232052. [CrossRef]
2. Li, Y.; Sun, Z.; Liu, D.; Gao, Y.; Wang, Y.; Bu, H.; Li, M.; Zhang, Y.; Gao, G.; Ding, S. A composite solid polymer electrolyte incorporating MnO_2 nanosheets with reinforced mechanical properties and electrochemical stability for lithium metal batteries. *J. Mater. Chem. A* **2020**, *8*, 2021–2032. [CrossRef]
3. Xu, W.; Wang, J.; Ding, F.; Chen, X.; Nasybulin, E.; Zhang, Y.; Zhang, J.-G. Lithium metal anodes for rechargeable batteries. *Energy Environ. Sci.* **2014**, *7*, 513–537. [CrossRef]
4. Zhang, J.-G.; Xu, W.; Xiao, J.; Cao, X.; Liu, J. Lithium Metal Anodes with Nonaqueous Electrolytes. *Chem. Rev.* **2020**, *120*, 13312–13348. [CrossRef]
5. Chu, S.; Majumdar, A. Opportunities and challenges for a sustainable energy future. *Nature* **2012**, *488*, 294–303. [CrossRef]
6. Dunn, B.; Kamath, H.; Tarascon, J.M. Electrical energy storage for the grid: A battery of choices. *Science* **2011**, *334*, 928–935. [CrossRef]
7. Hwang, J.; Matsumoto, K.; Chen, C.-Y.; Hagiwara, R. Pseudo-solid-state electrolytes utilizing the ionic liquid family for rechargeable batteries. *Energy Environ. Sci.* **2021**, *14*, 5834–5863. [CrossRef]
8. Sun, M.; Zeng, Z.; Hu, W.; Sheng, K.; Wang, Z.; Han, Z.; Peng, L.; Yu, C.; Cheng, S.; Fan, M.; et al. Scalable fabrication of solid-state batteries through high-energy electronic beam. *Chem. Eng. J.* **2022**, *431*, 134323. [CrossRef]
9. Zhang, X.; Wang, A.; Liu, X.; Luo, J. Dendrites in Lithium Metal Anodes: Suppression, Regulation, and Elimination. *Acc. Chem. Res.* **2019**, *52*, 3223–3232. [CrossRef]
10. Aruchamy, K.; Ramasundaram, S.; Divya, S.; Chandran, M.; Yun, K.; Oh, T.H. Gel Polymer Electrolytes: Advancing Solid-State Batteries for High-Performance Applications. *Gels* **2023**, *9*, 585. [CrossRef]
11. Yu, F.; Zhao, L.; Zhang, H.; Sun, Z.; Li, Y.; Hu, Q.; Chen, Y. Cathode/gel polymer electrolyte integration design based on continuous composition and preparation technique for high performance lithium ion batteries. *RSC Adv.* **2021**, *11*, 3854–3862. [CrossRef] [PubMed]
12. Zhang, Z.; Huang, Y.; Li, C.; Li, X. Metal–Organic Framework-Supported Poly(ethylene oxide) Composite Gel Polymer Electrolytes for High-Performance Lithium/Sodium Metal Batteries. *ACS Appl. Mater. Interfaces* **2021**, *13*, 37262–37272. [CrossRef] [PubMed]
13. Wang, S.; Zhou, L.; Tufail, M.K.; Yang, L.; Zhai, P.; Chen, R.; Yang, W. In-Situ synthesized Non-flammable gel polymer electrolyte enable highly safe and Dendrite-Free lithium metal batteries. *Chem. Eng. J.* **2021**, *415*, 128846. [CrossRef]
14. Wang, Q.; Liu, P.; Li, S.; Wang, X.; Li, F.; Ma, J.; Chai, J.; Zhang, J.; Xu, G.; Huang, Z.; et al. A Flame Retardant Ionic Conductor Additive for Safety-Reinforced Liquid Electrolyte of Lithium Batteries. *J. Electrochem. Soc.* **2017**, *164*, A1559. [CrossRef]
15. Hu, Z.; Wang, Y.; Huo, S.; Bao, W.; Fan, W.; Zhang, Y.; Jing, X.; Ahmad, N.; Cheng, H.; Zhang, Y. Integrated design of multifunctional all-in-one polymer electrolyte membranes with 3D crosslinking networks toward high-performance lithium metal batteries. *J. Membr. Sci.* **2023**, *677*, 121643. [CrossRef]
16. Hu, Z.; Chen, J.; Guo, Y.; Zhu, J.; Qu, X.; Niu, W.; Liu, X. Fire-resistant, high-performance gel polymer electrolytes derived from poly(ionic liquid)/P(VDF-HFP) composite membranes for lithium ion batteries. *J. Membr. Sci.* **2020**, *599*, 117827. [CrossRef]
17. Wróbel, P.; Kubisiak, P.; Eilmes, A. MeTFSI (Me = Li, Na) Solvation in Ethylene Carbonate and Fluorinated Ethylene Carbonate: A Molecular Dynamics Study. *J. Phys. Chem. B* **2021**, *125*, 1248–1258. [CrossRef]
18. Tasaki, K. Solvent Decompositions and Physical Properties of Decomposition Compounds in Li-Ion Battery Electrolytes Studied by DFT Calculations and Molecular Dynamics Simulations. *J. Phys. Chem. B* **2005**, *109*, 2920–2933. [CrossRef]
19. Haskins, J.B.; Bauschlicher, C.W., Jr.; Lawson, J.W. Ab Initio Simulations and Electronic Structure of Lithium-Doped Ionic Liquids: Structure, Transport, and Electrochemical Stability. *J. Phys. Chem. B* **2015**, *119*, 14705–14719. [CrossRef]
20. Chawla, N.; Bharti, N.; Singh, S. Recent Advances in Non-Flammable Electrolytes for Safer Lithium-Ion Batteries. *Batteries* **2019**, *5*, 19. [CrossRef]
21. Shi, P.; Zheng, H.; Liang, X.; Sun, Y.; Cheng, S.; Chen, C.; Xiang, H. A highly concentrated phosphate-based electrolyte for high-safety rechargeable lithium batteries. *Chem. Commun.* **2018**, *54*, 4453–4456. [CrossRef]
22. Cats, P.; Evans, R.; Härtel, A.; van Roij, R. Primitive model electrolytes in the near and far field: Decay lengths from DFT and simulations. *J. Chem. Phys.* **2021**, *154*, 124504. [CrossRef]
23. Diddens, D.; Heuer, A. Lithium Ion Transport Mechanism in Ternary Polymer Electrolyte-Ionic Liquid Mixtures: A Molecular Dynamics Simulation Study. *ACS Macro Lett.* **2013**, *2*, 322–326. [CrossRef] [PubMed]
24. Huang, M.; Abrams, C.F. HTPolyNet: A general system generator for all-atom molecular simulations of amorphous crosslinked polymers. *SoftwareX* **2023**, *21*, 101303. [CrossRef]
25. Wang, J.; Wolf, R.M.; Caldwell, J.W.; Kollman, P.A.; Case, D.A. Development and testing of a general amber force field. *J. Comput. Chem.* **2004**, *25*, 1157–1174. [CrossRef] [PubMed]

26. Berendsen, H.J.C.; van der Spoel, D.; van Drunen, R. GROMACS: A message-passing parallel molecular dynamics implementation. *Comput. Phys. Commun.* **1995**, *91*, 43–56. [CrossRef]
27. Abraham, M.J.; Murtola, T.; Schulz, R.; Páll, S.; Smith, J.C.; Hess, B.; Lindahl, E. GROMACS: High performance molecular simulations through multi-level parallelism from laptops to supercomputers. *SoftwareX* **2015**, *1–2*, 19–25. [CrossRef]
28. Sousa da Silva, A.W.; Vranken, W.F. ACPYPE—AnteChamber PYthon Parser interfacE. *BMC Res. Notes* **2012**, *5*, 367. [CrossRef]
29. Shen, X.; Hua, H.; Li, H.; Li, R.; Hu, T.; Wu, D.; Zhang, P.; Zhao, J. Synthesis and molecular dynamic simulation of a novel single ion conducting gel polymer electrolyte for lithium-ion batteries. *Polymer* **2020**, *201*, 122568. [CrossRef]
30. Levine, B.G.; Stone, J.E.; Kohlmeyer, A. Fast analysis of molecular dynamics trajectories with graphics processing units—Radial distribution function histogramming. *J. Comput. Phys.* **2011**, *230*, 3556–3569. [CrossRef]
31. Valiev, M.; Bylaska, E.J.; Govind, N.; Kowalski, K.; Straatsma, T.P.; Van Dam, H.J.J.; Wang, D.; Nieplocha, J.; Apra, E.; Windus, T.L.; et al. NWChem: A comprehensive and scalable open-source solution for large scale molecular simulations. *Comput. Phys. Commun.* **2010**, *181*, 1477–1489. [CrossRef]
32. Hertwig, R.H.; Koch, W. On the parameterization of the local correlation functional. What is Becke-3-LYP? *Chem. Phys. Lett.* **1997**, *268*, 345–351. [CrossRef]
33. Chen, X.; Yao, N.; Zeng, B.-S.; Zhang, Q. Ion–solvent chemistry in lithium battery electrolytes: From mono-solvent to multi-solvent complexes. *Fundam. Res.* **2021**, *1*, 393–398. [CrossRef]
34. Boys, S.F.; Bernardi, F. The calculation of small molecular interactions by the differences of separate total energies. *Some Proced. Reduc. Errors. Mol. Phys.* **1970**, *19*, 553–566.
35. Simon, S.; Duran, M.; Dannenberg, J.J. How does basis set superposition error change the potential surfaces for hydrogen-bonded dimers? *J. Chem. Phys.* **1996**, *105*, 11024–11031. [CrossRef]
36. Ademmer, M.; Su, P.-H.; Dodell, L.; Asenbauer, J.; Osenberg, M.; Hilger, A.; Chang, J.-K.; Manke, I.; Neumann, M.; Schmidt, V.; et al. Unveiling the Impact of Cross-Linking Redox-Active Polymers on Their Electrochemical Behavior by 3D Imaging and Statistical Microstructure Analysis. *J. Phys. Chem. C* **2023**, *127*, 19366–19377. [CrossRef]
37. Luque Di Salvo, J.; De Luca, G.; Cipollina, A.; Micale, G. Effect of ion exchange capacity and water uptake on hydroxide transport in PSU-TMA membranes: A DFT and molecular dynamics study. *J. Membr. Sci.* **2020**, *599*, 117837. [CrossRef]
38. Luque Di Salvo, J.; De Luca, G.; Cipollina, A.; Micale, G. A full-atom multiscale modelling for sodium chloride diffusion in anion exchange membranes. *J. Membr. Sci.* **2021**, *637*, 119646. [CrossRef]

batteries

MDPI

Review

Advancements in Vibration Testing: Effects on Thermal Performance and Degradation of Modern Batteries

Khursheed Sabeel, Maher Al-Greer * and Imran Bashir

School of Computing, Engineering and Digital Technologies, Teesside University, Middlesbrough TS1 3BX, UK; k.sabeel@tees.ac.uk (K.S.); i.bashir@tees.ac.uk (I.B.)
* Correspondence: m.al-greer@tees.ac.uk

Abstract: Lithium-ion cells are increasingly being used as central power storage systems for modern applications, i.e., e-bikes, electric vehicles (EVs), satellites, and spacecraft, and they face significant and constant vibrations. This review examines how these vibrations affect the batteries' mechanical, thermal, and electrical properties. Vibrations can cause structural issues, such as the separation of electrodes and the deformation of separators. These problems raise internal resistance and lead to localized heat generation. As a result, thermal management becomes more complicated, battery aging accelerates, and safety risks arise, including short circuits and thermal runaways. To tackle these challenges, we need more realistic testing protocols that consider the combined effects of vibrations, temperature, and mechanical stress. Improving thermal management systems (TMSs) using advanced cooling techniques and materials, e.g., phase change solutions, can help to alleviate these problems. It is also essential to design batteries with vibration-resistant materials and enhanced structural integrity to boost their durability. Moreover, vibrations play a significant role in various degradation mechanisms, including dendrite formation, self-discharge, and lithium plating, all of which can reduce battery capacity and lifespan. Our current research builds on these insights using a multiscale physics-based modeling approach to investigate how vibrations interact with thermal behavior and contribute to battery degradation. By combining computational models with experimental data, we aim to develop strategies and tools to enhance lithium-ion batteries' safety, reliability, and longevity in challenging environments.

Keywords: battery vibration; battery safety; battery management system; thermal management battery degradation; safe battery; thermal runaway

1. Introduction

Lithium-ion batteries in high-performance energy storage applications are subjected to continuous mechanical stress, particularly vibrations arising from real-world operating conditions. Unlike controlled laboratory settings, these environments expose batteries to fluctuating forces that gradually degrade their internal structure, leading to microfractures, electrode delamination, and separator deformation. Over time, these structural instabilities contribute to increased internal resistance, localized heating, and accelerated thermal degradation, compromising battery safety and performance. Traditional testing methods often fail to capture the full extent of vibration-induced degradation, so developing more advanced testing protocols that integrate mechanical, thermal, and electrochemical stress factors is essential. A deeper understanding of how vibrations influence battery failure mechanisms is crucial for improving thermal management strategies, enhancing material resilience, and extending battery lifespan in demanding applications.

 https://doi.org/10.3390/batteries11020082

1.1. Research Background

Lithium-ion (Li-ion) batteries have become the dominant energy storage technology in various applications, from consumer electronics, i.e., smartphones and laptops, to large-scale systems such as EVs and renewable energy storage [1]. Lithium-ion batteries have gained widespread popularity due to their numerous advantages, such as high energy density, long cycle life, low self-discharge rates, and minimal maintenance needs compared to rechargeable batteries such as nickel-cadmium or lead-acid. With the growing demand for clean energy solutions, these batteries have become essential in revolutionizing industries, especially transportation and power generation [2]. As the UK's Net Zero goal aims to achieve zero carbon by 2050, so battery energy storage technology is pivotal in this effort. For instance, EVs are a key strategy to reduce greenhouse gas emissions, and efficient energy storage is essential to harness the full potential of renewable energy sources, including wind and solar [3,4]. However, the rapid adoption of Li-ion batteries also brings challenges related to their performance, safety, and durability, especially when subjected to harsh environmental conditions or mechanical and electrical stresses [5,6]. The discrepancy in Figure 1 arises because estimated battery life is often based on ideal conditions—consistent moderate temperatures, optimal charging practices, and regular maintenance. By contrast, actual battery life reflects the impact of factors such as extreme weather, frequent rapid charging, and variable driving styles, all of which can accelerate battery wear and reduce longevity.

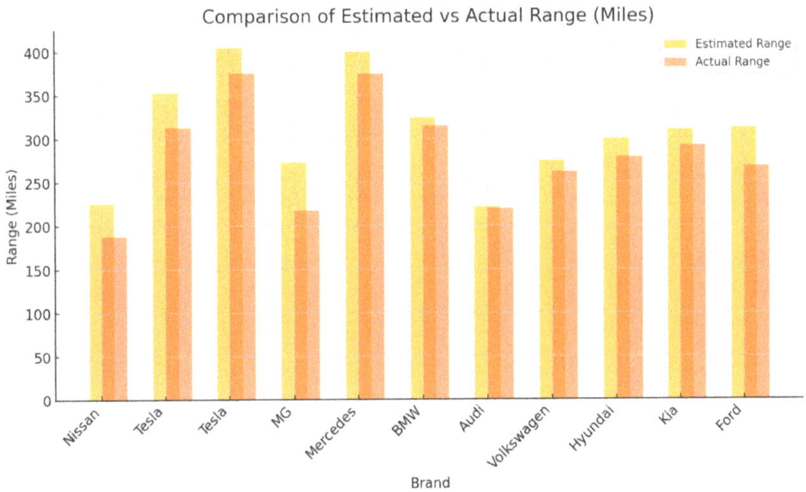

Figure 1. Comparison of estimated and actual range of EVs.

The difference between actual and estimated battery life in Figure 1 highlights how vibrations and factors, e.g., extreme weather and frequent rapid charging, contribute to battery degradation. This makes them critical factors in battery system design for EVs and other applications. This has driven significant efforts in recent years to improve battery modeling, enhance performance, and ensure safety while understanding how external factors impact battery life [7]. Li-ion batteries are sensitive to temperature, pressure, sudden electrical abuse, and other mechanical effects. Increased temperature can accelerate degradation and lead to thermal runaway, a dangerous situation during which the battery overheats and can catch fire or explode [8]. Proper TMSs are critical, especially in EVs. Brand et al. demonstrated that real-world vibrational loads could produce unique internal damage not typically seen in standardized testing. They emphasized the need for real-world vibration simulations to evaluate the true impact of mechanical stress on battery health [9]. It was found that prolonged vibrational stress leads to internal structural failures in lithium-ion batteries, such as short circuits and material displacement, with cylindrical cells especially at risk. He et al. developed an adaptive fuzzy inference model to assess lead-acid battery durability under vibrations, enabling accurate real-time detection without causing severe damage. This adaptable method offers potential for evaluating lithium-ion batteries, particularly as EVs face more complex mechanical environments [10]. Research on the robustness and safety of lithium-ion batteries across multiple applications emphasizes the importance of comprehensive testing for structural integrity and performance, especially under varied environmental stresses. Batteries undergo rigorous certification in aerospace and satellite applications, addressing potential failures due to vibration, short-circuiting, and extreme temperatures [11]. Vibration signal monitoring also serves as a novel method for pre-emptively diagnosing issues within lithium-ion cells, presenting a potential tool for battery management systems in high-stress applications [12]. This aligns with developments in transport shock and vibration monitoring systems, where MEMS technology is employed to enhance the sensitivity and precision of impact detection during transit [13]. The JARI Working Group introduced UN-standardized vibration testing, simulating multi-axial stress during transport. Their approach addressed large and small lithium-ion battery assemblies, setting the groundwork for consistent testing standards across different battery sizes and applications, from electric motorcycles to full-sized EVs [14]. Jung introduced numerical models to predict the performance of lithium-ion batteries with blended electrodes, demonstrating the application of these models in designing new batteries [15]. The team of researchers created a multistage vibration profile tailored for aviation, setting new standards for testing lithium-ion batteries in high-stress environments such as aircraft. Their research established a robust framework for evaluating battery performance, addressing criteria beyond typical vehicle application demands [16]. In an automotive context, pre-compliance vibration tests for LFP battery packs ensure long-term reliability by simulating real-world dynamic loads, thus helping to maintain structural stability during electric powertrain usage [17]. It was found that internal damage in 18650 lithium-ion cells, particularly in loosely designed mandrels, occurs without immediate performance decline under vibrations. Using computed tomography (CT) scans, researchers visualized subtle structural failures that became significant over time, emphasizing the need for robust internal cell design [18]. Electrochemical impedance spectroscopy (EIS) has become a pivotal tool in tracking internal resistance changes in cells subjected to vibrational forces, proving especially useful in EV battery health diagnostics [19]. Further research demonstrated that adding aluminum foam to TMS systems can reduce lithium-ion battery surface temperatures under mechanical stress, enhancing performance and lifespan. This method underscores the importance of temperature control in high-vibration environments such as EVs [20]. A comprehensive review examined temperature and vibrational effects on lithium-ion batteries, identifying a lack of studies combining these factors. The authors proposed a need for holistic thermal-mechanical coupled load testing, emphasizing that both temperature and vibrations are crucial for EV battery durability [21]. Sampath et al. presented a non-contact laser ultrasonic system for monitoring real-time state of charge (SOC). Their approach avoids the limitations of traditional contact-based testing, providing accurate SOC assessment while enhancing safety in EVs [22]. The Lithium-Ion Battery Analysis Guide advocated Fourier transform infrared (FTIR) techniques for monitoring chemical stability under long-term vibration, observing degradation in specific chemical bonds over time. This approach identifies early signs of failure, especially in high-frequency vibration environments [23]. Marco et al. documented unique failure patterns in motorcycle-specific vibration profiles, noting that these vibrations often cause higher battery failure rates than passenger vehicles. Their study provides insights for designing more durable battery systems tailored to two-wheeled transport needs [24]. High-impact research on six-degree-of-freedom vibrational stress revealed that battery aging accelerates under mechanical and thermal stresses, impacting internal resistance and capacity retention, which is critical for EV batteries [24,25]. Zhang et al. identified significant heat transfer improvements in PCM-based TMS systems when subjected to vibrations, suggesting enhanced cooling benefits in high-stress environments. This finding supports PCM's potential for managing EV battery temperatures efficiently without added energy costs [26]. Tian et al. examined dual-motor powertrain configurations in EVs, highlighting significant efficiency gains and reduced strain on battery systems. The dual-motor design enables more consistent power distribution, optimizing energy usage and extending battery life [27]. The 2023 Lithium-Ion Battery Analysis outlined QC protocols for lithium-ion batteries, including electrode material and separator testing. These methods ensure consistent battery quality, critical for safety and performance in large-scale EV deployments [28]. Further studies investigated the impact of vibrational stress on lithium battery cells' state of health (SOH), addressing its implications in applications such as EVs and aviation. Their findings suggest minor but noteworthy effects on charge capacity fade under vibrational conditions, pointing to mechanical vibrations as a factor that could influence ion transport within battery cells, impacting long-term performance. Furthermore, studies highlight the importance of environment-controlled testing setups for e-bike batteries, adhering to standards such as IEC 62133-2 and UN38.3 [29] for safe transportation and storage. Additional studies on the influence of operating conditions such as temperature and humidity further underline the need for robust seal designs to protect battery longevity, especially in flexible lithium-ion pouch cells [30]. Finally, feasibility analyses of lithium-ion batteries in unique environments, such as submarines, reveal that while these batteries provide high energy density and longevity, specific modifications are essential to meet stringent safety standards and functional requirements [31]. Zhao et al. proposed a method based on the vibration characteristics of over-discharged cells for fault detection in lithium-ion batteries. This method effectively distinguishes high-frequency vibration signals as indicators of potential over-discharge. It is suitable for integrating battery monitoring systems in EVs where safety is a concern due to high energy density and the potential for thermal runaway [32].

1.2. Research Hierarchy

Current gaps in lithium-ion battery testing pose significant risks for automakers, e.g., Tesla, Nissan, and Mercedes, as well as companies that build batteries for trucks, motorcycles, trains, cruise ships, and even space exploration.

The network visualization in Figure 2 illustrates the relationships and clustering of research topics related to lithium-ion batteries. The central themes "lithium-ion battery", "vibration", and "performance" connect to subtopics such as "safety", "EVs", and "durability". Each color represents a distinct cluster, showing how various concepts "temperature", "mechanical shock", and "battery packs" interlink across research areas. Current tests often fail to simulate the real-world effects on batteries' thermal behavior and degradation conditions, where batteries experience combined thermal and vibrational stresses that can accelerate degradation, compromise performance, and impact safety.

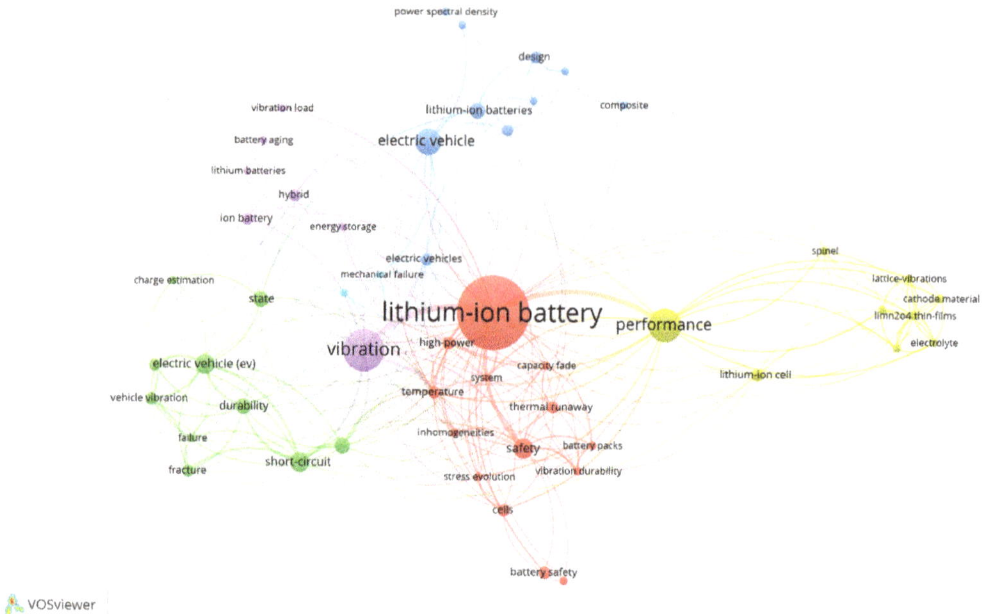

Figure 2. Literature focusing on lithium-ion battery vibration testing.

1.3. Research Motivation

Despite their growing market share and importance, there is limited comprehensive knowledge regarding how vibrations impact Li-ion batteries' mechanical integrity, safety, and electrical efficiency. Figure 3 depicts a traditional vibration testing method for batteries. It involves analyzing parameters such as voltage, current, state of charge (SOC), and temperature to monitor battery behavior. A vibration tuner sets and controls the vibration profile, which is applied to the battery using a vibrator. The battery tester evaluates the electrical and thermal responses of the battery under vibrations, such as capacity, resistance, and heating effects. The process includes a feedback loop where test data inform adjustments to improve accuracy and reliability.

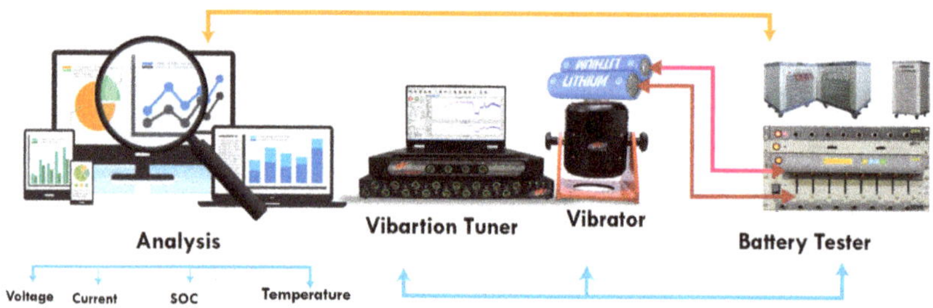

Figure 3. Traditional method used for battery vibration testing.

While a few studies have explored the fatigue and degradation of battery materials and pack structures under vibrational stress, the need to thoroughly assess and clarify the degradation mechanisms that compromise safety and performance remains critical [33].

1.4. Problem Statement

This review study aims to explore and synthesize the existing body of research on the thermal behavior and degradation mechanisms of lithium-ion batteries. It specifically focuses on the impact of vibrations, material behavior, and thermal properties on battery safety (BS) and performance, starting from modeling and examining its cycle performance under mechanical stress.

This study will explore simulation-based and hardware-based testing approaches utilized in the literature to model and assess the behavior of lithium-ion batteries under various mechanical and thermal stresses. This review will investigate the research landscape on thermal management and degradation in lithium-ion batteries, highlighting key findings related to the design of cells and packs and the thermal management strategies to mitigate these risks, as shown in Figure 4. A further application of [34] explores how mechanical and thermal factors contribute to the degradation processes observed in lithium-ion batteries. Furthermore, it will address these gaps by synthesizing industry and academia's latest computational, theoretical, and experimental research. It will offer valuable insights into how random vibrations and dynamic loads influence Li-ion batteries' mechanical and electrical characteristics, as shown in Figure 5 [35].

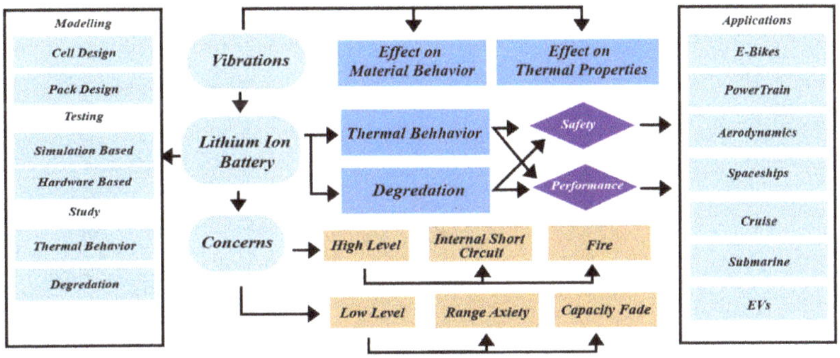

Figure 4. Methods of study for battery behavior under vibrations.

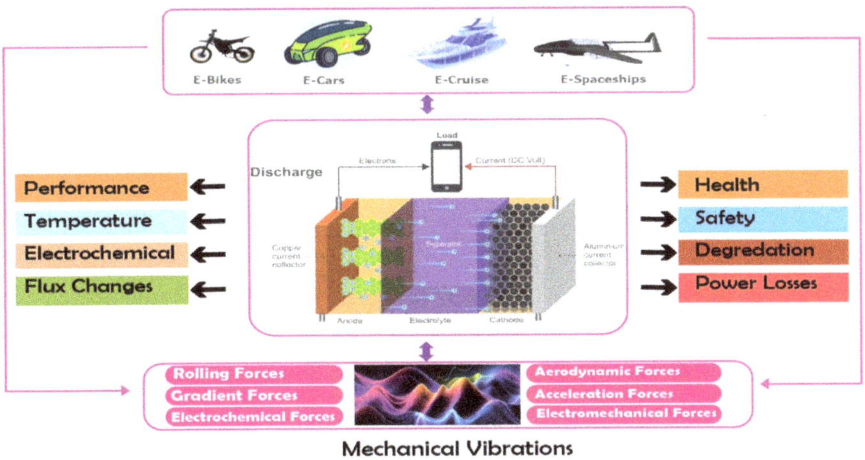

Figure 5. Importance of the study of vibrational impact on batteries.

Understanding these effects is crucial for advancing battery design and ensuring reliable performance in demanding environments.

1.5. Research Questions

The following research questions arise for this review study:

- What critical factors influence lithium-ion batteries' performance, thermal stability, and safety under vibrational conditions?
- To what extent do mechanical vibrations adversely impact the structural integrity, thermal behavior, and electrochemical performance of batteries?
- How do mechanical vibrations contribute to thermal runaway, electrochemical degradation, and long-term battery lifespan?
- Do vibrational forces compromise battery safety, operational range, and reliability in practical applications?

Figure 6 [36] highlights the importance of vibration studies and acoustic pressure in enhancing the performance of lithium-ion batteries for ships, airplanes, electric vehicles (EVs), and flying cars. Vibrations and noise can negatively affect battery performance, causing degradation or failure. Structural battery composites, which combine energy storage with structural support, must withstand these stresses. Vibration studies help to create batteries that endure mechanical forces and acoustic pressures in harsh environments. This research ensures that batteries are durable, efficient, and safe for ships, airplanes, and EVs. In flying electric vehicles, these studies are crucial for ensuring battery reliability during flight and ground operations, improving performance and safety across all applications.

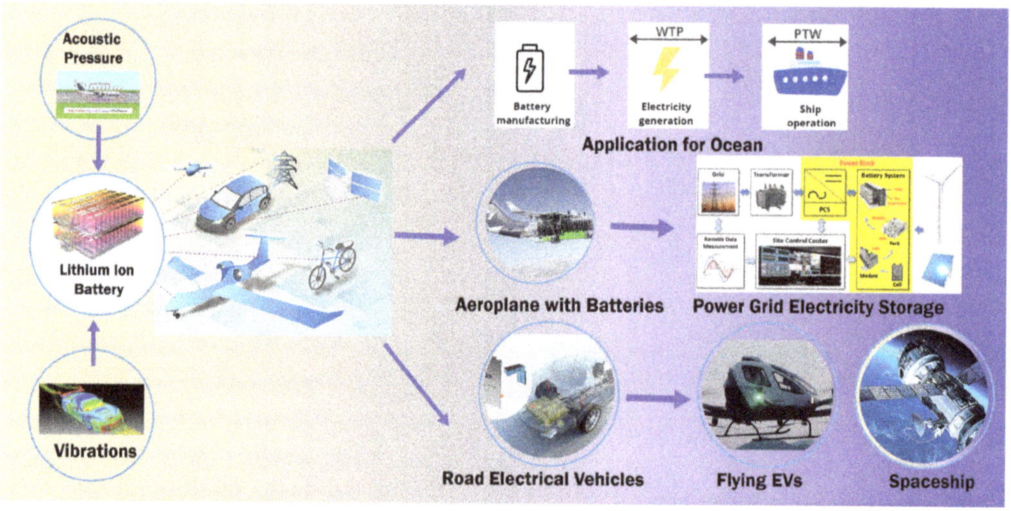

Figure 6. Application of battery vibration studies.

1.6. Research Aim and Objectives

This review highlights existing knowledge gaps and comprehensively understands how vibrations influence lithium-ion batteries' electrical and thermal behaviors in modern applications. This can be summarized as follows:

- Identify factors impacting battery performance and safety under vibrations.
- Assess the harmful effects of vibrations on battery systems.
- Investigate the effect of vibrations on the thermal behavior and degradation of batteries.
- Determine if vibrations compromise battery range, safety, and operational reliability.
- Review existing research on degradation mechanisms due to vibrations.

Section 2 explores advancements in battery modeling and vibration analysis, focusing on computational techniques like finite element analysis (FEA) and real-time monitoring to predict battery degradation. Section 3 examines how vibrations impact thermal behavior, leading to localized heating, electrolyte instability, and thermal runaway risks. Section 4 discusses degradation mechanisms, highlighting structural fatigue, electrode detachment, and electrolyte breakdown due to repeated mechanical stress. Section 5 links vibrations to thermal runaway, demonstrating how prolonged mechanical fatigue increases the risk of internal short circuits and heat propagation. Section 6 evaluates vibration-induced degradation in real-world applications, particularly its impact on charge cycles and energy retention in EVs and aerospace systems. Section 7 synthesizes the findings, identifying research gaps and assessing mitigation strategies. Section 8 recommends improvements in materials, structural reinforcements, and predictive modeling for enhanced battery resilience. Section 9 summarizes key insights, emphasizing the need for continued research to mitigate vibration-induced failures in lithium-ion batteries

2. Recent Advancements in Battery Modeling and Vibration Testing

Lithium-ion batteries are vital for energy storage in EVs and renewable systems, offering high energy density and long lifespans. However, real-world stresses and corresponding vibrations can cause structural damage, overheating, and accelerated degradation. Addressing these challenges requires advanced testing, improved designs, and modeling tools to enhance reliability and predict performance under various conditions [37].

2.1. *Advanced Battery Modeling Techniques*

Battery models now incorporate multiscale and multiphysics approaches, combining electrochemical, thermal, and mechanical factors to review Li-ion batteries' operation comprehensively [38]. Traditional models, starting from primary batteries and their equivalent circuit model (ECM), are still widely used for real-time applications because of their simplicity [39]. Still, more advanced models, like electrochemical-thermal models, have gained prominence due to their ability to simulate the behavior of batteries at a much deeper level [40]. These models can capture the interactions between the electrodes, electrolyte, and separator materials and the heat generation and mechanical stresses that occur during charging and discharging cycles.

Figure 7 provides an overview of the key considerations in the lifecycle of lithium-ion batteries, from raw materials to vehicle applications. It highlights critical aspects such as material sourcing, processing, and ethics; the development of precursors; electrode and component manufacturing; and cell formation and consistency. One of the breakthroughs in recent battery modeling is the development of physics-based models that can predict degradation mechanisms over time. This includes simulating solid–electrolyte interphase (SEI) layer growth, lithium plating, and capacity fade [41]. Models identical to these are invaluable for understanding long-term battery health and optimizing battery management systems (BMSs) that prolong battery life by controlling operational conditions as charging rates and temperature [42].

The bar graph in Figure 8 shows Europe's projected battery cell production capacities by format (cylindrical, prismatic, pouch, and unknown) from 2018 to 2030, reflecting the growing demand for EVs and energy storage. Moreover, machine learning and artificial intelligence advancements have been integrated into battery modeling to enhance predictive capabilities further. By analyzing large datasets generated from battery usage, machine learning algorithms can identify patterns and predict battery performance and failure modes with greater accuracy [43]. Integrating data-driven approaches with traditional models represents a significant step in improving battery design and lifecycle management.

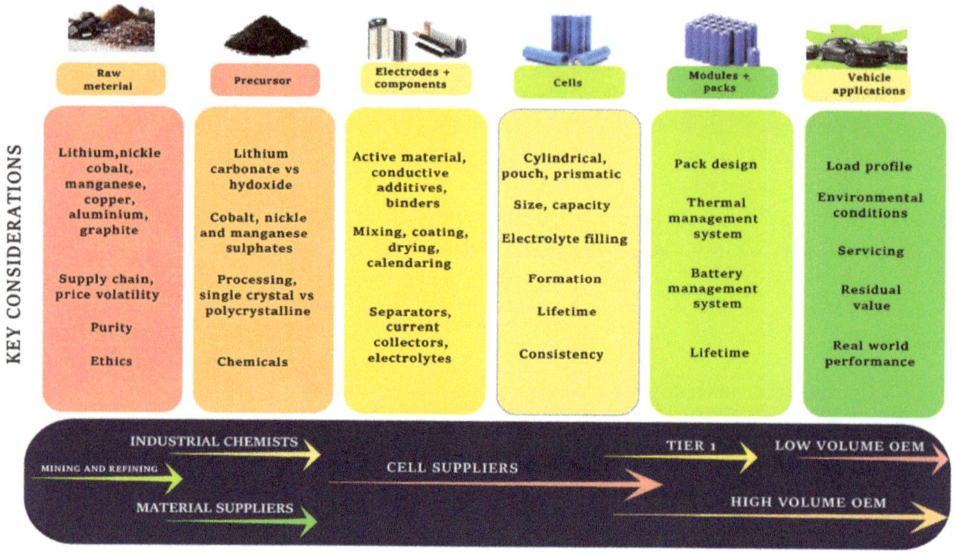

Figure 7. Battery cell modeling according to applications.

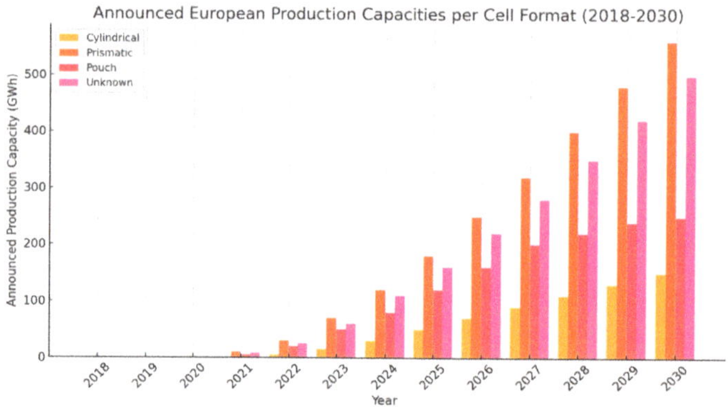

Figure 8. Battery cell design types and production used in different EVS models.

Recent advancements, as shown in Figure 9, in physics-based battery modeling have focused on integrating multiple domains, such as electrochemical, electromechanical, and electromagnetic physics, alongside AI-driven analytics to create highly accurate and predictive models. Electrochemical modeling captures charge transport and reaction kinetics, while electromechanical coupling accounts for stress and strain effects on battery components due to intercalation and thermal expansion. Electromagnetic analysis is used to study high-frequency effects and optimize current collection. This multidisciplinary approach offers enhanced insights into battery behavior, aiding in the design of safer, more efficient, and longer-lasting energy storage systems. Vibrational studies using COMSOL Multiphysics involve coupling mechanical, thermal, and electrochemical physics to analyze the effects of vibrations on battery performance and degradation. The process begins with a detailed geometry definition, including electrodes, separator, and casing, followed by assigning materials with properties such as Young's modulus, thermal conductivity, and diffusivity, which may be temperature or stress dependent. The Solid Mechanics interface is used to model vibrational loads through harmonic forces or sinusoidal displacements, with modal analysis identifying natural frequencies to avoid resonance. The Heat Transfer interface is included to model heat generation from electrochemical reactions and mechanical friction, while cooling mechanisms such as convection are applied to mimic thermal management. Coupling is achieved using the Thermal Stress multiphysics interface for thermal–mechanical interactions and manual linking of stress fields to degradation parameters in the Electrochemistry interface, accounting for stress-enhanced SEI layer growth or lithium diffusivity change.

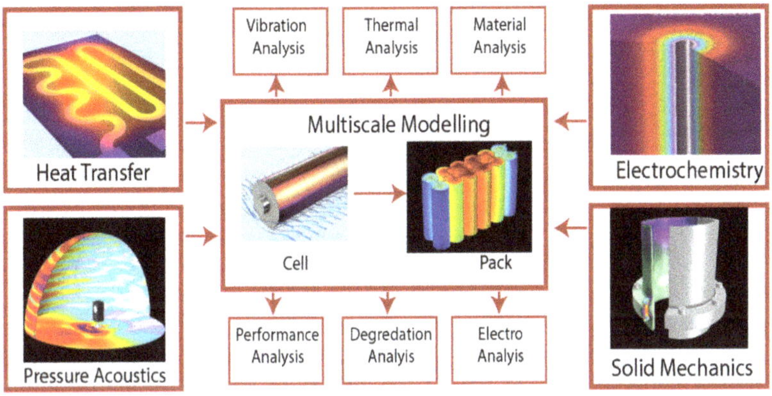

Figure 9. Physics-based battery modeling and analysis.

2.2. Battery Testing Procedures in COMSOL Multiphysics

COMSOL Multiphysics offers a range of methods and features to study vibrations inside systems such as batteries, combining advanced simulation capabilities in the Solid Mechanics, Heat Transfer, and Electrochemistry interfaces:

- Modal Analysis: Identifies natural frequencies and mode shapes to avoid resonance and localize vibration-sensitive areas [44,45].
- Frequency Domain Analysis: Models system responses to harmonic vibrations and identifies critical frequencies where thermal or mechanical failure occurs [46].
- Time-Dependent (Transient) Analysis: Captures real-time responses to shocks or impacts, such as stress wave propagation and transient heating [47].
- Multiphysics Coupling: Integrates the Solid Mechanics, Heat Transfer, and Electrochemistry interfaces for coupled thermal-stress effects, stress-enhanced degradation, and fatigue analysis [48].
- Parametric and Sensitivity Analysis: Explores effects of varying vibration amplitudes, frequencies, and material properties on battery behavior [49].
- Meshing and Solver Techniques: Includes adaptive meshing for stress hotspots and segregated solvers for large-scale multiphysics problems [50].

Vibration analysis is crucial for assessing lithium-ion battery performance, especially in high-stress applications like EVs and aerospace. The modal analysis identifies natural frequencies to prevent resonance-induced failures, while frequency domain analysis detects critical vibration points that may cause mechanical or thermal issues. Time-dependent analysis captures real-time responses to shocks and impacts, improving safety assessments. Multiphysics coupling integrates the Solid Mechanics, Heat Transfer, and Electrochemistry interfaces to study stress-induced degradation, enhancing lifespan predictions. Parametric analysis optimizes battery design by evaluating the effects of vibration amplitude, frequency, and material properties.

2.3. Battery Performance and Safety

One of the core challenges in battery performance is improving energy density, power density, choking time, and cycle life, which directly affects how much energy the battery can store relative to its weight or volume. Energy density (E) is defined as follows:

$$E = C \cdot V \quad (1)$$

where C is the battery capacity (in ampere hours), and V is the nominal voltage (in volts). While reliable, traditional lithium cobalt oxide (LCO) cathodes offer limited capacity, nickel-rich layered oxides, such as nickel-manganese-cobalt (NMC) cathodes, have demonstrated significantly higher capacities due to their ability to store more lithium ions [51]. Improving the performance of Li-ion batteries has been a primary focus of research and development, especially in the context of increasing energy density, enhancing charging speed, and extending cycle life [52]. Understanding the effects of mechanical vibrations on batteries is critical, particularly for drones used in agriculture. As shown in [53], agricultural drones experience varying operating conditions, making vibration analysis essential for assessing battery performance and reliability. Energy density remains one of the most critical performance metrics, as it determines how much energy the battery can store relative to its size and weight [54]. Advancements in materials science have led to the development of high-performance cathode materials, such as nickel-rich layered oxides, which offer higher energy capacities than traditional lithium cobalt oxide (LCO) cathodes. However, silicon's significant expansion (up to 300%) during lithiation presents challenges, as this expansion leads to mechanical stresses that degrade the anode's structural integrity over time. On the anode side, the shift from graphite to silicon-based anodes offers the potential for dramatic increases in energy density. Silicon's theoretical capacity (4200 mAh/g) is much higher than that of graphite (372 mAh/g) [55]. However, silicon's significant expansion (up to 300%) during lithiation presents challenges, as this expansion leads to mechanical stresses that degrade the anode's structural integrity over time. The following degradation rate model can describe the relationship between battery capacity and cycle life:

$$C_n = C_o\left(1 - \frac{n}{L}\right) \quad (2)$$

where C_n is the capacity of the battery after (n) discharge, and C_o is the capacity at the start. On the anode side, the shift from graphite to silicon-based anodes holds great promise for significantly increasing energy density. Silicon has a much higher theoretical capacity for lithium storage than graphite. Additionally, electrolyte improvements, such as developing solid-state electrolytes, enhance battery performance by addressing safety concerns associated with liquid electrolytes [56]. Solid-state batteries offer the potential for higher energy densities and improved safety by eliminating the flammable liquid electrolyte used in conventional Li-ion batteries.

Although still in the research and development phase, solid-state technologies are expected to transform battery performance in the coming years. Figure 10 [57] shows how the driving range of a new battery is limited by reserving a grace capacity. After approximately 900 cycles, the upper grace capacity starts to be utilized. Software adjustments can extend the battery's lifespan by adding more grace capacity, as depicted in the graph, but this comes at the cost of reducing the driving range. Battery safety remains a critical concern in developing Li-ion technology, mainly because of the risks associated with thermal runaway [58]. Lithium-ion batteries generate heat; this heat can degrade performance and shorten the battery's lifespan if not effectively managed. Heat generation can be modeled as follows:

$$Q_{heat} = I^2 R_{int} \quad (3)$$

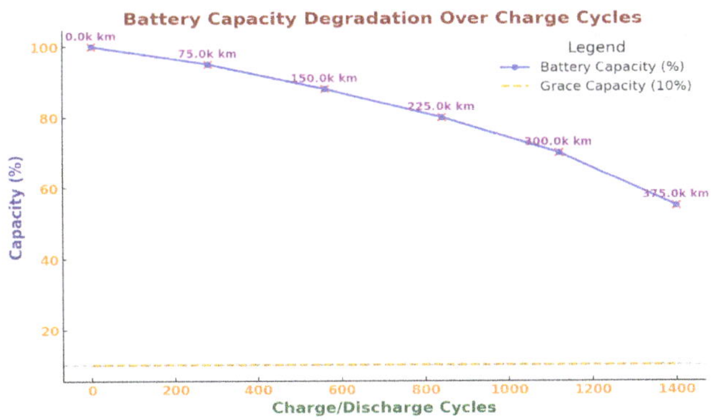

Figure 10. Energy band of an aging EV battery.

2.4. Battery Thermal Runaway

Thermal runaway is a self-propagating reaction when the battery's internal temperature rises uncontrollably due to internal short circuits, overcharging, or external heat exposure [47]. This process can lead to battery fires or explosions if not managed effectively. Battery safety has always been a key concern in developing and deploying Li-ion batteries, mainly because of the risks associated with thermal runaway due to the internal chemical process, as shown in Figure 11.

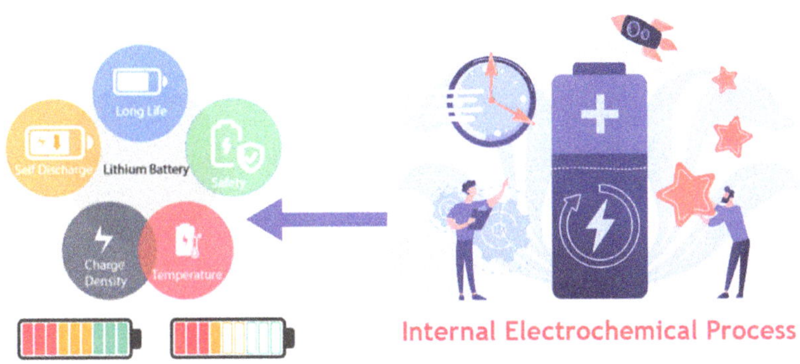

Figure 11. Safety and performance parameters of lithium batteries.

These safety features can lead to overheating, fire, or explosion if subjected to electrical or mechanical abuse [59]. Ensuring battery safety has become even more critical as the number of large battery packs in EVs and grid storage systems has increased. Recent advancements in safety mechanisms include integrating TMSs that actively regulate battery temperature during operation [60]. Lithium-ion batteries degrade over time, affecting their performance, as shown in Figure 11, due to an unknown self-internal process. The two main factors contributing to degradation in EVs are calendar aging (degradation over time, even when the battery is not in use) and cycle aging (degradation due to repeated charging and discharging).

$$k = Ae^{-\frac{E_a}{RT}} \quad (4)$$

The Arrhenius equation [61] is often used to describe how temperature accelerates the degradation rate, with higher temperatures leading to faster capacity loss. These systems use liquid cooling, phase change materials, or heat pipes to dissipate heat effectively and prevent hotspots from forming within the battery pack. BMSs also play a crucial role in safety by monitoring voltage, temperature, and current levels to detect and prevent abnormal conditions that could lead to failure. The graph in [62] shows how temperature affects battery capacity and life. Capacity is highest between 20 °C and 35 °C and decreases at extreme temperatures. Battery life drops significantly at temperatures above 30 °C and below −20 °C. Another area of focus in improving safety is the development of safer electrolyte formulations, including non-flammable electrolytes and additives that suppress dendrite formation—needle-like structures that can pierce the separator and cause short circuits [63]. These innovations and robust safety testing standards have significantly reduced the risk of catastrophic battery failure.

The section highlights how vibrations accelerate lithium-ion battery degradation, impacting structural integrity, thermal performance, and long-term reliability in EVs and aerospace applications. By integrating multiphysics modeling, advanced testing protocols, and AI-driven predictive tools, researchers can better assess failure risk and develop strategies to enhance battery safety and durability. Addressing these challenges requires vibration-resistant materials, improved thermal management systems, and optimized battery designs to ensure stable performance in high-vibration environments. The following section will focus on vibrations and their effects on thermal behavior and material properties. It will examine how mechanical stress influences heat generation, electrolyte stability, and structural fatigue, leading to performance loss and safety concerns in lithium-ion batteries.

3. Vibrations and Their Effects on Thermal Behavior and Material Properties

Vibrations refer to mechanical oscillations in systems due to dynamic forces, mechanical stress, or external excitations. In engineering systems, vibrations can arise from operational machinery, external environmental factors, or inherent material behavior [64]. Vibrations can significantly affect components' thermal behavior and material properties in automotive systems, aerospace, and battery technology applications. Heat generation due to vibrations can be explained using principles from thermodynamics and mechanics [65]. The following equation represents the heat generated in a material due to damping (energy dissipation) caused by vibrations:

$$Q_{vibration} = \frac{1}{2} m a^2 C_{damp} \tag{5}$$

where $Q_{vibration}$ is the heat generated by vibrations, m is the mass of a vibrating object, a is the amplitude of vibration, and C_{damp} is the damping coefficient. This equation shows that the heat generated is proportional to the vibration amplitude and frequency, meaning that higher vibrations (more significant a or f) lead to increased heat production.

Studies [66,67] have proven that an annular heat exchanger is effective for thermal energy storage systems operating at low frequencies. The phase change material (PCM) melts faster in this configuration than in a cubic one. The study demonstrated temperature contours and the solid–liquid melting interface, showing accelerated melting at low frequencies in the annular setup, enhancing heat transfer efficiency [67]. In systems, i.e., batteries, increased vibration-induced heat raises the internal temperature, which increases internal resistance (R_{int}). As the resistance increases, Joule heating generates more heat [68].

Figure 12 [69,70] presents how battery performance and safety issues affect vehicle types, such as 2-wheelers, 3-wheelers, and electric vehicles (EVs). Vibrational stress, thermal challenges, and other degradation mechanisms occur on uneven roads. This increase in temperature can accelerate material degradation, reduce efficiency, and lead to thermal runaway in extreme cases, especially in temperature-sensitive systems corresponding lithium-ion batteries. Prolonged exposure to vibrations can also lead to thermal fatigue [71].

As materials heat up due to vibrational energy, repeated thermal cycling (heating and cooling) can cause cracks or degradation, especially in materials that experience cyclic stress, such as metals and composites. Materials exposed to constant vibrations experience cyclic loading, which leads to mechanical fatigue. Fatigue is the weakening of a material caused by repeated loading and unloading cycles, leading to microcrack formation and, eventually, structural failure. By enhancing natural convection heat transfer, vibrations reduce temperature rise and create a more consistent thermal environment. Specifically, a vibration frequency of 10 Hz results in a marked cooling effect, while a frequency of 50 Hz optimizes temperature uniformity. Additionally, increased vibration amplitude further contributes to thermal stability, reducing the battery's peak temperatures and temperature differentials. Fatigue can be modeled using the S–N curve (stress–life curve) [72], which represents the relationship between the cyclic stress (σ) and the number of cycles to failure (N):

$$\sigma \cdot N^b = C \tag{6}$$

where σ is the applied cyclic stress (in Pa), N is the number of cycles to failure, b is a material-dependent fatigue exponent, and C is a constant related to the material's properties. The combined effects of vibrations and thermal cycling cause thermo-mechanical fatigue, a key failure mode in high-performance materials and systems such as turbines, engines, and battery systems. Thermo-mechanical fatigue can be modeled by combining the mechanical stress and thermal cycling effects, as follows:

$$\epsilon_{total} = \epsilon_{mechanical} + \epsilon_{thermal} + \epsilon_{vibration} \tag{7}$$

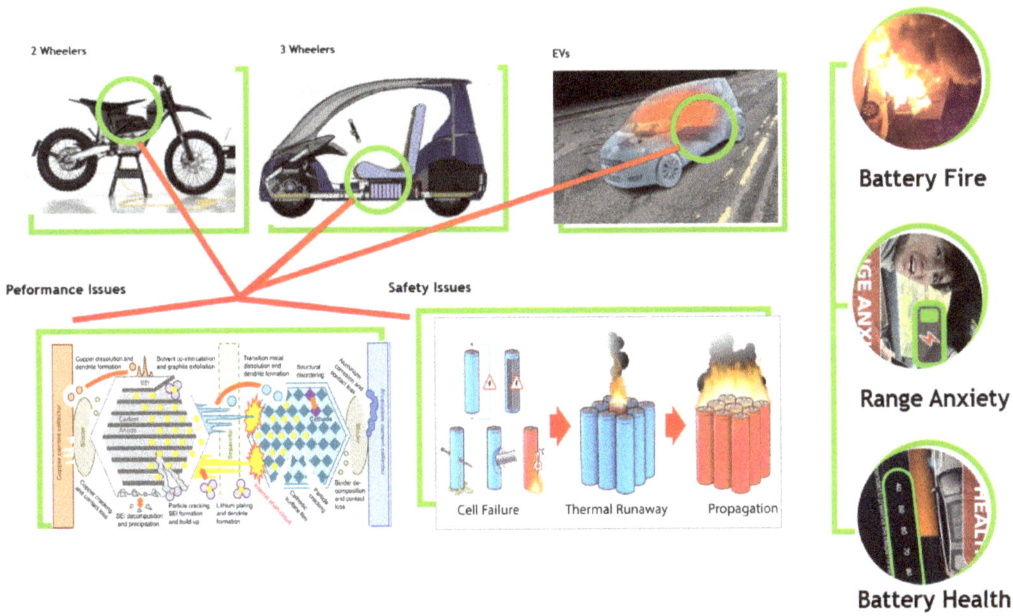

Figure 12. Effect of shocks and mechanical vibrations on batteries and related issues.

The interaction of these two factors accelerates the material degradation rate, particularly in temperature-sensitive components, such as lithium-ion battery cells and turbine blades. Repeated vibrations may cause plastic deformation in materials, particularly metals and alloys. Plastic deformation occurs when a material undergoes permanent strain after being subjected to stresses beyond its yield strength. Vibrations can lead to plastic deformation by causing localized yielding due to cyclic loading. The stress–strain relationship for plastic deformation can be modeled as follows:

$$\sigma = E \cdot \epsilon \tag{8}$$

where σ is the stress applied to the material (in Pa), E in Young's modulus (in Pa), and ϵ is the strain experienced by the material. As cyclic loading continues, the material may yield and deform plastically, leading to loss of structural integrity, lower strength, and failure.

3.1. *External Mechanical and Electrical Effects on Batteries*

Li-ion batteries, especially in demanding applications such as EVs and aerospace, are often exposed to external mechanical and electrical abuses. Mechanical abuse can come in the form of vibrations, which may cause physical damage to battery components, such as electrode cracking or separator degradation. These effects can lead to internal short circuits, capacity loss, and safety hazards. Studies have shown that batteries subjected to repeated mechanical stresses can experience accelerated degradation, reducing their effective lifespan [71]. Electrical abuse, such as overcharging, over-discharging, or operating the battery outside its recommended voltage window, also harms battery life. Overcharging can lead to lithium plating on the anode, which reduces capacity and increases the risk of internal short circuits, while over-discharging can cause irreversible damage to the cathode material, making it difficult for the battery to regain its original capacity [73]. Addressing these issues requires the development of more resilient battery designs and improved management systems that can protect against these external stresses. By integrating advanced sensors and control algorithms, battery management systems can detect and mitigate mechanical and electrical abuses before they cause considerable damage.

3.1.1. Thermal Effects on Lithium-Ion Battery Performance and Safety

Temperature plays a pivotal role in determining the electrochemical performance of lithium-ion batteries, directly influencing key metrics such as capacity, power output, charge rate, and efficiency. The operating temperature of a Li-ion battery has distinct effects depending on whether the temperature deviates toward higher or lower extremes.

3.1.2. Low-Temperature Performance Degradation

At low temperatures, particularly below 0 °C, the performance of Li-ion batteries is notably diminished. The reduced ionic conductivity of the electrolyte at these temperatures slows down lithium-ion mobility between the cathode and anode, decreasing overall capacity and power output [74]. Moreover, the increased internal impedance limits the discharge power, particularly in high-power applications such as EVs and energy storage systems (ESSs). Cold conditions also impair the efficiency of charging processes, as the sluggish kinetics of the intercalation reactions lead to poor charge acceptance. Critically, lithium plating on the anode can occur when fast charging is attempted under such conditions, causing irreversible capacity loss and contributing to safety hazards [75].

3.2. *High Temperature and Accelerated Degradation*

High-temperature operation (above 40 °C) introduces a separate set of challenges. Elevated temperatures accelerate the chemical reactions inside the battery, temporarily increasing ion transport and enhancing capacity in the short term [76]. However, prolonged exposure to high temperatures leads to faster degradation of the electrolyte, cathode, and anode materials. This accelerates the formation of the solid–electrolyte interphase (SEI) layer on the anode, consuming active lithium and reducing the amount available for reversible intercalation. Consequently, the battery's cycle life is shortened, as is its ability to maintain capacity over extended usage periods.

3.2.1. Thermal Effects on Safety

The safety of lithium-ion batteries is intimately tied to their thermal behavior, with extreme temperatures posing significant risks. Both high and low temperature extremes can initiate processes that compromise the structural integrity of the battery, increasing the possibility of failure modes such as thermal runaway, internal short circuits, and, in the worst case, fire or explosion [77]. These reactions generate further heat, leading to a cascading rise in temperature. The initial trigger for thermal runaway is often overheating, whether due to high ambient temperatures, excessive current loads, or internal short circuits [78]. As the temperature rises, the separator that physically isolates the cathode and anode may melt or become deformed, leading to internal shorting.

3.2.2. Impact of Overcharging and Temperature Synergy

Another significant safety concern arises from the synergistic effects of overcharging and high-temperature conditions. Overcharging leads to an excessive accumulation of lithium ions at the anode, which increases the possibility of lithium plating. This process reduces capacity and heightens the risk of internal short circuits [79]. This can trigger thermal runaway in combination with high temperatures. Moreover, high charging rates at elevated temperatures compound the risks by inducing localized hotspots within the cell, which can lead to the breakdown of materials and gas generation.

3.2.3. Long-Term Effects of Temperature on Cycle Life

The long-term impact of temperature on lithium-ion batteries is most evident in cycle life degradation. Elevated temperatures accelerate the deterioration of active materials and promote side reactions that consume electrolytes and lithium, reducing the battery's ability to maintain capacity over multiple charge–discharge cycles [80]. Conversely, while temporarily reducing capacity and power output, low temperatures can contribute to long-term degradation through lithium plating and increased internal resistance. Therefore, effective thermal management is essential for maintaining immediate performance and extending the usable life of lithium-ion batteries [81].

This section explored how vibrations impact lithium-ion batteries' thermal behavior and material properties. Vibrational stress increases internal resistance and Joule heating, leading to uneven heat distribution and accelerated thermal degradation. Prolonged exposure weakens material integrity, causing microcracks, electrode delamination, and separator damage, which contribute to capacity loss and thermal runaway risks. Additionally, cyclic loading and thermal fatigue further degrade structural components, reducing battery lifespan. Addressing these challenges requires vibration-resistant materials, optimized battery enclosures, and advanced cooling techniques to enhance battery durability. The following section will focus on degradation mechanisms in lithium-ion batteries under vibrations, examining how electrode deterioration, electrolyte instability, and mechanical fatigue contribute to long-term performance decline.

4. Mechanisms of Degradation in Lithium-Ion Batteries Under Vibrations

Degradation in lithium-ion batteries arises from several interacting mechanisms that gradually impair their performance and safety over time. These processes are typically divided into calendar aging (degradation over time regardless of use) and cycle aging (degradation resulting from repeated charge and discharge cycles).

4.1. Process of Degradation in Batteries

The primary degradation mechanisms include loss of active lithium, electrode material degradation, solid–electrolyte interphase (SEI) layer growth, lithium plating, and electrolyte decomposition. These mechanisms result in progressive capacity, power output, and efficiency decline while introducing significant safety risks [82].

Figure 13 presents the degradation mechanism in lithium-ion batteries, highlighting key processes that lead to capacity loss, safety risk, and failure. It shows that overcharging and high-rate charging cause excessive lithium deposition, leading to dendrite formation, which can penetrate the separator and cause internal short circuits. Low-temperature charging slows lithium-ion diffusion, increasing the likelihood of lithium plating, during which inactive lithium accumulates on the anode, further contributing to performance degradation [83]. The SEI (solid–electrolyte interphase) layer forms on the anode surface and is essential for battery stability. Still, its non-uniform growth or degradation leads to increased resistance, lithium loss, and reduced battery efficiency. Self-discharge and non-uniform discharging create local high-current regions, accelerating dendrite growth and separator damage. Temperature fluctuations further degrade the SEI layer and increase stress on the battery components, leading to structural instability and a higher risk of thermal runaway [84,85]. These interconnected degradation mechanisms significantly affect battery lifespan and safety, underscoring the importance of proper charging protocols, effective thermal management, and advanced materials to mitigate performance decline and improve the reliability of lithium-ion batteries.

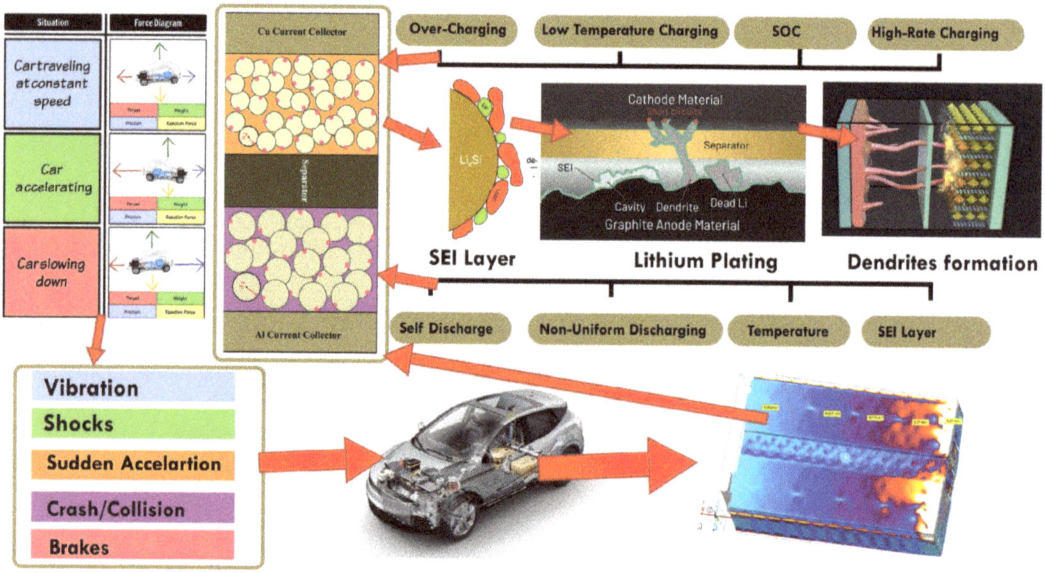

Figure 13. Mechanism of battery degradation.

4.1.1. Solid–Electrolyte Interphase (SEI) Layer Growth

One of the most critical degradation mechanisms is the formation and growth of the SEI layer on the anode, typically made of graphite. The SEI is initially formed during the first few charge–discharge cycles, acting as a protective barrier that allows lithium-ion transport while preventing further electrolyte decomposition [86]. However, over time and with repeated cycling, the SEI layer grows thicker as it continually consumes lithium ions and electrolyte components, resulting in the loss of active lithium. This process reduces the available lithium for electrochemical reactions, lowering the battery's capacity. Excessive SEI layer growth can increase internal resistance, reducing the battery's power delivery capabilities.

4.1.2. Electrode Material Degradation

Degradation of electrode materials, particularly the cathode, significantly reduces capacity and efficiency. Cathode materials, such as lithium cobalt oxide ($LiCoO_2$) or nickel manganese cobalt (NMC) oxide, undergo structural changes due to repeated lithium insertion and extraction during cycling. These changes include phase transformations, mechanical stress, and particle fracture, which reduce the cathode's structural integrity and limit its ability to store lithium ions effectively [87]. Anode materials, particularly those using graphite, can also experience mechanical degradation, especially in high-energy-density designs where large volumes of lithium are intercalated and de-intercalated. This mechanical stress may cause microcracks in the material, further accelerating performance loss.

4.1.3. Lithium Plating

Lithium plating occurs when lithium ions are deposited as metallic lithium on the anode surface rather than intercalating into the anode material. This degradation mechanism is particularly prominent at low temperatures or during high-rate charging, during which lithium deposition becomes kinetically favored [88]. Lithium plating not only leads to capacity loss but also poses a significant safety risk, as metallic lithium can form dendrites—needle-like structures that can grow and puncture the separator, causing an internal short circuit. This increases the chances of thermal runaway and, consequently, fire or explosion.

4.1.4. Electrolyte Decomposition

Decomposition of the electrolyte is another key factor in lithium-ion battery degradation. Electrolyte decomposition is often initiated by high operating temperatures, overcharging, or sustained cycling at high voltages, which promote unwanted side reactions that reduce electrolyte volume and increase the concentration of byproducts [89]. The decomposition products can coat the electrode surfaces, further increasing internal resistance and impeding lithium-ion transport. In addition to performance loss, the breakdown of electrolyte components can generate gases, raising the cell's internal pressure and creating safety risks, especially in sealed battery packs.

4.2. *Performance Decline Due to Degradation*

The degradation mechanisms outlined above contribute to several observable performance issues in lithium-ion batteries over time. These performance declines are typically reflected in capacity fade, increased internal resistance, reduced power output, and efficiency loss.

4.2.1. Capacity Fade

Capacity fade is the gradual reduction in the battery's charge over time. This decline is driven by the loss of active lithium, the degradation of electrode materials, and the increased impedance caused by SEI layer growth and electrolyte decomposition. As the battery loses capacity, its ability to power devices for extended periods diminishes, shortening the runtime between charges for consumer electronics and reducing the driving range of EVs.

4.2.2. Increased Internal Resistance and Power Loss

Internal resistance increases as the battery ages, due to thickening of the SEI layer, electrode degradation, and electrolyte decomposition. Higher internal resistance reduces the efficiency of charge and discharge cycles, as more energy is lost as heat within the cell. This results in lower power output, particularly in high-current applications, and can limit the battery's ability to deliver energy during periods of peak demand. This leads to slower acceleration and diminished overall performance in EVs, while in grid storage, it reduces the system's ability to respond to dynamic energy needs [90].

4.2.3. Reduced Coulombic Efficiency

Coulombic efficiency, the ratio of charge extracted during discharge to the charge input during charging, decreases as the battery degrades. Reduced efficiency is typically a result of parasitic side reactions, such as SEI formation, electrolyte oxidation, and gas generation, which consume lithium ions and reduce the amount available for electrochemical reactions. This degradation-induced loss of efficiency means that more energy is required to charge the battery for the same amount of usable output, contributing to overall inefficiencies in EVs and energy storage systems.

4.2.4. Safety Risks Arising from Degradation

As degradation progresses, the structural and chemical changes within the lithium-ion battery impair its performance and increase the prospect of safety failure. These risks are primarily associated with the breakdown of internal components, such as the SEI layer, electrolyte, and electrodes, which can lead to internal short circuits, thermal runaway, and other hazardous events.

4.2.5. Increased Risk of Thermal Runaway

The potential for thermal runaway increases significantly as the battery degrades. Thermal runaway occurs when the exothermic decomposition reactions within the cell exceed the battery's ability to dissipate heat, leading to a self-sustaining increase in temperature that can cause fires or explosions [91]. Degradation exacerbates this risk in several ways:

- SEI Layer Failure: If the SEI layer becomes unstable or grows excessively thick, it can crack, exposing the anode to the electrolyte and initiating exothermic reactions.
- Electrolyte Decomposition: Degraded electrolytes can produce flammable gases, increasing the cell's pressure and raising the fire risk if the gases are ignited.
- Lithium Plating and Dendrite Formation: Lithium plating can cause the formation of lithium dendrites, puncturing the separator and causing internal short circuits. These generate heat and further increase the probability of thermal runaway.

4.2.6. Short Circuit and Gas Generation

Internal short circuits resulting from dendrite formation, separator failure, or electrode material fracture are direct consequences of degradation. When short circuits occur, localized heating is produced, which can initiate further degradation and, in extreme cases, thermal runaway [92]. Additionally, gas generation from electrolyte breakdown products raises the internal pressure, potentially leading to cell venting or rupture.

4.2.7. Mechanical and Structural Instability

The mechanical degradation of electrodes, particularly in high-capacity designs, can lead to particle fracturing and loss of contact between active materials and current collectors. This reduces battery performance and increases the risk of structural failure within the cell, leading to electrical isolation of portions of the electrode or the development of hotspots due to uneven current distribution [93]. Mechanical instability is a key safety concern in high-power applications, where rapid cycling can exacerbate structural damage and lead to sudden performance failures.

This section examined how vibrations accelerate degradation in lithium-ion batteries, impacting performance, lifespan, and safety. Battery degradation occurs through calendar aging (time-dependent capacity loss) and cycle aging (repeated charge–discharge wear). Vibrations exacerbate key degradation mechanisms, including SEI layer growth, electrode material breakdown, lithium plating, and electrolyte decomposition, leading to higher internal resistance, capacity fade, and efficiency loss. Vibrational stress weakens electrode structures, causing particle detachment, microcracks, and separator damage, which increase the risk of internal short circuits and thermal runaway. Additionally, vibrations contribute to uneven lithium deposition, worsening dendrite formation and separator punctures, which can trigger electrical failures and safety hazards. As vibrations accelerate mechanical fatigue and heat generation, they further degrade the battery's thermal stability, increasing the likelihood of overheating, gas buildup, and ignition risks in high-vibration applications. The following section will focus on how vibrations influence thermal runaway in lithium-ion batteries, examining how mechanical stress intensifies heat generation, disrupts thermal regulation, and increases the probability of catastrophic failure.

5. Effects of Vibrations on Thermal Runaway in Lithium-Ion Batteries

Lithium-ion batteries, especially in dynamic applications such as EVs, aerospace, and marine industries, are exposed to mechanical vibrations during operation. Vibrations affect the structural integrity, electrical performance, and thermal behavior of the battery. Mechanical vibrations introduce complex mechanical–electrochemical interactions in lithium-ion batteries. The resulting micromovements and strain affect the battery's internal components, including the electrodes, separator, and current collectors. This leads to higher internal resistance and increased heat generation, as shown in Figure 14.

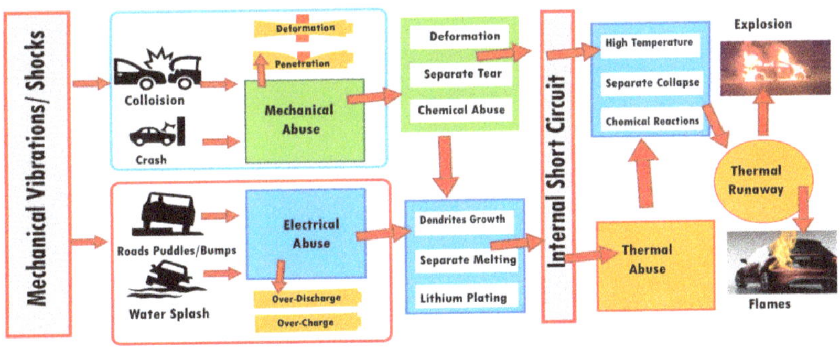

Figure 14. Mechanism of thermal runaway due to vibrations.

Mechanical vibrations and shocks play a significant role in the degradation and failure of lithium-ion batteries (LIBs), leading to internal short circuits and, in extreme cases, thermal runaway. Figure 14 illustrates the pathways through which mechanical, electrical, and thermal abuses contribute to battery failure, ultimately resulting in catastrophic outcomes such as explosions or vehicle fires.

5.1. Mechanical Abuse

Mechanical vibrations and external shocks arise from conditions such as road bumps, puddles, or vehicle collisions. These external stimuli lead to mechanical abuse, manifesting in penetration and deformation of battery components. Structural damage can further induce chemical abuse, causing separation tears and internal material failures [94,95]. Such deformations compromise the battery's structural integrity and increase the risk of short circuits.

5.2. *Electrical Abuse and Internal Short Circuits*

Vibrations also impact the electrochemical stability of LIBs. Road irregularities and water splashes can introduce electrical abuse conditions, including overcharge and over-discharge, accelerating dendritic lithium growth [95]. These lithium dendrites may penetrate the battery separator, causing internal short circuits that generate localized heating and intensify thermal stress [94]. Lithium plating and separator melting also contribute to further performance degradation and safety hazards.

5.3. *Thermal Abuse and Thermal Runaway*

Once an internal short circuit occurs, the battery experiences thermal abuse, leading to high temperature, separator collapse, and exothermic chemical reactions. These reactions trigger a thermal runaway event, wherein excessive heat buildup leads to flames and explosions [4]. This chain reaction is particularly hazardous in electric vehicles (EVs), where battery packs contain multiple interconnected cells. The interplay between mechanical vibrations, electrical instability, and thermal abuse highlights the necessity for robust battery management systems and vibration-resistant battery designs.

5.3.1. Heat Generation Mechanisms in Lithium-Ion Batteries

Heat generation in lithium-ion batteries primarily comes from three sources: Ohmic (Joule) heating due to resistance in internal components like electrodes and electrolytes, reversible heat generation associated with entropy changes during electrochemical reactions, and irreversible heat generation from side reactions such as solid–electrolyte interface (SEI) formation and degradation processes.

Vibrations can significantly exacerbate heat generation mechanisms in lithium-ion batteries. By introducing additional internal resistance and contributing to mechanical degradation, vibrations lead to higher internal energy loss, thereby increasing heat generation within the battery, as shown in Figure 15.

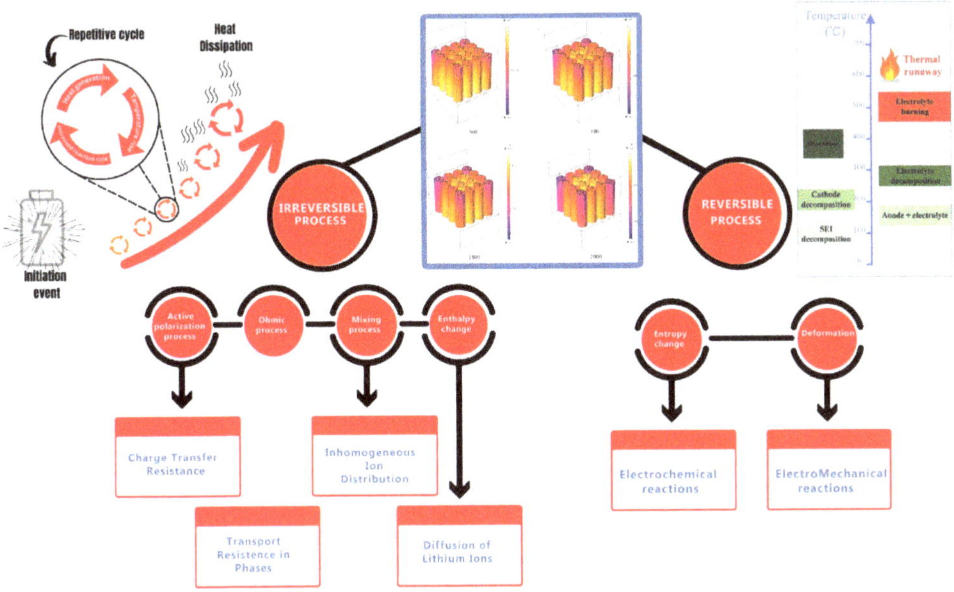

Figure 15. Thermal and mechanical processes in lithium-ion batteries: reversible and irreversible dynamics.

5.3.2. Vibration-Induced Internal Resistance Increase and Heat Generation

Vibrations have a direct impact on the internal structure of lithium-ion batteries. Introducing mechanical strain and deformation [96] increases internal resistance, affecting the battery's performance. Increased resistance (R_{int}) is a primary factor in heat generation within the cell. The relationship between internal resistance and heat generation can be modeled by Equation (1) under vibrations. Microscopic mechanical movements in the battery can cause poor contact between electrode particles, microcracks in the electrode materials, or slight misalignment of current collectors, which increase internal resistance and affect degradation, as shown in Figure 15. As R_{int} increases, the heat generation according to the equation for Q_{heat} increases with the current, making the battery more susceptible to overheating under vibrations at high discharge rates.

5.3.3. Impact of Vibrations on Electrochemical Reactions and Heat Generation

Vibrations can also disrupt the uniformity of lithium-ion diffusion within electrode materials. This uneven diffusion can lead to localized concentration gradients, causing non-uniform current densities and exacerbating heat generation in specific battery regions. The heat generated due to non-uniform current distribution is linked to the overpotential (η) in the battery, representing the difference between the actual operating and thermodynamically predicted voltage. The heat generated due to overpotential can be expressed as follows:

$$Q_{overpotential} = I \cdot \eta \tag{9}$$

where $Q_{overpotential}$ is the heat generated due to overpotential (in watts), I is the current (in amperes), and η is the overpotential (in volts). Vibrations cause mechanical fatigue in the electrodes, which can alter the local distribution of lithium ions and affect the kinetics of the electrochemical reactions. This leads to an increase in overpotential and, thus, more heat is generated.

5.3.4. Mechanical Strain and Heat Generation

Vibrations introduce mechanical strain in battery components, leading to material deformation and failure modes such as microcracking of the electrodes and delamination of electrode coatings from the current collectors [97]. The mechanical energy imparted by vibrations is partially converted to heat due to frictional loss and material deformation. This heat generation can be modeled as follows:

$$Q_{vibrations} = \frac{1}{2}a^2 m f C_{damp} \tag{10}$$

where $Q_{vibrations}$ is the heat generated due to mechanical strain (in watts), m is the mass of the battery (in kilograms), a is the acceleration caused by the vibrations (in meters per second squared), f is the frequency of vibrations (in hertz), and C_{damp} is the damping coefficient, which represents the material's ability to dissipate vibrational energy as heat.

In practical applications, such as automotive systems, vibrations typically occur in the frequency range of 5–200 Hz, and the damping coefficient depends on the specific materials used in battery construction [98]. Prolonged exposure to these vibrations increases the battery's temperature, further accelerating degradation processes.

5.3.5. Thermal Feedback and Heat Accumulation

Vibrations increase internal resistance and mechanical strain and contribute to a thermal feedback loop. The battery's temperature rises as heat is generated from mechanical vibrations and increased resistance. Elevated temperatures, in turn, increase internal resistance, further enhancing heat generation according to the $\boldsymbol{i^2 R_{int}}$ relationship. The overall heat balance in a battery under vibrations can be represented as follows:

$$Q_{total} = i^2 R_{int} + I \cdot \eta + Q_{vibrations} - Q_{dissipated} \tag{11}$$

where Q_{total} is the total heat generated in the battery (in watts); $i^2 R_{int}$ represents the heat generated due to internal resistance of the battery as current flows through it; $I \cdot \eta$ is the heat generated due to over potential, as previously discussed; $Q_{vibrations}$ is the heat generated due to mechanical strain from vibrations; and $Q_{dissipated}$ is the heat dissipates through cooling mechanisms, such as convection, conduction, or radiation (in watts). If the generated heat exceeds the dissipation capacity of the battery's thermal management system, the battery's temperature can rise uncontrollably, leading to thermal runaway in extreme cases [99].

Table 1 provides a detailed overview of how different types of mechanical vibrations and shocks impact lithium-ion battery performance, particularly in terms of heat generation and the risk of thermal runaway. Low-frequency vibrations, such as those from road unevenness, cause gradual SEI layer growth and minor electrode misalignment, leading to a moderate increase in internal resistance and heat buildup, with a low to moderate risk of thermal runaway. These findings highlight the need for improved battery designs, advanced thermal management strategies, and robust battery management systems to mitigate the effects of mechanical vibrations and shocks. Future studies should focus on real-time monitoring solutions and innovative materials that can withstand mechanical stress while maintaining optimal thermal stability.

Table 1. Impact of vibrations on thermal runaway.

Type of Shock/Vibration	Effect on Battery Components	Heat Generation Mechanism	Risk of Thermal Runaway
Low-Frequency Vibrations (Road Unevenness)	Gradual SEI layer growth, minor electrode misalignment	Increased internal resistance leading to moderate heat buildup	Low to Moderate
High-Frequency Vibrations (Motor Operation)	Accelerated electrode wear, increased lithium plating	Localized heating due to lithium plating and resistance rise	Moderate
Sudden Impacts (Potholes, Speed Bumps)	Internal microcracks, separator weakening	Sudden temperature spikes from localized internal stress	Moderate to High
Crash/Collision Shocks	Severe structural damage, high risk of short circuit	Extreme heat generation due to short circuits and thermal runaway	Very High
Continuous Micro-Vibrations (High-Speed Driving)	Increased internal resistance, gradual heat buildup	Progressive heat accumulation from increased resistance and loss	Moderate to High

5.4. Battery Risks in EV Crashes

EVs are equipped with advanced braking, power electronics, and battery systems to ensure safety during sudden stops and collisions. However, sudden braking, vehicle crashes, and impacts with stationary objects (e.g., buildings, barriers, or poles) can introduce vibrational stress, mechanical deformation, and electrical failures in EVs. These factors can significantly affect battery integrity, thermal stability, and overall system performance, posing risks such as thermal runaway, high-voltage failures, and post-crash hazards, as shown in Figure 16.

The safety of EVs depends on the ability of the battery system to withstand sudden mechanical shocks and maintain its thermal and electrical stability under extreme conditions. As the adoption of EVs continues to grow, improving battery pack designs, integrating intelligent battery management systems, and developing more robust crash protection mechanisms will be crucial in preventing fire hazards and ensuring long-term vehicle reliability.

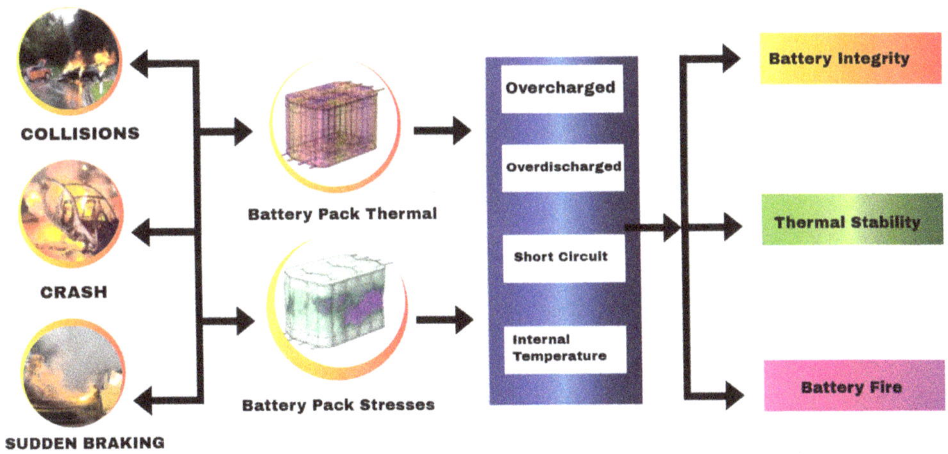

Figure 16. Battery failure analysis under adverse conditions.

5.4.1. Sudden Brake Vibrations

Unlike conventional vehicles, EVs primarily use regenerative braking supplemented by mechanical friction braking. During hard braking or emergency stops, oscillatory forces act on battery modules, generating high-frequency vibrations that stress interconnects, terminals, and cooling plates. These vibrations have been shown to reduce battery efficiency and induce microstructural damage in lithium-ion cells. Studies indicate that regenerative braking also leads to voltage fluctuations, which may trigger false fault detections in the battery management system and result in unnecessary power limitations or shutdowns.

5.4.2. Collision Vibrations

Collisions and exceptionally high-impact crashes introduce additional risks for battery packs and EV safety systems. Lithium-ion cells are prone to short internal circuits when subjected to mechanical deformation, which can trigger thermal runaway. Side impact collisions are particularly concerning, as they can cause direct battery intrusion, leading to module displacement and cooling system failures. If the cooling plates are punctured or disconnected due to the impact, thermal regulation becomes ineffective, accelerating internal heating. Battery enclosure crumpling is another significant factor, especially in underbody impacts, where road debris or structural elements can penetrate the pack. Studies analyzing EV crashworthiness have found that structural deformation exceeding 12% of battery enclosure space significantly raises the likelihood of thermal events. Post-collision fire hazards remain one of the most critical concerns in EV crashes. Thermal runaway can occur immediately upon impact or be delayed for hours or even days due to residual stress within the cells.

5.4.3. Collision with Buildings, Poles, and Barriers: Unique Challenges

Collisions with stationary objects such as buildings, barriers, and poles introduce unique challenges in EV safety. Unlike vehicle-to-vehicle impacts, where energy dissipation is distributed across both vehicles, crashes with stationary objects cause localized force concentrations. This increases pressure on specific vehicle sections, such as the front-end, underbody, or battery casing. Front-end collisions may compromise the power electronics, affecting inverter and BMS operation, while side impacts often pose direct threats to battery safety due to potential penetration into the enclosure.

5.4.4. Unique Challenges

Collisions, crashes, and sudden braking in electric vehicles (EVs) create significant risks for battery safety due to thermal and mechanical stress on the battery pack. These events can lead to structural deformation, electrical failures, and temperature imbalances, increasing the chances of thermal runaway and fire hazards. The impact of crashes or sudden braking generates excessive force on the battery pack, causing internal mechanical stresses that weaken the structural integrity of battery modules.

5.4.5. Failure Analysis Testing

Failure analysis of electric vehicle (EV) batteries under collisions, sudden braking, and vibrational stress requires mechanical, thermal, electrical, and computational testing. Crash impact and vibration testing assesses structural integrity, identifying microcracks, weld failures, and module displacement. Thermal runaway and fire suppression tests evaluate heat propagation and containment strategies, ensuring post-crash safety. Electrical safety tests, including short circuit analysis and insulation resistance measurement, detect high-voltage failures and electrocution risks. Computational simulations such as finite element analysis (FEA) and AI-driven battery monitoring predict failure risks by analyzing stress distribution, thermal anomalies, and voltage inconsistencies. Integrating these methods can enhance battery resilience, enabling the development of safer, crash-resistant EV systems.

This section examined how vibrations contribute to thermal runaway in lithium-ion batteries. Mechanical stress causes electrode misalignment, separator damage, and increased internal resistance, leading to localized heating and performance degradation. Sudden shocks, collisions, and road irregularities accelerate lithium dendrite formation and short circuits, triggering uncontrolled heat buildup and thermal runaway risks. Vibrations also disrupt lithium diffusion, increase overpotential, and amplify heat generation, creating a thermal feedback loop that can overwhelm battery cooling systems. These effects underscore the need for vibration-resistant designs, advanced thermal management, and failure detection systems to enhance battery safety and durability. The next section will explore vibration-induced degradation in lithium-ion batteries, focusing on structural wear, capacity fade, and long-term performance loss.

6. Vibration-Induced Degradation in Lithium-Ion Batteries

Vibrations play a significant role in the degradation of lithium-ion batteries, particularly in applications such as electric vehicles, aviation, and marine systems, where mechanical vibrations are unavoidable. These vibrations cause mechanical stress, material fatigue, and structural failures in battery components, accelerating key degradation processes.

Mechanical vibrations and shocks are inevitable in electric vehicles (EVs) due to road irregularities, acceleration, braking, and other dynamic conditions. These external mechanical stimuli significantly impact lithium-ion batteries' electrochemical and structural integrity (LIBs), leading to performance degradation, capacity loss, and inaccuracies in state of charge (SOC) estimation. Figure 17 [100] illustrates how mechanical vibrations affect battery components and the resulting implications for battery health and EV range anxiety. Mechanical stress increases reactivity within the battery, accelerating unwanted side reactions at the electrode–electrolyte interface. One of the most critical consequences is the accelerated formation of the solid–electrolyte interphase (SEI) layer. While the SEI layer is essential for stabilizing battery chemistry, excessive growth increases internal resistance, reduces charge transfer efficiency, and contributes to long-term capacity fade. The mechanical agitation of electrode materials also leads to microcrack formation, exposing fresh electrode surfaces to the electrolyte and further increasing SEI formation. As a result, repeated exposure to mechanical shocks results in reduced Coulombic efficiency and progressive loss of usable capacity over multiple charge–discharge cycles [101]. Beyond electrochemical effects, mechanical vibrations induce structural damage within the battery. Stress on the cathode, anode, and separator can weaken the physical integrity of these components, leading to separator deformation or misalignment of active materials. Damage to the current collectors, particularly copper and aluminum foils, increases the risk of internal short circuits, leading to localized heating and, in severe cases, thermal runaway [102]. These degradation mechanisms compromise battery safety and impair the accuracy of SOC estimation, leading to discrepancies between the apparent SOC reported by the battery management system (BMS) and the actual remaining charge. This inconsistency causes EV users to experience range anxiety, as the displayed battery level may not accurately reflect the actual available energy [103]. As the figure demonstrates, these degradation pathways ultimately affect battery health and operational reliability. The growing mismatch between the apparent and actual SOC results in premature charging, inefficient energy utilization, and increased operational costs for EV users. Additionally, excessive battery degradation can shorten the overall lifespan of the battery pack, increasing the frequency of replacement and raising sustainability concerns. Addressing these challenges requires advanced diagnostic tools capable of detecting early-stage mechanical degradation and improved battery designs that can withstand vibrational stress without significant performance loss. The impact of mechanical vibrations on battery behavior underscores the need for further research into resilient electrode materials, robust separators, and advanced BMS algorithms to mitigate performance loss.

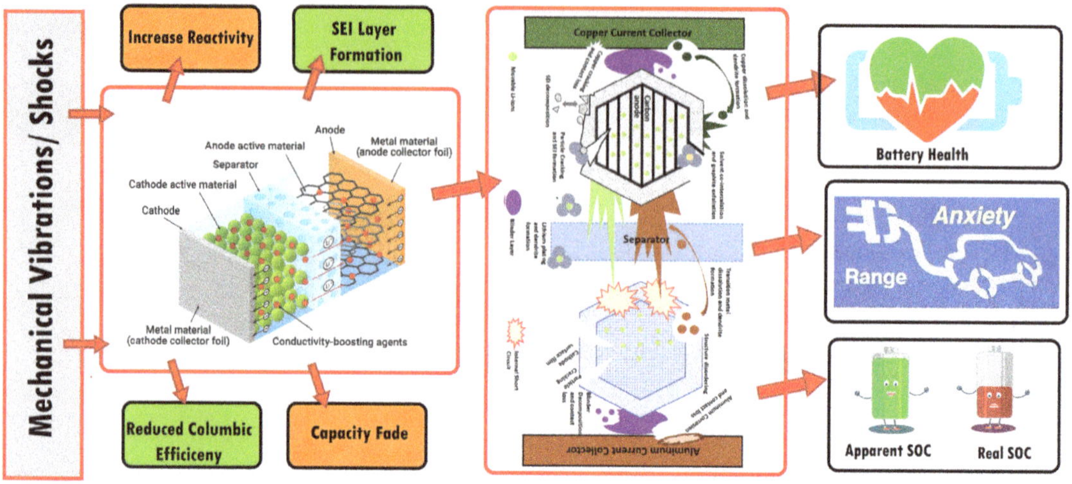

Figure 17. Mechanism of degradation due to vibrations.

6.1. Mechanisms of Degradation Under Vibrations

The primary degradation mechanisms influenced by vibrations are:

Mechanical Fatigue: Repeated mechanical stress causes the battery's material components (electrodes, separators, and current collectors) to undergo fatigue and microstructural changes.

Electrode Delamination: Vibrations can cause the active material to delaminate from the current collector, increasing internal resistance and reducing active material utilization.

Separator Damage: Vibrations can induce physical strain on the separator, which increases the prospect of short circuits and accelerates degradation.

Increased Resistance and Heat Generation: Vibrational stress can disrupt the internal contact between electrode particles, increasing internal resistance and further accelerating ageing through excessive heat generation.

Figure 18 [104] provides an overview of the effects of vibrations on lithium-ion batteries and the application of multiphysics analysis to understand their behavior as an effect of thermal behavior. It highlights the key processes and mechanisms within the battery—such as structural degradation, thermal dynamics, and stress distribution—caused by vibrations during operation. The battery's performance under real-world conditions is simulated using tools as a vibration controller and vibrating machine. Multiphysics thermal and stress analysis enables the visualization of heat distribution and mechanical deformation, ensuring better design and safety, which is in high demand. These mechanical and electrochemical processes interact to degrade battery performance. Mathematical models can describe how vibrations contribute to these mechanisms.

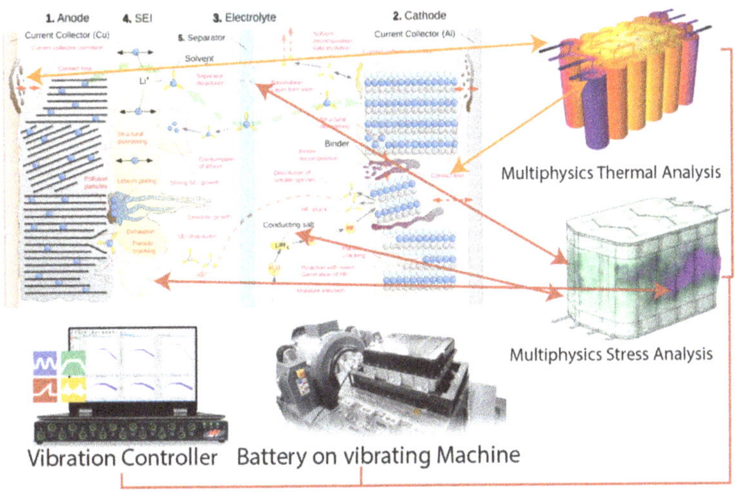

Figure 18. Battery degradation mechanisms under vibrations.

6.2. Electrode Fatigue and Delamination Due to Vibrations

Electrode fatigue refers to the weakening of electrode materials due to repeated mechanical loading caused by vibrations. As vibrations propagate through the battery, they induce cyclic strain in the active materials [105,106]. This cyclic loading creates microcracks in the electrodes, reducing the available surface area for lithium intercalation and increasing internal resistance. The fatigue behavior of materials under cyclic loading is described by S–N curves (stress–life curves), which relate the stress amplitude (σ) to the number of cycles to failure (N), as follows:

$$\sigma \cdot N^b = C \tag{12}$$

where σ is the applied stress (in Pa), N Is the number of cycles to failure, N^b is a material-dependent fatigue exponent, and C is a constant that depends on the material's properties. As σ increases due to higher vibration amplitudes, the cycle number before failure decreases, meaning that the battery degrades more quickly under stronger or more frequent vibrations. Electrode delamination arises when vibrational forces degrade the adhesion between the active material and the current collector. This mechanical stress causes the electrode coating to detach from the collector, diminishing electrical conductivity and impairing active material utilization. As a result, internal resistance increases, and the battery's capacity is compromised. The delamination rate is closely tied to the amplitude and frequency of vibrations, which inject strain energy into the system, further accelerating the deterioration of the battery. The strain energy density (U_{strain}) can be modeled using the energy associated with this delamination, as follows:

$$U_{strain} = \frac{1}{2} 2\sigma \cdot \epsilon \tag{13}$$

where U_{strain} is the strain energy density (in J/m^3), σ is the applied stress (in Pa), and ϵ is the strain induced by vibrations. As strain increases due to vibrations, delamination progresses more rapidly, further contributing to the battery's degradation.

6.3. Separator Damage and Short Circuit Risk

The separator plays a vital role in preventing direct contact between the anode and cathode while still permitting the flow of lithium ions. However, vibrations can cause mechanical stress on the separator, resulting in fatigue and the formation of small tears. Any damage to the separator raises the risk of internal short circuits, which can drastically speed up degradation and potentially cause catastrophic failure, such as thermal runaway [107]. The probability of separator failure due to vibrations can be modeled using fatigue failure models based on the accumulated strain (ϵ) over time. The relationship between strain and failure probability is often described by the Coffin–Manson equation, which is used to predict fatigue life in cyclically loaded materials, as follows:

$$N_f = \frac{\Delta\epsilon}{C_{sep}} \tag{14}$$

where N_f is the number of cycles to failure for the separator, $\Delta\epsilon$ is the vibration-imposed strain range, and C_{sep} is a constant that depends on the separator material's properties. As the vibration frequency and amplitude increase, $\Delta\epsilon$ increases, leading to a decrease in N_f (i.e., faster separator degradation). Internal shorts may develop once the separator is compromised, leading to safety issues and rapid capacity loss.

6.4. Vibration-Induced Resistance Increase and Heat Generation

As mechanical vibrations disrupt the battery's internal structure, internal resistance increases due to poor contact between electrode particles and current collectors. Increased resistance causes more heat generation, as described by Joule's law in the equation. i^2R_{int}. The increase in internal resistance (R_{int}) can be modeled as a function of the vibration amplitude (A) and frequency (f), as follows:

$$R_{int}(A,F) = R_o + K_1A + K_1f \tag{15}$$

where R_o is the baseline internal resistance (in ohms), A is the vibration amplitude (in meters), f is the vibration frequency (in hertz), and $K_1$and K_2 are constants related to the battery's material and structural properties. As vibrations increase in amplitude and frequency, internal resistance rises, generating more heat [108]. This heat further accelerates degradation mechanisms such as electrolyte decomposition and SEI layer growth.

Table 2 highlights the impact of various shocks and vibrations on battery heat generation. Low-frequency vibrations, such as those from road unevenness, gradually increase SEI layer thickness and internal resistance, leading to moderate heat buildup. More severe shocks, such as those from crashes or collisions, can cause extreme structural damage, increasing the risks of internal short circuits and thermal runaway. Continuous micro-vibrations from high-speed driving progressively degrade battery components, causing steady heat accumulation and increasing thermal instability. Understanding the effects of mechanical shocks and vibrations on battery heat generation is crucial for improving lithium-ion battery safety and performance in electric vehicles. Addressing these challenges requires enhanced battery designs, advanced materials, and real-time monitoring systems to mitigate thermal risks and extend battery lifespan.

Table 2. Impact of vibrations on battery materials, safety, and performance.

Type of Vibration	Effect on Electrode	Effect on Separator	Effect on Battery	Effect on Internal Resistance	Effect on Thermal Runaway	References
Low-Frequency Vibrations (Road Unevenness)	Minor electrode misalignment, SEI layer thickening	Minimal impact, slight deformation over time	Slight capacity fade over time	Slight increase over time due to SEI layer growth	Low risk, but prolonged stress can contribute to localized heating	[1,9,15,21]
High-Frequency Vibrations (Motor Operation)	Accelerated electrode wear, increased contact resistance	Increased mechanical stress, risk of separator thinning	Increased resistance, reduced cycle life	Noticeable increase, leading to efficiency loss	Increased heat generation, reduced thermal stability	[6,10,18,25]
Sudden Shocks (Potholes, Speed Bumps)	Microcracks in electrode material, loss of active material	Potential small tears, localized damage	Accelerated degradation, loss of efficiency	Sharp increase in resistance due to material fractures	Sudden energy release may trigger localized overheating	[6,9,17,24]
Crash/Collision Impact	Severe structural damage, detachment of active material	High risk of complete separator failure	Severe capacity loss, potential safety hazard	Extreme increase, leading to power loss and safety failure	There is a high likelihood of thermal runaway due to short circuits and heat buildup	[6,14,20,27]
Continuous Micro-Vibrations (High-Speed Driving)	Gradual electrode degradation, uneven lithium deposition	Gradual wear leading to increased resistance	Progressive performance decline, reduced lifespan	Gradual rise in resistance, affecting efficiency over time	Long-term heat accumulation can lead to self-heating and instability	[9,15,19,26]

This section explored how vibrations accelerate the degradation of lithium-ion batteries, where mechanical stress is unavoidable. Vibrations induce mechanical fatigue, electrode delamination, and separator damage, leading to increased internal resistance, capacity loss, and heat generation. Repeated mechanical stress accelerates SEI layer growth, disrupts lithium diffusion, and weakens electrode materials, causing progressive efficiency to decline and potential short circuits. Key degradation mechanisms include electrode fatigue, separator wear, and increased Joule heating, which raise thermal instability and the likelihood of thermal runaway. Vibrations also affect state of charge accuracy, creating inconsistencies that contribute to range anxiety and inefficient battery utilization in EVs. Resilient electrode materials, advanced separators, and enhanced battery management systems are essential to mitigate vibration-induced degradation.

7. Discussion

This review of studies on lithium-ion battery performance under vibrational conditions reveals the multifaceted impact of mechanical vibrations on battery safety, range, and operational reliability. The following discussion summarizes the findings, evaluates how well the review's objectives were met, and highlights emerging research gaps. The studies indicate that thermal management, structural resilience, and environmental conditions are crucial factors affecting battery performance under vibrational stress. Research shows thermal management, structural resilience, and environmental conditions influence battery performance under vibrational stress. High temperatures and mechanical vibrations significantly degrade battery performance and reduce lifespan. Studies have consistently reported that increased internal resistance and capacity fade are primary degradation mechanisms caused by thermal and mechanical stress. These findings highlight the critical relationship between external abuse and battery performance decline. However, while the individual effects of these factors are well documented, there are limited data on how combined stressors, such as temperature, pressure, and vibrations, interact to impact battery performance.

Table 3 summarizes the studies investigating the impact of vibrations on lithium-ion batteries, focusing on thermal behavior, degradation mechanisms, and experiment types. Thermal investigations are often secondary to mechanical durability or state of health evaluations, with only a few studies directly addressing heat generation under vibrational stress. Studies like [11,17,24] identified the risks of overheating during short circuits or high vibration amplitudes, while others like [13,15] did not explore thermal behavior, concentrating instead on structural integrity or compliance with transportation safety standards. Degradation mechanisms, however, are more comprehensively studied, with many papers [18,19,24] linking vibrations to increased internal resistance, capacity fade, and structural damage. Vibrations cause delamination, separator thinning, and electrode degradation, contributing to performance loss over time. While practical experiments dominate this area of research, simulation studies like [21,31] focused more on state of charge (SOC) estimation or feasibility analysis in extreme environments.

Table 3. Technical overview of thermal and degradation studies on lithium-ion batteries under vibrations.

Ref.	Thermal Investigation Results	Degradation Investigation Results	Experiment Type
[11]	Risk of overheating during short circuit and overcharge tests.	Vibration and shorting tests revealed overheating risk.	Practical
[8]	Focused on mechanical effects on batteries rather than thermal aspects.	Vibrations and shocks impact cell performance.	Practical
[15]	Concentrated on SOH under vibrations, no thermal evaluation.	Minimal capacity fade was observed across different vibrational conditions.	Practical
[18]	Resistance buildup with vibrations suggests heat increase.	Increased resistance and reduced capacitance were observed under vibrations.	Practical
[17]	High heat release under vibrational stress.	Structural integrity is compromised under severe vibrational conditions.	Practical
[24]	Higher heat output under vibrations.	Increased resistance and capacity fade.	Practical
[22]	Studied thermal issues in various stress conditions.	Investigated degradation rates under different environmental conditions.	Practical
[13]	Focused on safe battery transport protocols.	No degradation focus, the study emphasized vibration testing for transportation.	Practical
[29]	Monitored temperature for compliance.	Ensured safety under different stress tests.	Practical
[19]	Effects of temperature and vibrations on battery lifespan.	Degradation rates are higher under stress conditions.	Practical
[21]	No thermal analysis, SOC estimation study.	SOC accurately tracked under various operating conditions.	Simulation
[23]	No thermal analysis, focused on durability in motorcycle vibrations.	Tested unique motorcycle vibration patterns on battery durability.	Practical
[11]	Mechanical vibrational effects were explored without thermal aspects.	Linked abnormal vibration patterns to potential battery degradation.	Practical
[16]	No thermal investigation, focused on vibration testing in powertrains.	Vibration resistance was checked in prototype testing.	Practical
[32]	High-frequency signals indicated over-discharge faults.	Demonstrated over-discharge risks through vibration signals.	Practical
[12]	Shock and vibration monitoring for transportation, thermal impact not studied.	The MEMS system is adequate for transport monitoring.	Practical
[30]	No thermal effect was noted, the environmental impact on battery performance was emphasized.	Assessed the influence of pressure, humidity, and temperature on battery performance.	Practical
[26]	A thermal study was not conducted, the focus was on unique powertrain vibrations.	Evaluated performance under dynamic load conditions.	Practical
[20]	Review paper on performance, no new thermal testing.	Mechanical and electrical performance parameters summarized.	Review
[14]	Vibrational effects on electrical performance but no thermal analysis.	Performance impact observed under continuous vibration.	Practical
[9]	Thermal aspects are secondary, focused on lead-acid battery vibrations.	Performance evaluated under vibrational effects.	Practical
[33]	Dynamic loads but thermal effects were not included.	Studied dynamic load impact on battery life.	Practical
[27]	Nondestructive testing focus, no thermal study.	SOC monitored using structural changes.	Practical
[25]	Phase change materials evaluated without thermal assessment.	Evaluated thermal management through the use of phase change materials.	Simulation
[28]	SOC under vibrations, no thermal focus.	Analysis under variable vibrational stress levels.	Practical
[6]	Comparative study, thermal focus secondary to mechanical evaluation.	Failure mechanisms under mechanical abuse were explored.	Practical
[31]	Thermal management was explored under vibrations for submarine use.	Analyzed feasibility of battery use in high-stress environments.	Simulation

Several studies highlight that vibrations exacerbate internal resistance buildup, causing localized heat generation and further accelerating degradation. However, some investigations [20,28] lack direct thermal evaluation and instead focus on environmental or electrical performance impacts. Simulation-based studies tend to overlook the combined thermal and mechanical effects seen in real-world conditions, whereas practical studies excel in identifying specific risks but often lack detailed thermal models.

7.1. Harmful Effects of Vibrations on Battery Systems

The studies support that vibrations are harmful to lithium-ion battery systems. For instance, prolonged exposure to mechanical stress results in physical damage to internal components, such as electrode delamination and separator deformation. Vibration-induced structural damage increases the risk of internal short circuits, which poses significant safety hazards and accelerates capacity loss. The review successfully assessed the harmful effects, confirming that vibrations are a risk factor for battery degradation and safety. However, further investigation is needed to quantify the specific vibration thresholds that delineate safe operating limits. These could help to establish guidelines to minimize battery damage, particularly in high-vibration environments such as EVs or aerospace applications.

Table 4 highlights the effects of different vibrational conditions on lithium-ion batteries. Low-frequency vibrations cause minor capacity fade and negligible safety concerns, while high-frequency vibrations result in specified hotspots, increased resistance, and heightened thermal runaway risks. Six-degree-of-freedom vibrations lead to moderate heat rises, capacity loss, and safety risks from separator damage. Shock vibrations cause electrode cracking, severe capacity fade, and high safety risks from short circuits. Random vibrations show minimal heat effects but significantly increase resistance and structural risks over time. Coupled cycling and vibrations accelerates SEI layer growth, heat rise, and capacity fade, increasing safety risks. Over-discharge with vibrations triggers severe overheating, capacity loss, and thermal runaway risks. Marine vibrations show negligible thermal and performance impacts with minimal safety concerns, particularly in LFP batteries. Studies such as [1,8,17] revealed internal resistance increases of up to 257.82% in 18,650 cells subjected to random vibrations, attributed primarily to tab loosening and structural deformation. These findings provide valuable insights into mechanical degradation but lack material-level analyses. Incorporating advanced diagnostic tools like SEM or XPS could have helped to uncover critical degradation mechanisms, such as cathode cracking and SEI instability, to better predict long-term performance and safety. High-frequency vibrations were extensively studied in [15,76,80], showing localized hotspots of 2–3 °C, impedance growth of up to 40%, and 4–6% capacity fade, mainly due to cracking in NMC cathode materials. These studies effectively link mechanical stress to electrochemical performance loss. However, they fail to address the compounding effects of gas generation or prolonged electrolyte decomposition, which are critical safety concerns in extended operations. Future work should explore these prolonged degradation pathways and their impact on thermal runaway risk. In [16,18,24], six-degree-of-freedom vibrations led to 3–4 °C uniform thermal rises, 18.8% ohmic resistance increase, and capacity losses of up to 0.78 Ah after 200 cycles. While these studies successfully demonstrate the connection between mechanical stress and electrochemical degradation, they lack evaluations under high C-rate cycling conditions typical of electric vehicles. Such data are critical to optimizing battery design for demanding applications. Other papers [13,15,76] reported that shock vibrations caused 2–4 °C heat rises, separator deformation, electrode cracking, and capacity losses of 5–10%. These degradation mechanisms introduce high safety risks, such as internal short circuits and gas generation, which can lead to thermal runaway. However, critical evaluations of real-world thermal and mechanical stress interactions remain unexplored.

Some studies [5,17,80] focused on low-amplitude vibrations in marine applications, demonstrating minimal capacity fade (<1%) and negligible temperature rise (<1 °C). LFP batteries consistently outperformed NMC batteries in durability under these conditions. However, the effects of prolonged exposure to environmental factors such as salinity or humidity were not addressed, leaving a gap in the understanding of long-term marine battery performance. Research on PCM use for thermal management under vibrations [16,57,76] showed that PCM reduced the temperature rise by 2–4 °C, but uneven stress distribution within PCM layers was noted. These studies highlight PCM's potential for enhancing thermal performance but lack insights into its long-term durability under combined vibrational and cycling stress. Significant findings in [1,19,24] indicated 3–6 °C thermal increases and 5–12% capacity fade under coupled cycling and vibrations. These impacts were driven by accelerated SEI layer growth and lithium plating, which pose performance and safety concerns.

Table 4. Vibrational effects on lithium-ion battery performance and safety.

Vibrational Conditions	Thermal Impact	Performance Impact	Safety Impact	Citations
Low frequency: 5–20 Hz	Temperature rise of up to 3 °C (some papers do not confirm thermal impact)	Minor capacity fade (<2%); increased SEI layer growth and minor lithium loss observed in long-term vibration tests.	Low safety concerns: no critical deformation observed under low-stress conditions.	[1,7,8,24]
High frequency: 100–1000 Hz	Localized hotspots; temperature increase of 2–3 °C (validated)	Resistance increased by 10–40%; cathode cracking and SEI layer thickening caused a 4–6% capacity fade.	Increased internal pressure from electrode damage raises thermal runaway risks under high-stress conditions.	[18,19,76,80]
6-DOF Vibrational Stress: 20–200 Hz	Heat rise not explicitly discussed in some studies, 3–4 °C validated in others	Capacity loss of 0.78 Ah after 200 cycles; ohmic resistance increased by 18.8%.	Safety concerns arise from mechanical fatigue leading to separator damage, increasing short circuit risks.	[9,16,24,80]
Shock Vibrations: 30–100 Hz, Amplitude > 2 mm	Temperature increase of up to 4 °C (verified where noted)	Electrode cracking, separator deformation, and polarization resistance increase caused 5–10% capacity loss.	High safety risk from internal short circuits and gas generation during severe shocks.	[13,15,18,76]
Random Vibrations (SAE J2380)	No significant heat generation; temperature rise < 1 °C (validated)	Internal resistance increased up to 257.82%, especially in cylindrical 18650 cells.	Safety risks include tab loosening and structural deformation, which could lead to electrical faults over time.	[8,17,24]
Cycling and Vibrations Coupling	Heat rise of 3–6 °C validated	SEI layer growth accelerated; capacity fade between 5–12% over extended cycling with vibrations.	Increased safety risk from accelerated degradation and gas generation under prolonged stress.	[8,24,80]
Over-discharge Coupled Vibrations	Overheating observed; temperature rise to 55 °C validated	Severe capacity fade of 10–15%, driven by anode degradation and electrolyte decomposition.	High safety risk from thermal runaway triggered by anode collapse and rapid heat generation.	[18,19,76]
Marine Vibrations (low amplitude)	No measurable heat rise; confirmed negligible thermal impact	Negligible capacity fade (<1%); stable performance with no structural damage observed.	Minimal safety concerns; LFP batteries demonstrated high resilience under low-amplitude vibrations.	[5,17,80]

Despite identifying key degradation mechanisms, these studies fail to propose mitigation strategies, such as developing vibration-damping materials or optimized battery enclosures. Railway applications, analyzed in [5,15,18], showed controlled thermal impacts (<1 °C) and capacity fade of up to 12%, with LFP batteries showing superior vibration resistance compared to NMC batteries.

However, actionable recommendations such as advanced anti-vibration mounts or structural reinforcements are not proposed, leaving a practical gap in improving long-term performance. Aerospace-specific research [10,18,24] showed that long-term vibration exposure resulted in minor capacity fade and structural deformation in 18650 cells. However, these studies lack real-time monitoring of impedance and temperature, which are critical for ensuring safety during satellite missions. Incorporating advanced sensing systems could significantly enhance reliability evaluations. Innovative fault detection methods, explored in [7,11,12], focused on abnormalities such as tab loosening and electrode misalignment. While these methods demonstrate high sensitivity, they are not scalable or validated in dynamic field conditions for large modules. Future work could focus on the scalability and field implementation of these techniques. Papers such as [18,19,76] highlight SEI layer thickening and increased charge transfer resistance under sinusoidal vibrations. However, they fail to establish direct correlations between these impedance changes and physical damage mechanisms, such as electrolyte decomposition or electrode fractures. Coupling EIS with imaging techniques (e.g., SEM, TEM) could provide a more comprehensive understanding of the degradation pathways.

7.2. Effects of Vibrations on Battery Thermal Behavior and Degradation

Vibrations were found to increase the thermal load within batteries, as evidenced by studies showing up to a 5 °C rise in surface temperature under vibrational stress. This thermal response exacerbates degradation, as higher temperatures accelerate aging mechanisms such as electrolyte decomposition and electrode damage. Additionally, the review found that vibrational stress amplifies the need for robust thermal management systems, especially for high-energy-demand applications. This insight aligns with investigating vibration effects on thermal behavior and highlights that managing heat in vibrating environments is critical to prolonging battery life. Still, a gap remains in identifying optimal thermal management solutions tailored to mitigate the dual challenges of thermal and vibrational stress. Studies examining the efficacy of phase change materials, enhanced cooling channels, or advanced TMSs under vibrations could provide valuable insights.

7.3. Compromises to Safety, Range, and Operational Reliability

The studies confirm that mechanical vibrations can compromise battery safety and reliability, directly impacting range and operational stability. For example, vibrations induce capacity fade and increased internal resistance, diminishing energy output and reducing the range of EVs. Additionally, vibration-induced thermal effects exacerbate safety concerns by increasing the chances of thermal runaway under extreme conditions. This aligns with determining if vibrations compromise battery range and safety. However, existing research primarily focuses on immediate impacts, while long-term studies examining the cumulative effects of low-level, chronic vibrations on battery range and reliability are limited. Addressing this research gap could enhance predictions of battery lifespan in practical applications. Table 5 summarizes the factors influencing battery performance under various conditions. Vibrations cause structural failures, increased resistance, and degradation, which are analyzed using simulations and CT scans and are critical for applications such as EVs and aerospace. High temperatures accelerate degradation and thermal runaway, necessitating advanced thermal management for safer EV design.

Table 5. Factors affecting battery performance: analysis, methods, and applications.

Factor Affecting Battery Performance	Discovery	Testing/Analysis Methods	Application/Implication
Vibrations	Cause battery degradation, internal structural failure, and increased resistance.	Real-world vibration simulations, MEMS-based monitoring, CT scans.	EVs, aerospace, and transportation applications.
Temperature	High temperatures accelerate degradation and risk of thermal runaway.	Thermal management systems, temperature monitoring.	It is essential in EV battery design to prevent overheating and improve safety.
Mechanical Stress	It can lead to short circuits, material displacement, and internal damage in cylindrical cells.	Six-degree-of-freedom vibration tests, mechanical abuse simulations.	Impacts EV, motorcycle, and aerospace batteries due to structural concerns.
Environmental Conditions (Humidity, Pressure)	They influence battery life and require robust seals in lithium-ion cells.	Environment-controlled tests.	Critical for durability in EVs and e-bike battery transport.
Over-discharge	High-frequency vibration signals can indicate over-discharge, a potential safety concern.	Vibration signal monitoring for pre-emptive diagnostics.	Useful in high-energy-density applications like EVs to detect faults early.
Battery Health Monitoring	Innovations like laser ultrasonics and FTIR for SOC and degradation tracking.	Non-contact SOC monitoring, FTIR for chemical stability.	Enhances safety and longevity, especially in high-stress applications.
Phase Change Materials (PCMs)	Vibration-enhanced PCM-based TMSs show potential for improved cooling.	PCM-based thermal management testing under vibrations.	Applicable in EVs for temperature control without high energy costs.
Dual-Motor Powertrains	Reduce strain on battery systems, enabling consistent power distribution.	Real-time performance assessment.	Optimizes energy usage in EVs, extending battery life.
Fourier Transform Infrared (FTIR)	Identifies chemical stability changes over time due to vibrations.	FTIR techniques for initial failure signs detection.	It is essential for detecting early degradation signs in high-stress setups.
Standards and Testing Protocols (IEC 62133-2) [29]	Establishes consistency across battery applications, especially for large battery assemblies.	UN-standardized vibration testing protocols.	Critical for safety across industries like automotive and aerospace.
Structural Integrity	CT scans reveal hidden structural failures in batteries under vibrations.	Computed tomography scans.	Strong internal design in cylindrical cells (e.g., 18650).

7.4. Review of Degradation Mechanisms Due to Vibrations

The studies comprehensively examine degradation mechanisms, including capacity fade, impedance increases, and structural damage under vibrational stress. Various techniques, such as EIS and incremental capacity analysis, provide valuable insights into these mechanisms. The review meets its objective by identifying specific degradation pathways, confirming that vibrational stress can lead to mechanical failures that, in turn, impair electrochemical performance. However, despite these insights, there is a gap in understanding how different vibration frequencies and intensities impact degradation. A more granular exploration into how varying vibration profiles influence degradation mechanisms across different battery chemistries could improve the design of vibration-resistant battery systems.

8. Recommendations for Future Research

As lithium-ion batteries continue to play a pivotal role in modern energy storage solutions across industries such as EVs, drones, aerospace, and renewable energy systems, there remains a demanding need to address the challenges posed by their performance, safety, and longevity under diverse operating conditions. Vibrations, temperature extremes, mechanical stress, and environmental influences significantly impact battery reliability and efficiency. While significant progress has been made, gaps in understanding and technological innovation persist. Future research must focus on developing advanced diagnostics, robust materials, and innovative designs to optimize battery performance, ensure safety, and meet the demands of next-generation applications. This section highlights key areas for further exploration to advance lithium-ion battery technology.

8.1. Real-World Application Testing

Existing research has yet to fully simulate real-world vibration profiles under various operational conditions. For example:

Electric Vehicles (EVs): Studies like [24,80] highlight the need to replicate high-frequency vibrations caused by uneven road surfaces and braking scenarios. Future studies should integrate vibration profiles with dynamic cycling to reflect actual EV operations.

Aerospace Vibration Profiles: Research in [10] shows insights into long-term structural impacts of vibrations in satellite missions. However, it lacks real-time monitoring of impedance or thermal responses. Future work could include combined low-pressure and vibrational stress scenarios.

Marine Applications: Low-amplitude marine vibrations were studied in [5,17], showing minimal thermal and performance impacts. However, the influence of salinity and long-term exposure to humidity remain unexplored.

Such real-world simulations would ensure battery designs meet the durability and safety requirements of automotive, aerospace, and marine environments.

8.2. Material-Level Diagnostics

Microstructural and chemical degradation mechanisms remain underexplored. For example:

Advanced Imaging Techniques: Incorporating scanning electron microscopy (SEM) and transmission electron microscopy (TEM), as used in [18,76], can help to visualize microcracks, cathode fractures, or separator deformation under vibrations.

Electrochemical Impedance Spectroscopy (EIS): Studies like [19,29] revealed increased charge transfer resistance under sinusoidal vibrations. However, coupling EIS with X-ray photoelectron spectroscopy (XPS) could better correlate impedance growth with SEI changes.

These advanced diagnostics would provide a comprehensive understanding of vibration-induced material degradation.

8.3. Vibration-Tolerant Battery Design

Enhanced battery designs are necessary to mitigate vibrational impacts. Examples include:

Vibration-Damping Materials: Studies like [16,76] suggested using elastomers or composites to absorb vibrational energy. However, further exploration is needed to optimize such materials for different frequency ranges.

Structural Optimization: As highlighted in [13,15], evenly distributing mechanical stress across battery modules reduces localized damage. Innovative designs could integrate optimized enclosures and anti-vibration mounts.

Phase Change Materials (PCMs): Research in [57,76] demonstrated that PCMs reduced temperature rises by 2–4 °C during vibration. Future work should focus on their long-term stability and effectiveness under combined cycling and vibrational stress.

Improved designs will enhance both performance and safety in vibration-prone applications.

8.4. Combined Thermal and Mechanical Stress Testing

Few studies simultaneous address thermal and mechanical stresses, although their coupled effects are critical in real-world conditions. For instance:

Simulating Coupled Stresses: Studies like [24,80] highlighted significant heat rises (3–6 °C) during cycling and vibration. Extending these studies to include high C-rate cycling or high-frequency vibrations could provide valuable insights.

Thermal Runaway Risks: High-temperature scenarios combined with vibrations, as discussed in [29,76], increase the risk of catastrophic failure. Real-time thermal and structural monitoring systems should be developed to capture early warning signs.

Understanding the interplay between thermal and mechanical stresses will guide safer battery operation and design.

8.5. Predictive Modeling and AI Integration

Machine learning can revolutionize vibration research. For example:

Predictive Models: AI-based modeling, as suggested in [12,80], can predict degradation patterns under vibrations, reducing testing time and costs.

Digital Twins: Virtual replicas of battery systems could simulate vibration and stress scenarios, identifying weak points without extensive physical testing.

Such tools will streamline research and development while improving battery reliability.

8.6. Addressing Long-Term Safety Concerns

Safety concerns from vibration-induced damage require more attention. For example:

Gas Generation Analysis: Research in [19,76] identified gas formation as a significant risk under high-frequency vibrations. Future studies should quantify gas buildup and its impact on thermal runaway risks.

Real-Time Monitoring: Real-time systems discussed in [8,29] could detect early signs of mechanical damage, such as tab loosening or impedance growth. Standardized protocols addressing these concerns will significantly enhance battery safety under vibrational stress.

8.7. Gaps and Future Directions

- Thermal Runaway Risks: Most studies fail to investigate vibration-induced thermal runaway under combined thermal and mechanical stress.
- Advanced Diagnostics: Limited use of advanced characterization techniques (e.g., SEM, TEM, XPS) to correlate microstructural changes with performance degradation.
- Real-World Scenarios: Few studies simulate real-world conditions (e.g., EV dynamics, aerospace vibration profiles), leaving applicability gaps.
- Mitigation Strategies: The research lacks proposals for effective vibration-damping materials or optimized cell designs for enhanced durability.

9. Conclusions

This review explores how vibrations affect lithium-ion batteries' performance, degradation, and thermal behavior, which is crucial for technologies such as EVs, aerospace systems, and marine applications. Vibrations can cause structural damage, such as electrode delamination and separator deformation, which increases internal resistance and generates hotspots. These effects make thermal management more challenging and accelerate battery aging. Vibrations also significantly contribute to dendrite formation, self-discharge, and lithium plating. These processes reduce the battery's capacity and lifespan and increase safety risks such as short circuits and thermal runaway. To address these challenges, there is a pressing need for more realistic testing protocols that replicate the combined effects of vibrations, temperature, and mechanical stress. Enhancing TMSs with advanced cooling techniques and materials, such as phase change solutions, can help to mitigate these issues. Additionally, designing batteries with vibration-resistant materials and improved structural integrity will be crucial for increasing their durability. Long-term studies focused on the cumulative effects of vibrations and their role in promoting dendrite growth, self-discharge, and lithium plating are essential for better predicting battery life and reliability. Building on this, our current research uses a multiscale physics-based modeling approach to explore how vibrations interact with thermal behavior and contribute to battery degradation. At the same time, we are developing an electromechanical system to analyze the impact of mechanical degradation on thermal and electrical performance. This work focuses on understanding how vibrational stresses influence key processes such as internal resistance rise, SEI layer growth, dendrite formation, and lithium plating. It also examines how they affect the battery's structural and thermal stability. By combining computational models with experimental data, this research aims to develop tools and strategies to improve lithium-ion batteries' safety, reliability, and lifespan in demanding environments.

Author Contributions: Conceptualization, K.S. and M.A.-G.; formal analysis, K.S.; Investigation, K.S.; writing—original draft preparation, K.S.; writing—review and editing, K.S., M.A.-G., and I.B.; Visualization, K.S. and M.A.-G.; Supervision, M.A.-G. and I.B. All authors have read and agreed to the published version of the manuscript.

Funding: This research received no external funding.

Conflicts of Interest: The authors declare no conflicts of interest.

References

1. Bajolle, H.; Lagadic, M.; Louvet, N. The future of lithium-ion batteries: Exploring expert conceptions, market trends, and price scenarios. *Energy Res. Soc. Sci.* **2022**, *93*, 102850. [CrossRef]
2. Bridge, G.; Faigen, E. Lithium, Brexit and Global Britain: Onshoring battery production networks in the UK. *Extr. Ind. Soc.* **2023**, *16*, 101328. [CrossRef]
3. Bridge, G.; Faigen, E. Towards the lithium-ion battery production network: Thinking beyond mineral supply chains. *Energy Res. Soc. Sci.* **2022**, *89*, 102659. [CrossRef]
4. Tingey, M.; Webb, J. *Net Zero Localities: Ambition & Value in UK Local Authority Investment*; University of Strathclyde Publishing: Glasgow, UK, 2020.
5. Wu, J.; Wu, Y.; Wang, L.; Ye, H.; Lu, J.; Li, Y. Challenges and Advances in Rechargeable Batteries for Extreme-Condition Applications. *Adv. Mater.* **2024**, *36*, 2308193. [CrossRef] [PubMed]
6. Wang, R.; Liu, G.; Wang, C.; Ji, Z.; Yu, Q. A comparative study on mechanical-electrical-thermal characteristics and failure mechanism of LFP/NMC/LTO batteries under mechanical abuse. *eTransportation* **2024**, *22*, 100359. [CrossRef]
7. Alkhedher, M.; Al Tahhan, A.B.; Yousaf, J.; Ghazal, M.; Shahbazian-Yassar, R.; Ramadan, M. Electrochemical and thermal modeling of lithium-ion batteries: A review of coupled approaches for improved thermal performance and safety lithium-ion batteries. *J. Energy Storage* **2024**, *86*, 111172. [CrossRef]
8. Krishna, T.; Kumar, S.V.S.V.P.D.; Rao, S.S.; Chang, L. Powering the Future: Advanced Battery Management Systems (BMS) for Electric Vehicles. *Energies* **2024**, *17*, 3360. [CrossRef]
9. Brand, M.J.; Schuster, S.F.; Bach, T.; Fleder, E.; Stelz, M.; Gläser, S.; Müller, J.; Sextl, G.; Jossen, A. Effects of vibrations and shocks on lithium-ion cells. *J. Power Sources* **2015**, *288*, 62–69. [CrossRef]
10. He, B.; Wang, H.; He, X. Vibration test methods and their experimental research on the performance of the lead-acid battery. *J. Power Sources* **2014**, *268*, 326–330. [CrossRef]

11. Cameron, Z.; Kulkarni, C.S.; Luna, A.G.; Goebel, K.; Poll, S. A battery certification testbed for small satellite missions. In Proceedings of the 2015 IEEE AUTOTESTCON, National Harbor, MD, USA, 2–5 November 2015.
12. Liu, G.; Li, C.; Ma, H.; Xu, J.; Chen, L.; Peng, X. Research progress and prospect of vibration signal of li-ion battery. In Proceedings of the 2020 12th IEEE PES Asia-Pacific Power and Energy Engineering Conference (APPEEC), Nanjing, China, 20–23 September 2020.
13. Huang, S.; Geng, Y.; Han, Y.; Gu, M. Design of shocks & vibrations monitoring system for long distance transportation in confined spaces. In Proceedings of the 2023 China Automation Congress (CAC), Chongqing, China, 17–19 November 2023.
14. JARI and Battery Transportation Working Group. Vibration Test for Large Lithium-ion Battery Assemblies on UN Transportation Manual of Tests & Criteria. 2009. Available online: https://www.prba.org/wp-content/uploads/JARI_presentation_UN_LIBWG_20090420-1.pdf (accessed on 13 January 2025).
15. Jung, S. Mathematical model of lithium-ion batteries with blended-electrode system. *J. Power Sources* **2014**, *264*, 184–194. [CrossRef]
16. Caposciutti, G.; Bandini, G.; Marracci, M.; Buffi, A.; Tellini, B. Analysis of SoH for lithium battery cells operating under vibration stress. In Proceedings of the 2020 IEEE 7th International Workshop on Metrology for AeroSpace (MetroAeroSpace), Pisa, Italy, 22–24 June 2020.
17. Eidinejad, H.; Madaro, F.; Brugo, T.M.; Rossi, C.; Rivola, A.; Troncossi, M.; Martini, A. Pre-compliance vibration testing of a LFP battery pack prototype for electric powertrains. In Proceedings of the 2024 IEEE International Workshop on Metrology for Automotive (MetroAutomotive), Bologna, Italy, 26–28 June 2024.
18. Berg, P.; Spielbauer, M.; Tillinger, M.; Merkel, M.; Schoenfuss, M.; Bohlen, O.; Jossen, A. Durability of lithium-ion 18650 cells under random vibration load concerning the inner cell design. *J. Energy Storage* **2020**, *31*, 101499. [CrossRef]
19. Bangal, O.A.; Chaturvedi, V.; Babu, P.K.A.; Shelke, M.V. Impedance analysis and equivalent circuit modelling of cells subjected to sinusoidal vibration test using electrochemical impedance spectroscopy. In Proceedings of the 2019 IEEE Transportation Electrification Conference (ITEC-India), Bengaluru, India, 17–19 December 2019.
20. Siddique, A.R.M.; Tasnim, S.; Mahmud, S.; Van Heyst, B. Thermal field investigation of lithium-ion battery with porous medium under vibration. *J. Energy Storage* **2022**, *47*, 103632. [CrossRef]
21. Rojas, O.E.; Khan, M.A. A review on electrical and mechanical performance parameters in lithium-ion battery packs. *J. Clean. Prod.* **2022**, *378*, 134381. [CrossRef]
22. Sampath, S.; Yin, X.; Tham, Z.W.; Chen, Y.F.; Zhang, L. Real-time and non-contact estimation of state of charge for lithium-ion battery using laser ultrasonics. *J. Power Sources* **2024**, *605*, 234544. [CrossRef]
23. Chen, Y.; Kang, Y.; Zhao, Y.; Wang, L.; Liu, J.; Li, Y.; Liang, Z.; He, X.; Li, X.; Tavajohi, N.; et al. A review of lithium-ion battery safety concerns: The issues, strategies, and testing standards. *J. Energy Chem.* **2021**, *59*, 83–99. [CrossRef]
24. Hooper, J.M.; Williams, D.; Roberts-Bee, K.; McGordon, A.; Whiffin, P.; Marco, J. Defining a vibration test profile for assessing the durability of electric motorcycle battery assemblies. *J. Power Sources* **2023**, *557*, 232541. [CrossRef]
25. Li, W.; Jiao, Z.; Xiao, Q.; Meng, J.; Mu, Y.; Jia, H.; Teodorescu, R.; Blaabjerg, F. A study on performance characterization considering six-degree-of-freedom vibration stress and aging stress for electric vehicle battery under driving conditions. *IEEE Access* **2019**, *7*, 112180–112190. [CrossRef]
26. Cai, S.; Zhang, X.; Ji, J. Recent advances in phase change materials-based battery thermal management systems for electric vehicles. *J. Energy Storage* **2023**, *72*, 108750. [CrossRef]
27. Tian, Y.; Wang, Z.; Ji, X.; Ma, L.; Zhang, L.; Hong, X.; Zhang, N. A concept dual-motor powertrain for battery electric vehicles: Principle, modeling and mode-shift. *Mech. Mach. Theory* **2023**, *185*, 105330. [CrossRef]
28. Lambert, S.M.; Armstrong, M.; Attidekou, P.S.; Christensen, P.A.; Widmer, J.D.; Wang, C.; Scott, K. Rapid nondestructive-testing technique for in-line quality control of Li-ion batteries. *IEEE Trans. Ind. Electron.* **2016**, *64*, 4017–4026. [CrossRef]
29. Savrum, M.; Poyrazoglu, G. Safety test setup installation for e-bike battery systems. In Proceedings of the 2022 4th Global Power, Energy and Communication Conference (GPECOM), Nevsehir, Turkey, 14–17 June 2022.
30. Liao, C.; Wang, J. Study on the influence of operating environment conditions on the performance of power lithium battery system. In Proceedings of the 2022 IEEE International Conference on Electrical Engineering, Big Data and Algorithms (EEBDA), Changchun, China, 25–27 February 2022.
31. Wang, F.; Gao, J. Feasibility analysis of new energy batteries equipped on conventional submarines. In Proceedings of the 2022 7th International Conference on Integrated Circuits and Microsystems (ICICM), Xi'an, China, 28–31 October 2022.
32. Zhao, X.; Xiong, Q.; Huang, X.; Zhang, J.; Li, J.; Ji, S. An overdischarge fault detection method for lithium-ion battery module based on vibration characteristics. In Proceedings of the 2023 IEEE 6th International Electrical and Energy Conference (CIEEC), Hefei, China, 12–14 May 2023.
33. Hua, X.; Thomas, A. Effect of dynamic loads and vibrations on lithium-ion batteries. *J. Low Freq. Noise Vib. Act. Control* **2021**, *40*, 1927–1934. [CrossRef]
34. Xu, L.; Xiao, Y.; Yu, Z.X.; Yang, Y.; Yan, C.; Huang, J.Q. Revisiting the Electrochemical Impedance Spectroscopy of Porous Electrodes in Li-ion Batteries by Employing Reference Electrode. *Angew. Chem.* **2024**, *136*, e202406054. [CrossRef]

35. Zhang, M.; Liu, Y.; Li, D.; Cui, X.; Wang, L.; Li, L.; Wang, K. Electrochemical impedance spectroscopy: A new chapter in the fast and accurate estimation of the state of health for lithium-ion batteries. *Energies* **2023**, *16*, 1599. [CrossRef]
36. Choi, D.; Shamim, N.; Crawford, A.; Huang, Q.; Vartanian, C.K.; Viswanathan, V.V.; Paiss, M.D.; Alam, M.J.E.; Reed, D.M.; Sprenkle, V.L. Li-ion battery technology for grid application. *J. Power Sources* **2021**, *511*, 230419. [CrossRef]
37. Zhao, J.; Lv, Z.; Li, D.; Feng, X.; Wang, Z.; Wu, Y.; Shi, D.; Fowler, M.; Burke, A.F. Battery engineering safety technologies (BEST): M5 framework of mechanisms, modes, metrics, modeling, and mitigation. *eTransportation* **2024**, *22*, 100364. [CrossRef]
38. Goswami, B.R.D.; Mastrogiorgio, M.; Ragone, M.; Jabbari, V.; Shahbazian-Yassar, R.; Mashayek, F.; Yurkiv, V. A combined multiphysics modeling and deep learning framework to predict thermal runaway in cylindrical Li-ion batteries. *J. Power Sources* **2024**, *595*, 234065. [CrossRef]
39. Turner, J.M. *Charged: A History of Batteries and Lessons for a Clean Energy Future*; University of Washington Press: Seattle, WA, USA, 2022.
40. Cai, C.; Gong, Y.; Fotouhi, A.; Auger, D.J. A novel hybrid electrochemical equivalent circuit model for online battery management systems. *J. Energy Storage* **2024**, *99*, 113142. [CrossRef]
41. Mehta, R.; Gupta, A. Mathematical modelling of electrochemical, thermal and degradation processes in lithium-ion cells—A comprehensive review. *Renew. Sustain. Energy Rev.* **2024**, *192*, 114264. [CrossRef]
42. Goswami, B.R.D.; Abdisobbouhi, Y.; Du, H.; Mashayek, F.; Kingston, T.A.; Yurkiv, V. Advancing battery safety: Integrating multiphysics and machine learning for thermal runaway prediction in lithium-ion battery module. *J. Power Sources* **2024**, *614*, 235015. [CrossRef]
43. Franco, A.A.; Vijay, U.; Saravanan, S. Combining physics-based modeling and artificial intelligence for the optimization of battery manufacturing processes. In Proceedings of the ASME 2024 19th International Manufacturing Science and Engineering Conference, Knoxville, TN, USA, 17–21 June 2024.
44. Hu, R.; Zhou, D.; Jia, Y.; Chen, Y.; Zhang, C. Dynamic mechanical behaviors of load-bearing battery structure upon low-velocity impact loading in electric vehicles. *eTransportation* **2024**, *21*, 100334. [CrossRef]
45. Lee, Y.K.; Song, J.; Park, J. Multi-scale coupled mechanical-electrochemical modeling for study on stress generation and its impact on multi-layered electrodes in lithium-ion batteries. *Electrochim. Acta* **2021**, *389*, 138682. [CrossRef]
46. Spotnitz, R.; Franklin, J. Abuse behavior of high-power, lithium-ion cells. *J. Power Sources* **2003**, *113*, 81–100. [CrossRef]
47. Mallarapu, A.; Kim, J.; Carney, K.; DuBois, P.; Santhanagopalan, S. Modeling extreme deformations in lithium ion batteries. *eTransportation* **2020**, *4*, 100065. [CrossRef]
48. Castro, M.T.; Del Rosario, J.A.D.; Chong, M.N.; Chuang, P.-Y.A.; Lee, J.; Ocon, J.D. Multiphysics modeling of lithium-ion, lead-acid, and vanadium redox flow batteries. *J. Energy Storage* **2021**, *42*, 102982. [CrossRef]
49. Li, W.; Cao, D.; Jöst, D.; Ringbeck, F.; Kuipers, M.; Frie, F.; Sauer, D.U. Parameter sensitivity analysis of electrochemical model-based battery management systems for lithium-ion batteries. *Appl. Energy* **2020**, *269*, 115104. [CrossRef]
50. Di Mauro, G.; Guida, M.; Gomez, L.M.; Olivares, G. Multiphysics modeling for safe batteries using LS-dyna. In Proceedings of the ICAS Conference, Florence, Italy, 9–13 September 2024.
51. Zheng, X.; Cai, Z.; Sun, J.; He, J.; Rao, W.; Wang, J.; Zhang, Y.; Gao, Q.; Han, B.; Xia, K.; et al. Nickel-rich layered oxide cathodes for lithium-ion batteries: Failure mechanisms and modification strategies. *J. Energy Storage* **2023**, *58*, 106405. [CrossRef]
52. Placke, T.; Kloepsch, R.; Dühnen, S.; Winter, M. Lithium ion, lithium metal, and alternative rechargeable battery technologies: The odyssey for high energy density. *J. Solid State Electrochem.* **2017**, *21*, 1939–1964. [CrossRef]
53. Rejeb, A.; Abdollahi, A.; Rejeb, K.; Treiblmaier, H. Drones in agriculture: A review and bibliometric analysis. *Comput. Electron. Agric.* **2022**, *198*, 107017. [CrossRef]
54. Niu, H.; Zhang, N.; Lu, Y.; Zhang, Z.; Li, M.; Liu, J.; Song, W.; Zhao, Y.; Miao, Z. Strategies toward developing high-energy-density lithium batteries. *J. Energy Storage* **2024**, *88*, 111666. [CrossRef]
55. Wang, T.; Wang, Z.; Li, H.; Cheng, L.; Wu, Y.; Liu, X.; Meng, L.; Zhang, Y.; Jiang, S. Recent status, key strategies, and challenging prospects for fast charging silicon-based anodes for lithium-ion batteries. *Carbon* **2024**, *230*, 119615. [CrossRef]

56. Machín, A.; Morant, C.; Márquez, F. Advancements and challenges in solid-state battery technology: An in-depth review of solid electrolytes and anode innovations. *Batteries* **2024**, *10*, 29. [CrossRef]
57. Battery University. BU-1003a: Battery Aging in an Electric Vehicle (EV). 2021. Available online: https://batteryuniversity.com/article/bu-1003a-battery-aging-in-an-electric-vehicle-ev (accessed on 13 January 2025).
58. Chombo, P.V.; Laoonual, Y. A review of safety strategies of a Li-ion battery. *J. Power Sources* **2020**, *478*, 228649. [CrossRef]
59. Wang, K.; Ouyang, D.; Qian, X.; Yuan, S.; Chang, C.; Zhang, J.; Liu, Y. Early warning method and fire extinguishing technology of lithium-ion battery thermal runaway: A review. *Energies* **2023**, *16*, 2960. [CrossRef]
60. Osmani, K.; Alkhedher, M.; Ramadan, M.; Choi, D.S.; Li, L.K.; Doranehgard, M.H.; Olabi, A.-G. Recent progress in the thermal management of lithium-ion batteries. *J. Clean. Prod.* **2023**, *389*, 136024. [CrossRef]
61. Kucinskis, G.; Bozorgchenani, M.; Feinauer, M.; Kasper, M.; Wohlfahrt-Mehrens, M.; Waldmann, T. Arrhenius plots for Li-ion battery ageing as a function of temperature, C-rate, and ageing state—An experimental study. *J. Power Sources* **2022**, *549*, 232129. [CrossRef]
62. Temperature Effects on Battery Capacity and Service Life. Available online: https://www.prostarsolar.net/article/temperature-effects-on-battery-capacity-and-service-life.html (accessed on 13 March 2019).
63. Molaiyan, P.; Bhattacharyya, S.; dos Reis, G.S.; Sliz, R.; Paolella, A.; Lassi, U. Towards greener batteries: Sustainable components and materials for next-generation batteries. *Green Chem.* **2024**, *26*, 7508–7531. [CrossRef]
64. Esmaeili, M.H.; Haft-Cheshmeh, Y.M. Dynamic characteristics of elastomeric materials used as railway vibration mitigation measures considering the effect of shape factor. *Measurement* **2024**, *236*, 115058. [CrossRef]
65. Wu, G.; Chen, S.; Zeng, S. Effects of mechanical vibration on melting behaviour of phase change material during charging process. *Appl. Therm. Eng.* **2021**, *192*, 116914. [CrossRef]
66. Gao, P.; Xiang, C.; Liu, H.; Walker, P.; Zhang, N. Design of the frequency tuning scheme for a semi-active vibration absorber. *Mech. Mach. Theory* **2019**, *140*, 641–653. [CrossRef]
67. Zhou, W.; Mohammed, H.I.; Chen, S.; Luo, M.; Wu, Y. Effects of mechanical vibration on the heat transfer performance of shell-and-tube latent heat thermal storage units during charging process. *Appl. Therm. Eng.* **2022**, *216*, 119133. [CrossRef]
68. Zhou, Z.; Chen, S.; Luo, M.; Du, W.; Wu, Y.; Yu, Y. Effect of mechanical vibration on phase change material based thermal management system for a cylindrical lithium-ion battery at high ambient temperature and discharge rate. *Int. J. Heat Mass Transf.* **2023**, *211*, 124255. [CrossRef]
69. Shahjalal, M.; Shams, T.; Islam, E.; Alam, W.; Modak, M.; Bin Hossain, S.; Ramadesigan, V.; Ahmed, R.; Ahmed, H.; Iqbal, A. A review of thermal management for Li-ion batteries: Prospects, challenges, and issues. *J. Energy Storage* **2021**, *39*, 102518. [CrossRef]
70. Ma, R.; Liu, J.; Wang, S.; Rao, Z.; Cai, Y.; Wu, W. Progress on thermal runaway propagation characteristics and prevention strategies of lithium-ion batteries. *Chin. Sci. Bull.* **2021**, *66*, 2991–3004. [CrossRef]
71. Du, W.; Chen, S. Effect of mechanical vibration on phase change material based thermal management module of a lithium-ion battery at high ambient temperature. *J. Energy Storage* **2023**, *59*, 106465. [CrossRef]
72. Murakami, Y.; Takagi, T.; Wada, K.; Matsunaga, H. Essential structure of S-N curve: Prediction of fatigue life and fatigue limit of defective materials and nature of scatter. *Int. J. Fatigue* **2021**, *146*, 106138. [CrossRef]
73. Zhang, G.; Wei, X.; Chen, S.; Zhu, J.; Han, G.; Dai, H. Revealing the impact of slight electrical abuse on the thermal safety characteristics for lithium-ion batteries. *ACS Appl. Energy Mater.* **2021**, *4*, 12858–12870. [CrossRef]
74. Huang, R.; Wei, G.; Zhou, X.; Zhu, J.; Pan, X.; Wang, X.; Jiang, B.; Wei, X.; Dai, H. Targeting the low-temperature performance degradation of lithium-ion batteries: A non-destructive bidirectional pulse current heating framework. *Energy Storage Mater.* **2024**, *65*, 103173. [CrossRef]
75. Tomaszewska, A.; Chu, Z.; Feng, X.; O'Kane, S.; Liu, X.; Chen, J.; Ji, C.; Endler, E.; Li, R.; Liu, L.; et al. Lithium-ion battery fast charging: A review. *eTransportation* **2019**, *1*, 100011. [CrossRef]
76. Diao, W.; Xing, Y.; Saxena, S.; Pecht, M. Evaluation of present accelerated temperature testing and modeling of batteries. *Appl. Sci.* **2018**, *8*, 1786. [CrossRef]
77. Ma, S.; Jiang, M.; Tao, P.; Song, C.; Wu, J.; Wang, J.; Deng, T.; Shang, W. Temperature effect and thermal impact in lithium-ion batteries: A review. *Prog. Nat. Sci. Mater. Int.* **2018**, *28*, 653–666. [CrossRef]
78. Lyu, P.; Liu, X.; Qu, J.; Zhao, J.; Huo, Y.; Qu, Z.; Rao, Z. Recent advances of thermal safety of lithium ion battery for energy storage. *Energy Storage Mater.* **2020**, *31*, 195–220. [CrossRef]
79. Zhang, Q.; Li, P.; Liu, C.; Wei, F.; Wang, M.; Li, J.; Zhu, S.; Shao, G.; Mao, J. Heat generation and temperature rise characteristics of single overcharged lithium-ion batteries. *J. Electrochem. Soc.* **2022**, *169*, 060502. [CrossRef]
80. Capron, O.; Jaguemont, J.; Gopalakrishnan, R.; Bossche, P.V.D.; Omar, N.; Van Mierlo, J. Impact of the temperature in the evaluation of battery performances during long-term cycling—Characterisation and modelling. *Appl. Sci.* **2018**, *8*, 1364. [CrossRef]

81. Stadler, R.; Maurer, A. Methods for durability testing and lifetime estimation of thermal interface materials in batteries. *Batteries* **2019**, *5*, 34. [CrossRef]
82. Guo, J.; Li, Y.; Pedersen, K.; Stroe, D.-I. Lithium-ion battery operation, degradation, and aging mechanism in electric vehicles: An overview. *Energies* **2021**, *14*, 5220. [CrossRef]
83. Rahman, T.; Alharbi, T. Exploring lithium-Ion battery degradation: A concise review of critical factors, impacts, data-driven degradation estimation techniques, and sustainable directions for energy storage systems. *Batteries* **2024**, *10*, 220. [CrossRef]
84. Menye, J.S.; Camara, M.; Dakyo, B. Lithium Battery Degradation and Failure Mechanisms: A State-of-the-Art Review. *Energies* **2025**, *18*, 342. [CrossRef]
85. Shchurov, N.I.; Dedov, S.I.; Malozyomov, B.V.; Shtang, A.A.; Martyushev, N.V.; Klyuev, R.V.; Andriashin, S.N. Degradation of lithium-ion batteries in an electric transport complex. *Energies* **2021**, *14*, 8072. [CrossRef]
86. Sun, S.; Zhang, X.; Wang, Y.; Li, J.; Zheng, Z.; Huang, J. Understanding the transport mechanism of lithium ions in solid-electrolyte interphase in lithium metal batteries with liquid electrolytes. *Mater. Today* **2024**, *77*, 39–65. [CrossRef]
87. Jia, K.; Yang, G.; He, Y.; Cao, Z.; Gao, J.; Zhao, H.; Piao, Z.; Wang, J.; Abdelkader, A.M.; Liang, Z.; et al. Degradation Mechanisms of Electrodes Promotes Direct Regeneration of Spent Li-Ion Batteries: A Review. *Adv. Mater.* **2024**, *36*, 2313273. [CrossRef]
88. Gao, J.; Wang, S.; Hao, F. A Review of Non-Destructive Testing for Lithium Batteries. *Energies* **2024**, *17*, 4030. [CrossRef]
89. Hu, C.; Geng, M.; Yang, H.; Fan, M.; Sun, Z.; Yu, R.; Wei, B. A Review of Capacity Fade Mechanism and Promotion Strategies for Lithium Iron Phosphate Batteries. *Coatings* **2024**, *14*, 832. [CrossRef]
90. Zhang, L.; Ning, Z.; Peng, H.; Mu, Z.; Sun, C. Effects of vibration on the electrical performance of lithium-ion cells based on mathematical statistics. *Appl. Sci.* **2017**, *7*, 802. [CrossRef]
91. Hu, X.; Gao, F.; Xiao, Y.; Wang, D.; Gao, Z.; Huang, Z.; Ren, S.; Jiang, N.; Wu, S. Advancements in the safety of Lithium-Ion Battery: The Trigger, consequence and mitigation method of thermal runaway. *Chem. Eng. J.* **2024**, *481*, 148450. [CrossRef]
92. Jung, J.; Lim, M.-G.; Kim, J.-Y.; Han, B.-G.; Kim, B.; Rho, D.-S. Safety assessment for external short circuit of Li-ion battery in ESS application based on operation and environment factors. *Energies* **2022**, *15*, 5052. [CrossRef]
93. Matsumoto, F.; Fukunishi, M. Review of Current Collector-, Binder-, Conductive Additive-Free, and Freestanding Electrodes in Lithium and Related Batteries. *Batteries* **2024**, *10*, 330. [CrossRef]
94. Zhang, G.; Wei, X.; Zhu, J.; Chen, S.; Han, G.; Dai, H. Revealing the failure mechanisms of lithium-ion batteries during dynamic overcharge. *J. Power Sources* **2022**, *543*, 231867. [CrossRef]
95. Cortada-Torbellino, M.; Elvira, D.G.; El Aroudi, A.; Valderrama-Blavi, H. Review of Lithium-Ion Battery Internal Changes Due to Mechanical Loading. *Batteries* **2024**, *10*, 258. [CrossRef]
96. Wang, Z.; Zhao, Q.; Yu, X.; An, W.; Shi, B. Impacts of vibration and cycling on electrochemical characteristics of batteries. *J. Power Sources* **2024**, *601*, 234274. [CrossRef]
97. Xiong, R.; Yu, Y.; Chen, S.; Li, M.; Li, L.; Zhou, M.; Zhang, W.; Yan, B.; Li, D.; Yang, H.; et al. Overpotential decomposition enabled decoupling of complex kinetic processes in battery electrodes. *J. Power Sources* **2022**, *553*, 232296. [CrossRef]
98. Heinzen, T.; Plaumann, B.; Kaatz, M. Influences on Vibration Load Testing Levels for BEV Automotive Battery Packs. *Vehicles* **2023**, *5*, 446–463. [CrossRef]
99. Liu, S.; Zhang, G.; Wang, C. Challenges and Innovations of Lithium-Ion Battery Thermal Management Under Extreme Conditions: A Review. *ASME J. Heat Mass Transf.* **2024**, *145*, 080801. [CrossRef]
100. Birkl, C.R.; Roberts, M.R.; McTurk, E.; Bruce, P.G.; Howey, D.A. Degradation diagnostics for lithium ion cells. *J. Power Sources* **2017**, *341*, 373–386. [CrossRef]
101. Li, Y.; Guo, W.; Stroe, D.-I.; Zhao, H.; Kristensen, P.K.; Jensen, L.R.; Pedersen, K.; Gurevich, L. Evolution of aging mechanisms and performance degradation of lithium-ion battery from moderate to severe capacity loss scenarios. *Chem. Eng. J.* **2024**, *498*, 155588. [CrossRef]
102. Ramasubramanian, B.; Reddy, M.V.; Zaghib, K.; Armand, M.; Ramakrishna, S. Growth mechanism of micro/nano metal dendrites and cumulative strategies for countering its impacts in metal ion batteries: A review. *Nanomaterials* **2021**, *11*, 2476. [CrossRef]
103. Soyoye, B.D. State of Charge and State of Health Estimation in Electric Vehicles: Challenges, Approaches and Future Directions. *Batteries* **2025**, *11*, 32. [CrossRef]
104. Schlasza, C.; Ostertag, P.; Chrenko, D.; Kriesten, R.; Bouquain, D. Review on the aging mechanisms in li-ion batteries for electric vehicles based on the FMEA method. In Proceedings of the 2014 IEEE Transportation Electrification Conference and Expo (ITEC), Dearborn, MI, USA, 15–18 June 2014.
105. Júnior, C.A.R.; Sanseverino, E.R.; Gallo, P.; Amaral, M.M.; Koch, D.; Kotak, Y.; Diel, S.; Walter, G.; Schweiger, H.-G.; Zanin, H. Unraveling the Degradation Mechanisms of Lithium-Ion Batteries. *Energies* **2024**, *17*, 3372. [CrossRef]
106. Tan, J.; Chen, X.; Bu, Y.; Wang, F.; Wang, J.; Huang, X.; Hu, Z.; Liu, L.; Lin, C.; Meng, C.; et al. Incorporating FFTA based safety assessment of lithium-ion battery energy storage systems in multi-objective optimization for integrated energy systems. *Appl. Energy* **2024**, *367*, 123472. [CrossRef]

107. An, Z.; Shi, T.; Du, X.; An, X.; Zhang, D.; Bai, J. Experimental study on the internal short circuit and failure mechanism of lithium-ion batteries under mechanical abuse conditions. *J. Energy Storage* **2024**, *89*, 111819. [CrossRef]
108. Xin, K.; Lee, G. Effect of Cell-to-Cell Internal Resistance Variations on the Thermal Performance of Lithium-Ion Batteries for Urban Air Mobility. *World Electr. Veh. J.* **2024**, *15*, 423. [CrossRef]

batteries

MDPI

Article

Comparative Study of Thermal Runaway Propagation and Material Barrier Effect of Lithium-Ion Batteries

Yikai Mao †, Yaoyu Chen †, Yanglin Ye, Yin Chen * and Mingyi Chen *

School of the Environment and Safety Engineering, Jiangsu University, Zhenjiang 212013, China; 2222309078@stmail.ujs.edu.cn (Y.M.); 3220904013@stmail.ujs.edu.cn (Y.C.); 2222309080@stmail.ujs.edu.cn (Y.Y.)

* Correspondence: chenyin@ujs.edu.cn (Y.C.); chenmy@ujs.edu.cn (M.C.)

† These authors contributed equally to this work.

Abstract: Battery thermal runaway (TR) is usually accompanied by a large amount of heat release, as well as a jet of flame. This not only causes harm to the surrounding environment but even exacerbates thermal runaway propagation (TRP). At this stage, many types of materials are used to suppress TRP, and people tend to focus on improving one characteristic of the material while ignoring other properties of the material. This may leave potential pitfalls for TRP suppression, suggesting the need to study multiple properties of multiple materials. In order to better weigh the advantages and disadvantages of different types of materials when suppressing TRP, we compared three typical materials for suppressing TRP behavior in lithium-ion batteries (LIBs). These materials are phase change materials (PCM), ceramic fibers, and glass fibers. They are all available in two different thicknesses, 2 mm and 3 mm. The experiments started with a comparative analysis of the TR experimental phenomena in the presence of the different materials. Then, the temperature and mass loss of the battery module during TR were analyzed separately and comparatively. The 3 mm glass fiber showed the best inhibition effect, which extended the TR interval between cells 1 and 2 to 894 s and successfully inhibited the TR of cell 3. Compared with the blank group, the total mass loss decreased from 194.3 g to 182.2 g, which is a 6.2% reduction. Subsequently, we comprehensively analyzed the performance of the three materials in suppressing TRP by combining their suppressing mechanisms. The experimental results show that glass fiber has the best effect in suppressing TRP due to its excellent thermal insulation and mechanical properties. This study may provide new insights into how to trade-off material properties for TRP suppression in the future.

Keywords: lithium-ion battery; thermal runaway; overheating; comparative analysis

1. Introduction

In recent years, lithium-ion batteries (LIBs) have been widely used in electric vehicles (EVs) and energy storage plants due to their high energy density, environmental friendliness, and long life [1–3]. However, thermal runaway (TR) can occur in batteries for a variety of reasons, and TR can create thermal runaway propagation (TRP) in battery modules [4,5]. The flames and toxic gases produced during the TRP process are extremely hazardous to the surrounding environment [6]. Therefore, it is necessary to improve the safety of battery modules by adding materials that can suppress TRP between battery modules when carrying out the design of battery modules.

There are many causes of battery TR, including thermal, electrical, and mechanical abuse [7]. Therefore, it is better to prevent these three dimensions when we choose materials for the design of the battery module. From the thermal point of view, continuous and discontinuous heating affects the behavior of the battery TR. Zhou et al. [8] investigated the effects of continuous and discontinuous heating on battery TR. They found that continuous heating leads to a faster accumulation of heat inside the battery, which makes the battery TR more drastic, emphasizing the importance of heat dissipation and insulation in the battery module. From an electrical point of view, once an internal short circuit occurs in a battery, it is highly likely to cause TR [9,10]. This emphasizes the importance of the insulating properties of the material. Many times, batteries are short-circuited from external shocks [11]. Zhang et al. [12] illustrated the relationship between the pinning depth and the TR due to a short circuit inside the battery. The results, in conjunction with the battery structure, point out that the faster the pinning speed is, the more the battery is damaged and the higher the TRP rate is. Therefore, the mechanical properties of the material must also be taken into account. There is a lot of existing research on materials to suppress TRP, which will be further described below.

There are many studies on mitigating thermal behavior of LIBs under TR conditions by using phase change materials (PCMs). PCMs, with an excellent ability of absorbing heat, are widely used in the field of energy storage [13]. Chen et al. [14] conducted research on suppressing TR using different kinds of PCMs. Their findings reveal that, compared with natural convection cooling, PCM cooling can effectively decrease a battery's surface temperature. The combination of aerogel felt (AEGF) and PCM can greatly mitigate TR behavior. Unlike the mechanism of PCM on suppressing TR, the AEGF is remarkable for its good thermal insulation characteristic [15]. Thus, plenty of scholars put thermal insulation materials, like AEGF, into use to suppress TR. Niu et al. [16] prepared ultra-light hollow glass microsphere (HGM) plates with low thermal conductivity to suppress TRP. It was found that the plates, characterized by low thermal conductivity and excellent fireproof performance, can slow down the TRP rate, thereby enhancing the safety of high-energy battery systems. Zou et al. [17] investigated the TRP behavior of battery modules under conditions with three types of thermal insulation materials and without insulation materials. Thet found that all of the materials suppressed TRP, although they did not inhibit TRP completely. By analyzing the thermophysical properties of these materials, it was revealed that the thermal diffusion coefficient and maximum service temperature are critical parameters for the selection of thermal insulation materials. Some research has considered various factors to prepare barrier material. Bausch et al. [18] presented an eco-friendly endothermic barrier, characterized for its low conductivity and high compressive strength, to suppress TRP. The study shows that the thermal barrier material can not only delay TRP but also can avoid damage to the cells adjacent to the cell processing TR. However, at this stage, there are very limited comparative studies on TRP suppression by differently characterized materials. Based on this, we will carry out a comparative research work on TRP suppression with different materials and elaborate the mechanisms behind it.

This study addresses a key research gap by studying the suppression effect of three typical materials with advantages and disadvantages, respectively. With the analysis of battery surface temperature, TRP interval, and absolute mass loss, we deeply clarify the mechanism of these materials' functions. After analyzing the mechanism by which these materials suppress TRP, we used a radar chart to elaborate, in detail, how to weigh the advantages and disadvantages of materials that suppress TRP from multiple indicators. It is demonstrated that glass fiber has the best ability to suppress TRP, as it not only has a good command of thermal insulation but is also remarkable for its excellent mechanical properties. By comparing the differences underlying TRP suppression while systematically examining the critical balance among various excellent material properties, this paper provides valuable insights for the selection of materials for battery thermal management systems.

2. Methodology

2.1. Battery Sample

The battery sample chosen for this experiment is a square pouch battery with a cathode made of $LiNi_xCo_yMn_{1-x-y}O_2$ and an anode composed of graphite. The battery is designed to be able to cycle over 1000 times, and the electrolyte of the battery consists of lithium salts and a polymer matrix. The nominal battery voltage is 3.7 V, the charge cut-off voltage is 4.2 V, and the discharge limit voltage is 2.8 V. Its overall dimensions are $132 \times 65 \times 13.5$ mm. The physical diagram of the battery is as shown in Figure 1.

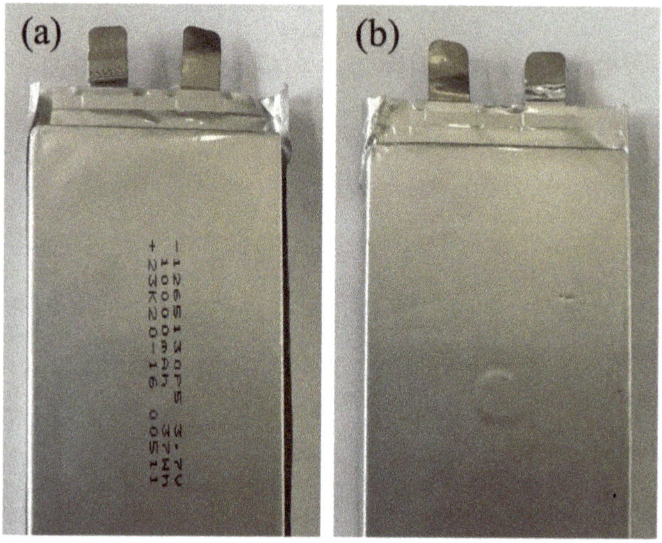

Figure 1. Physical battery diagram: (**a**) front view, (**b**) reverse view.

2.2. Material Characterization

2.2.1. PCM Sample

The PCM material used in this study is a new type of solid–solid phase change material (SSPCM), aiming to achieve strong mechanical properties and leak resistance. Hexamethylene diisocyanate (HDI) and polyethylene glycol (PEG) were used as precursors. Meanwhile, a hybrid heat transfer network was designed, including carbon nanotubes (CNT), expanded graphite (EG), and advanced ceramic materials, such as boron nitride (BN) and silicon carbide (SiC). HDI and a dibutyltin dilaurate (DBTDL) catalyst were purchased from Aikang Biotechnology Development Co., Ltd. (Zhenjiang, China). PEG and CNT (specification: 98 wt%, outer diameter 20 nm, length 10–30 μm) were purchased from Heowns Opde Technologies Co., Ltd. (Tianjin, China). H-BN (specification: 99 wt%, 1–2 μm) was purchased from Merck Biochemical Co., Ltd. (Shanghai, China). EG (purity: 99 wt%) was purchased from Tengshengda Carbon Graphite Co., Ltd. (Qingdao, China). SiC (specification: 99 wt%, 1–3 μm) was purchased from Dewei Chemical Co., Ltd. (Zhenjiang, China). The specific synthesis process can be found in a previous work [19].

2.2.2. Ceramic Fiber Sample

The ceramic fiber was purchased from Hebei Hualuo Sealing Materials Co., Ltd. (Cangzhou, China). The main component of the ceramic fiber is aluminum silicate (with an alumina content of 45–46% and a silica content of 51–52%). It also contains a small amount of zirconia. Its bulk density is approximately 200 kg/m^3, and the thermal conductivity is about 0.12 W/mK.

2.2.3. Glass Fiber Sample

The glass fiber material was purchased from Hebei Caike Thermal Insulation Materials Co., Ltd. (Langfang, China). The glass fiber is a non-woven fabric composed of high-purity silica (SiO_2, ~52–54 wt%), alumina (Al_2O_3, ~14–16 wt%), calcium oxide (CaO, ~17–19 wt%), and trace amounts of magnesium oxide (MgO) and boron oxide (B_2O_3). The fibers exhibit an average diameter of 8–12 μm.

2.3. Experimental Setup

The layout of the experiment is shown in Figure 2a. An electronic balance with 0.1 g accuracy was used to record the mass change during the TR process of the battery module, and a heat baffle between it and the battery module was used to protect the electronic balance. A high-definition video camera was placed at the front side of the TR battery module to record the experimental phenomena during the TR process. The exact layout of the battery module is shown in Figure 2b, with the corresponding TRP-suppressing material sandwiched between the cells. A heater was used to heat the first cell to TR. The heater was separated from the sandwich plate by a heat shield. Thermocouples (accuracy: 0.5 °C, temperature measurement range: 0–1300 °C) were used to record the surface temperatures at the center of the cells on the left and right sides, and the data were transferred to a computer via an ICPCON 7018 module. The TRP test was conducted in winter, when the ambient temperature was approximately (10 ± 5) °C. The air pressure was about one standard atmosphere. The battery modules were arranged horizontally, with the heating plate tightly attached to the leftmost battery's side. The heating plate was turned off immediately after the battery was heated to TR. During the test, the weather was either sunny or cloudy, and the battery was kept dry to prevent accidental short circuits.

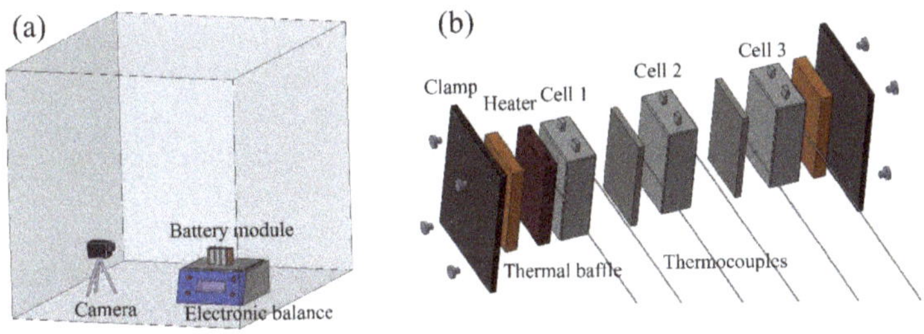

Figure 2. Layout of the experiment: (**a**) overall schematic; (**b**) specific layout in TR test.

2.4. Experimental Arrangement

The materials used to suppress TRP in the experiment were PCMs, glass fiber, and ceramic fiber. They are composed of two thicknesses, 2 mm and 3 mm. A multifunctional thermal conductivity tester (Xiangtan Dra-III type, accuracy: ±3%) (Dre-III, Xiangtan Xiangyi Instruments Co., Ltd., Xiangtan, China) was used to measure the thermal conductivity of the PCM, which is around 0.61 W/m·K. A Q500 thermogravimetric analyzer (TGA, Q500, TA Instruments Inc., Shanghai, China) was used to analyze the thermal stability of the PCM. Details can be found in the previous work of our subject group [19]. The physical properties of the glass and ceramic fibers are provided by the merchant. Their thermal conductivity is around 0.018 W/m·K and 0.12 W/m·K, respectively. The specific arrangement of the experiment is shown in Table 1, which is divided into a blank group and six other groups, with different thicknesses and materials. The purpose of this arrangement is to better compare the effect of TRP suppression between different materials. By setting up different thicknesses of the same materials, the internal mechanism of the corresponding materials to suppress TRP can be better explored.

Table 1. Specific arrangements of experiment.

Test	Materials	Thickness/mm
1	Blank	\
2	PCM	2
3	PCM	3
4	Ceramic fiber	2
5	Ceramic fiber	3
6	Glass fiber	2
7	Glass fiber	3

3. Experimental Results and Analyses

3.1. Comparative Analysis of TR Behavior Through Experimental Phenomenon

The three materials selected in the experiment all have the potential to suppress TRP, and their mechanisms for suppressing TRP are not completely the same. For the PCM, during the phase change process, it can absorb a large amount of heat, effectively suppressing the rise in battery temperature and thus achieving the purpose of suppressing TRP. Ceramic fiber is an excellent thermal insulation material. It can effectively prevent the heat of the TR battery from spreading to the surrounding batteries, thereby suppressing TRP. Glass fiber also has good thermal insulation performance. At the same time, due to its high porosity, while suppressing heat conduction, its structure with high porosity restrains the gas flow, and the low thermal conductivity reduces the heat transfer in the gas, effectively suppressing the occurrence of thermal convection. All these characteristics endow glass fiber with the potential to suppress TRP. In order to compare the effect of TRP suppression by the three materials more intuitively, we compared the experimental phenomena during the TRP suppression process. Figure 3 shows the typical experimental phenomena of cell 1, cell 2, and cell 3 during the TRP process of the blank group, as well as the three materials of 3 mm, respectively. It can be clearly observed that only the 3 mm glass fiber successfully inhibited the TRP of cell 3. The other three groups did not inhibit the TRP of cell 3. Overall, there was no significant difference in the flame behavior of the cells, except for cell 3 of the 3 mm glass fiber set. This is because the three materials themselves cannot directly act on the flame; they can only play a role as a flame retardant but cannot mitigate the intensity of the flame directly. A PCM can absorb heat and dissipate heat in the process of phase change, but due to its inability to insulate the heat, it is unable to successfully inhibit the TRP at a thickness of 3 mm. Ceramic fibers and glass fibers have a good ability to insulate heat. However, from the experimental phenomenon, compared to the glass fiber, the ceramic fiber failed to inhibit the violent flame jet of cell 3 as successfully as the glass fiber. The success of the glass fiber in inhibiting the TR of cell 3 may be attributed to the better mechanical and insulating properties of glass fiber. This allowed the glass fiber to not only act as a thermal insulator but also provide some cushioning and insulation during the expansion process of the pouch cells. This avoided the occurrence of short-circuiting between neighboring cells, which prevented the TR of cell 3 from bursting into flames, resulting in only a small fire emanating from cell 3.

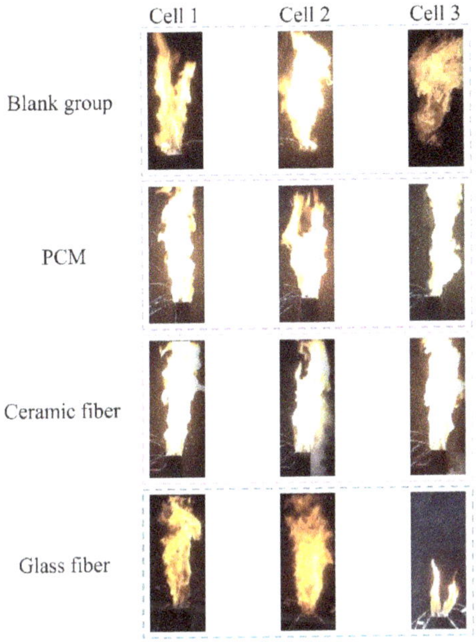

Figure 3. Experimental phenomena of TR under blank group and different material suppression.

3.2. Comparative Analysis of TR Behavior by Battery Temperature Characteristic

The temperature–time images of the seven groups of the TRP tests are shown in Figure 4. First of all, compared to the blank group, regardless of the PCM, ceramic fiber or glass fiber, they all effectively reduced the maximum temperature of the cell surface and prolonged the TRP interval. It is noteworthy that only the 3 mm glass fibers successfully inhibited the TRP of cell 3. Considering the material thickness dimension, it is obvious that, regardless of the material, the 3 mm material further prolonged the interval between the TRP of the cell compared to the 2 mm one and further reduced the maximum temperature of the cell surface. This is attributed to better heat dissipation and insulation, respectively. For the PCM, except for the material at 3 mm, it better suppressed the temperature rise of the battery compared to the insulating material, ceramic fiber. However, the total time it suppressed TRP was not as good as the ceramic fiber in both thickness cases. It can be concluded that although PCM material has certain heat absorption and dissipation ability, it can be well used in the thermal management of battery modules. However, in suppressing TRP, PCM alone does not suppress heat diffusion well. For glass fiber, with a thickness of 3 mm, it successfully blocks the TR of cell 3. Glass fiber has a strong balance of mechanical strength and flexibility, and its tensile strength is much greater than aluminum silicate ceramic fiber [20,21]. As a thermal insulation material, it has better insulation and mechanical properties than ceramic fiber, which may be the reason why it is more prominent than ceramic fiber in suppressing TRP. Glass fiber has good thermal and mechanical properties. Many scholars use it to improve the thermal and mechanical properties of materials. Chen et al. [22] incorporated glass fibers into graphene oxide, and the results show that glass fibers play an important role in enhancing strength and toughness. Feng et al. [23] used glass fibers to improve the mechanical and thermal properties of thermal insulation materials. The results show that the glass fibers effectively dispersed the external stresses of the material and significantly improved the thermal insulation ability of the material at high temperatures. The PCM used in this paper is a PCM with a low leakage rate and high mechanical strength, synthesized in the previous work of our group [19]. This PCM was synthesized via a polyurethane crosslinking reaction, using polyethylene glycol (PEG) and hexamethylene diisocyanate (HDI) as precursors. This approach chemically integrates the phase change molecular chains into the thermoset polymer matrix via PEG, enhancing the thermal stability and mechanical properties of the material while effectively mitigating the leakage problem commonly found in PCM composites. Nonetheless, the experimental results show that with increasing temperature, this PCM undergoes a slight decomposition at around 250 °C, which corresponds to the rupture of the polyurethane bonds. Subsequently, the PEG chain breakage rapidly degraded in the temperature range of 340 °C to 450 °C. This corresponds to a rapid phase transition of the PCM, with the maximum rate of mass loss occurring at about 400 °C. This is much lower than the surface temperature of the cell at TR, when the PCM is already undergoing phase transition. At the same time, the large amount of heat released by the cell makes it impossible to achieve a rapid temperature control and cooling of the already phase-transformed PCM. Therefore, as the results of this experiment show, the PCM alone cannot successfully block TRP at a relatively thin thickness (2 mm or 3 mm). From this, we can conclude that simple heat absorbing materials, such as PCMs, cannot successfully inhibit TRP at a relatively thin thickness, and how to design a heat absorbing material that can effectively inhibit TRP at a thinner thickness is a breakthrough point worthy of research in the future. As for the thermal insulation material, when choosing the material, not only should its thermal conductivity be taken into account but also its own mechanical properties and insulating properties are also very important.

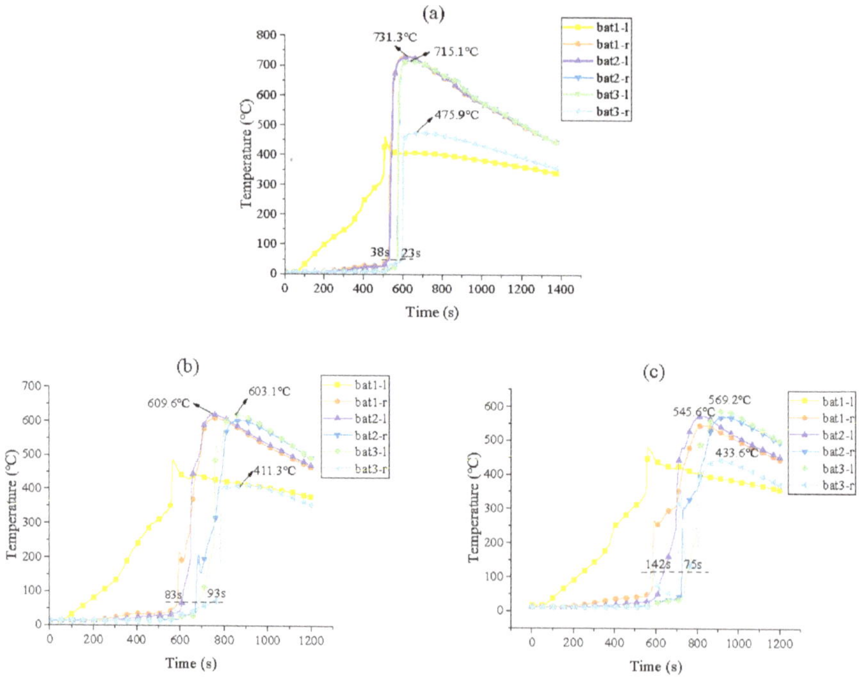

Figure 4. *Cont.*

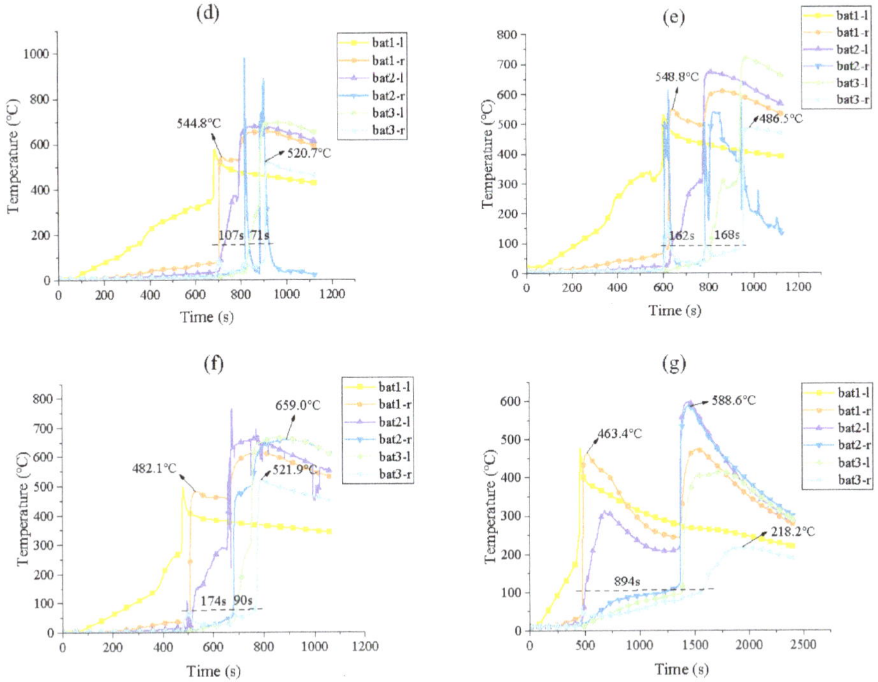

Figure 4. Comparison of battery module temperatures: (**a**) blank group, (**b**) 2 mm PCM, (**c**) 3 mm PCM, (**d**) 2 mm ceramic fiber, (**e**) 3 mm ceramic fiber, (**f**) 2 mm glass fiber, (**g**) 3 mm glass fiber.

3.3. Comparative Analysis of TR Behavior by Mass Loss of Battery Module

We have analyzed the mass loss of the battery module and discuss it in the following section. Figure 5 summarizes the total mass loss of the battery module for several cases. Compared with the 194.3 g of the blank group, the mass loss of the PCM group increased to 233.9 g and 226.2 g, respectively, which may be due to the loss of PCM by phase change melting when encountering high temperatures. The 3 mm thick PCM provides relatively better heat absorption and insulation, reducing the intensity of TR, which makes the total mass loss of the battery module slightly lower than that of the 2 mm PCM group at this time. The ceramic fiber mass loss in both thickness cases is higher than that of the blank group, 199.8 g and 209.3 g, respectively, even though it is more effective in suppressing the TRP of the battery module than both PCM groups. This may be due to the unstable chemical composition of the selected ceramic fiber materials, which led to the thermal decomposition of the materials and increased mass loss during TR. Through a thorough analysis of the experimental process and results, it was found that the layered stripping of aluminum silicate ceramic fiber paper at high temperatures is one of the main reasons for the decline and even failure of its barrier performance. When the ceramic fiber paper is exposed to a high-temperature environment, the internal fiber structure will undergo a series of changes. Under the action of high temperatures, the binding force between the fibers inside the ceramic fiber paper gradually weakens, and the originally closely arranged fiber layer begins to separate, forming the phenomenon of layered peeling. This is because ceramic fiber paper is usually made of fibers that are combined with each other by physical or chemical interactions. At high temperatures, these binding forces are disrupted and the friction between the fibers decreases, leading to easy separation of the fiber layer. Layer stripping will make the structure of ceramic fiber paper become loose, forming many gaps inside, thus greatly reducing its ability to block heat and gas. Heat can be transferred more easily through these voids, and the gas generated by TR can penetrate the ceramic fiber paper more smoothly, resulting in the TR transmission being unable to be effectively suppressed. The total mass loss of both glass fiber groups is smaller than that of the blank group, 190.6 g and 182.2 g. On the one hand, this is attributed to the good thermal stability of the glass fiber material itself. On the other hand, the good thermal insulation of the glass fibers reduced the intensity of TR, even at 3 mm, inhibiting the fierce ejection of cell 3. This resulted in a minimum mass loss of 182.2 g in the battery module with 3 mm glass fiber.

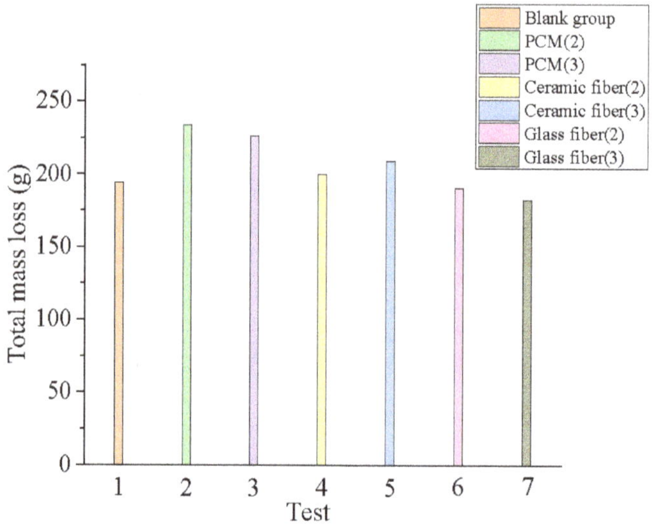

Figure 5. Comparison of total mass loss of battery packs in different cases.

3.4. Comprehensive Comparative Analysis of the Inhibition Effect of Different Materials

In the case of ceramic fibers, the internal fiber structure is destroyed at high temperatures, consequently resulting in a weakened blocking effect on conductive heat, as well as radiant heat from neighboring cells. At the same time, the relatively weak mechanical properties make it further extruded to the neighboring soft pack cells. Thus, a good TRP inhibition effect cannot be achieved in both thicknesses.

PCMs have relatively high thermal conductivity, which enhances the longitudinal heat transfer in the TR process. When a lithium-ion battery is experiencing TR, a lot of heat is generated inside the battery, which needs to be transmitted through the material inside the battery and the surrounding barrier materials. Due to the relatively high thermal conductivity of the PCM, the heat can be quickly transferred along the axial direction inside the PCM, that is, the longitudinal transmission. Inside the battery, when TR occurs in a certain part, the heat will be quickly transferred to other parts through the PCM, resulting in the rapid spread of TR inside the battery. The relatively high thermal conductivity of the PCM also enables the rapid transfer of heat to adjacent batteries or other components, further exacerbating TRP. Although the PCM has a strong heat dissipation capability, it cannot balance the heat that is quickly transferred to neighboring cells by its own strong thermal conductivity. Therefore, the effect of simply using thinner PCMs to inhibit TRP is not significant.

Glass fiber has extremely high porosity, up to 92%. This porous structure allows for the formation of many tiny pores inside the glass fiber wool, which are connected with each other and constitute a complex network of gas channels. During the process of TR, the heat is mainly transmitted through heat conduction, heat convection, and thermal radiation. For glass fiber, the tiny pores inside it can greatly hinder the progress of heat conduction. As the thermal conductivity of gas is far lower than that of solid materials, the heat needs to be continuously reflected and scattered in many tiny pores when passing through glass fiber, which greatly increases the path and difficulty of heat transfer and effectively slows down the speed of heat conduction. These tiny pores are also able to inhibit the occurrence of thermal convection. Glass fiber also has very low thermal conductivity of only around 0.018 W/mK. Thermal conductivity is an important index to measure the thermal insulation ability of a material. The lower the thermal conductivity, the better the thermal insulation ability of the material, and the stronger the ability to block heat transfer. There is a synergistic effect between the porosity and the low thermal conductivity to inhibit thermal diffusion. The structure of high porosity provides a good basis for the performance of low thermal conductivity, which results in the obstruction of heat transfer inside the glass fiber. During the heat conduction process, the transfer of heat is required not only in the solid skeleton but also in a large number of gas pores, which makes heat conduction much more difficult. The low thermal conductivity further limits the heat transfer speed in the solid skeleton and gas pores, so that the glass fiber can inhibit the heat diffusion more effectively. In terms of thermal convection, the structure of high porosity suppresses the gas flow, while the low thermal conductivity reduces the transfer of heat in the gas, which effectively prevents the occurrence of thermal convection. Therefore, the porosity and low thermal conductivity of glass fiber cooperate with each other, making it show excellent performance in blocking TRP.

4. Evaluation of the Effectiveness of Different Materials for TRP Inhibition

Based on the analysis above, we evaluated the effectiveness of TRP inhibition in the blank group, as well as in different materials with different thicknesses. This will help relevant personnel to better weigh the benefits of the materials, as well as the material properties. First of all, the higher the effectiveness of the material to inhibit TRP, the lower the risk of the battery module. In this paper, the effectiveness of materials to inhibit TRP is qualitatively evaluated from the perspective of inhibition effectiveness, and five key indicators are selected. They are battery TR trigger temperature T_{onset}, maximum temperature within the battery module T_{max}, total interval time for TR propagation, maximum temperature rise rate (MTR) of the battery, and mass loss rate (MLR) of the battery.

T_{max}: peak temperature of the battery after TR. The higher the maximum temperature of the battery after TR, the higher the risk of potential harm to other batteries [24], i.e., the lower the effectiveness of material inhibition to suppress TRP.

TR interval: the longer the time interval of TRP of the battery, the lower the harm of TR, and the higher the effectiveness of material inhibition to suppress TRP. In order to maintain consistency with the data above, the inverse of the TR interval is taken as the observation index in this paper.

MTR: the maximum temperature rise rate of the battery. The greater the maximum temperature rise rate of the battery, the greater the potential harm to the surrounding environment [24], that is, the less effective the material is in suppressing TRP. In order to make the results of MTR more representative, we chose the right-hand side temperature rise rate of battery 3 as the observation index.

MLR: the mass loss rate of the battery. The larger the mass loss rate of the battery module during the whole TRP process, the more completely it represents the loss of the battery, and the more harmful it is to the surrounding environment [25], that is, the lower the effectiveness of material to inhibit TRP.

T_{onset}: The higher the temperature at which the battery triggers TRP, the lower the risk of TRP in the battery [26], that is, the higher the effectiveness of material to inhibit TRP. In order to keep consistency with the data above, the inverse of T_{onset} is taken as the observation index in this paper.

The evaluation model data are shown in Figure 6a. We can clearly see that the total area of 3 mm glass fiber is the smallest. This indicates that the riskiness of the battery module under the action of 3 mm glass fibers is the smallest, i.e., the effectiveness of 3 mm glass fibers in suppressing TRP is the best. In order to visualize these areas, we use the value of the product of the five metrics as the final metric that qualitatively represents the effectiveness of the material to inhibit TRP. The smaller the product, the more effective the material is in inhibiting TRP. As we can see from Figure 6b, all the materials have smaller indices than the blank group, proving that the inhibition is all positive. Especially, the value of 3 mm glass fiber is only 0.00233, which is much smaller than the other materials. It is proved that the 3 mm glass fiber group attained the best ability in inhibiting TRP. Overall, the effectiveness of TRP inhibition is, in descending order, 3 mm glass fiber, 3 mm ceramic fiber, 2 mm PCM, 3 mm PCM, 2 mm glass fiber, and 2 mm ceramic fiber. In practical applications, the cost and weight of the materials also need to be taken into account. As can be known from the above evaluation, the glass fiber has the best effect in inhibiting TRP, and its manufacturing cost is lower than that of the aluminum silicate ceramic fiber, but it is heavier than the aluminum silicate ceramic fiber. Therefore, it is suitable for application in fixed-energy storage systems. For mobile battery packs, such as the battery packs in electric vehicles, considering the factor of weight, ceramic fiber is more suitable for application in this scenario. However, compared with glass fiber, the effect of ceramic fiber in inhibiting TRP is not good, and it is necessary to make a trade-off between safety and cost at this time. For the PCM, its thermal management performance is the most excellent. However, it has a high cost, a high density, and the effect of inhibiting TRP is not good. Therefore, when choosing it, it is recommended to use it in combination with thermal insulation materials.

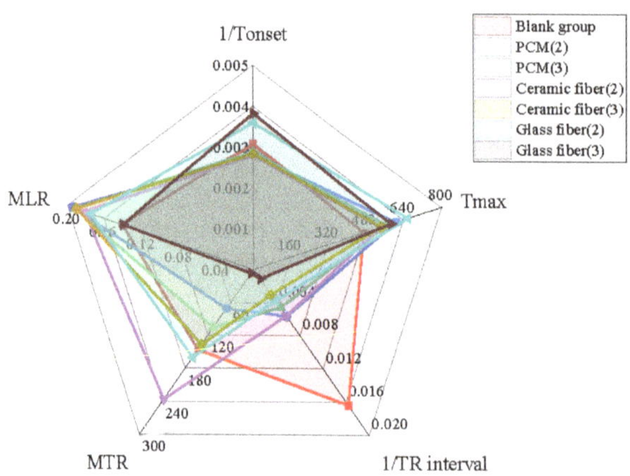

(a)

Figure 6. *Cont.*

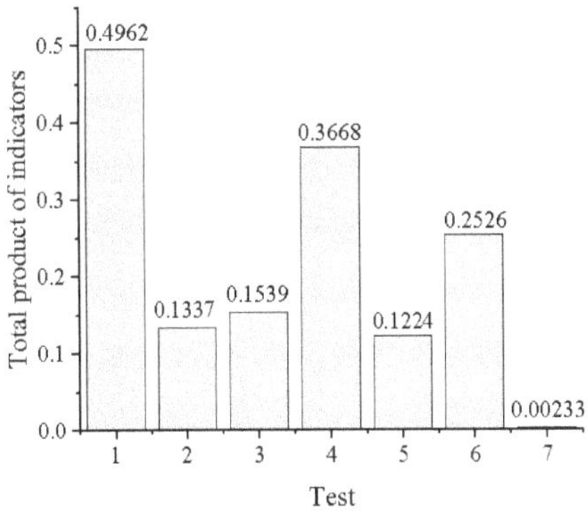

(b)

Figure 6. (**a**) A radar chart for the evaluation of the effectiveness of material suppression of TRP; (**b**) total product of indicators.

5. Conclusions

This project takes LIBs as the object, compares and analyzes the influence of different isolation layer materials on the TRP of LIBs, analyzes the influence law of different isolation layer materials on the TRP, and analyzes the blocking effect and influencing factors of different materials. The test results show that glass fiber has the best inhibiting effect on TRP. This is due to its high thermal stability, insulating properties, and good mechanical properties, along with its strong insulating ability. In the case of purely heat absorbing materials, such as PCMs, it is not possible to block TRP at a relatively thin thickness. In the case of ceramic fiber, despite its good insulating ability, its thermal stability is not up to the mark, making it ultimately unsuccessful in blocking TRP. Overall, the physical properties, chemical properties, structural properties, and use environment of the material will have an impact on the barrier effect of the material. Based on the comparison of the phenomena, temperatures, and mass losses of different materials in suppressing the TRP of battery packs, this study innovatively conducts a comprehensive evaluation of relevant indicators in a radar chart. In combination with practical situations, it is explained that when people select materials for suppressing TRP, they should not only focus on a single property of the material. Instead, they need to consider the application scenarios and make trade-offs among various advantages and disadvantages of the materials, such as thermal insulation performance, mechanical properties, heat dissipation performance, cost, and weight. The research results of this project will have important guiding significance for the selection and application of TR protection materials for LIBs.

Author Contributions: Investigation, Y.C. (Yaoyu Chen); resources, Y.C. (Yin Chen); data curation, Y.Y.; writing—original draft, Y.M. and Y.C. (Yaoyu Chen); writing—review & editing, Y.M. and M.C.; supervision, M.C. All authors have read and agreed to the published version of the manuscript.

Funding: This research was funded by the National Natural Science Foundation of China (52204213).

Data Availability Statement: The original contributions presented in this study are included in the article. Further inquiries can be directed to the corresponding authors.

Conflicts of Interest: The authors declare that they have no known competing financial interests or personal relationships that could have appeared to influence the work reported in this paper.

References

1. Ali, S.; Ahmad, T.; Tahir, M.Y.; Usman, M.; Chhattal, M.; Hussain, I.; Khan, S.; Hassan, A.M.; Assiri, M.A.; Rosaiah, P.; et al. The emergence of density functional theory for supercapacitors: Recent progress and advances. *J. Energy Storage* **2023**, *73*, 109100. [CrossRef]
2. Mao, Y.; Chen, Y.; Chen, M. Review of Flame Behavior and Its Suppression during Thermal Runaway in Lithium-Ion Batteries. *Batteries* **2024**, *10*, 307. [CrossRef]
3. Jiangyi, H.; Fan, W. Design and testing of a small orchard tractor driven by a power battery. *Eng. Agríc.* **2023**, *43*, e20220195. [CrossRef]
4. Börger, A.; Mertens, J.; Wenzl, H. Thermal runaway and thermal runaway propagation in batteries: What do we talk about? *J. Energy Storage* **2019**, *24*, 100649. [CrossRef]
5. Quan, Y.-Z.; Liu, Q.-S.; Liu, M.-C.; Zhu, G.-R.; Wu, G.; Wang, X.-L.; Wang, Y.-Z. Flame-retardant oligomeric electrolyte additive for self-extinguishing and highly-stable lithium-ion batteries: Beyond small molecules. *J. Energy Chem.* **2023**, *84*, 374–384. [CrossRef]
6. Sun, P.; Bisschop, R.; Niu, H.; Huang, X. A Review of Battery Fires in Electric Vehicles. *Fire Technol.* **2020**, *56*, 1361–1410. [CrossRef]
7. Feng, X.; Ren, D.; He, X.; Ouyang, M. Mitigating Thermal Runaway of Lithium-Ion Batteries. *Joule* **2020**, *4*, 743–770. [CrossRef]
8. Zhou, G.; Yang, S.; Liu, Y.; Zhang, Q.; Li, Y.; Lu, H.; Wei, Z.; Huang, Q.; Zhou, W. Thermal runaway response and flame eruption dynamics of lithium-ion batteries under continuous/intermittent heating mode of dual heat sources. *Appl. Therm. Eng.* **2024**, *249*, 123363. [CrossRef]
9. Zhang, G.; Wei, X.; Tang, X.; Zhu, J.; Chen, S.; Dai, H. Internal short circuit mechanisms, experimental approaches and detection methods of lithium-ion batteries for electric vehicles: A review. *Renew. Sustain. Energy Rev.* **2021**, *141*, 110790. [CrossRef]
10. Salloum, R.; Rabuel, F.; Abada, S.; Morcrette, M. Investigating the internal short-circuit in 18650 cells under thermal abuse conditions. *J. Power Sources* **2025**, *628*, 235905. [CrossRef]
11. Li, Y.; Zhang, N.; Jiang, L.; Wei, Z.; Zhang, Y.; Yu, Y.; Song, L.; Wang, L.; Duan, Q.; Sun, J.; et al. Assessment of the complete chain evolution process of LIBs from micro internal short circuit failure to thermal runaway under mechanical abuse conditions. *Process Saf. Environ. Prot.* **2024**, *185*, 296–306. [CrossRef]
12. Zhang, L.; Liu, Y.; Huang, X.; Huang, X. Intra-cell thermal runaway propagation within a cylindrical battery induced by nail penetration. *Int. J. Therm. Sci.* **2025**, *210*, 109633. [CrossRef]
13. Chen, M.; Yu, Y.; Ouyang, D.; Weng, J.; Zhao, L.; Wang, J.; Chen, Y. Research progress of enhancing battery safety with phase change materials. *Renew. Sustain. Energy Rev.* **2024**, *189*, 113921. [CrossRef]
14. Chen, M.; Zhu, M.; Zhang, S.; Ouyang, D.; Weng, J.; Wei, R.; Chen, Y.; Zhao, L.; Wang, J. Experimental investigation on mitigation of thermal runaway propagation of lithium-ion battery module with flame retardant phase change materials. *Appl. Therm. Eng.* **2023**, *235*, 121401. [CrossRef]
15. Ren, M.; Fakayode, O.A.; Kong, F.; Zhou, C.; Chen, L.; Fan, X.; Liang, J.; Li, H. Characterization of cellulose nanocrystals prepared by different delignification methods and application of ultra-light, hydrophobic aerogels as oil absorbent in food systems. *Ind. Crops Prod.* **2023**, *197*, 116653. [CrossRef]
16. Niu, H.; Chen, C.; Liu, Y.; Li, L.; Li, Z.; Ji, D.; Huang, X. Mitigating thermal runaway propagation of NCM 811 prismatic batteries via hollow glass microspheres plates. *Process Saf. Environ. Prot.* **2022**, *162*, 672–683. [CrossRef]
17. Zou, K.; Xu, J.; Zhao, M.; Lu, S. Effects and mechanism of thermal insulation materials on thermal runaway propagation in large-format pouch lithium-ion batteries. *Process Saf. Environ. Prot.* **2024**, *185*, 1352–1361. [CrossRef]
18. Bausch, B.; Frankl, S.; Becher, D.; Menz, F.; Baier, T.; Bauer, M.; Böse, O.; Hölzle, M. Naturally-derived thermal barrier based on fiber-reinforced hydrogel for the prevention of thermal runaway propagation in high-energetic lithium-ion battery packs. *J. Energy Storage* **2023**, *61*, 106841. [CrossRef]
19. Chen, M.; Gong, Y.; Zhao, L.; Chen, Y. Phase change material with outstanding thermal stability and mechanical strength for battery thermal management. *J. Energy Storage* **2024**, *104*, 114565. [CrossRef]
20. Gupta, A.K.; Kiran, R.; Zafar, S.; Pathak, H. Effect of hygrothermal ageing on the mechanical properties of glass fiber reinforced polymer composite: Experimental and numerical approaches. *Mater. Today Commun.* **2024**, *41*, 111060. [CrossRef]
21. Xu, C.; Liu, X.; Wang, L.; Gan, X.; Li, T.; Feng, X.; Yue, X.; Lv, Y.; Wang, X. Structural morphology evolution and characterization of electrospun magnesium aluminum silicate ceramic micro-nanofibers: A study. *Mater. Today Commun.* **2024**, *39*, 108816. [CrossRef]
22. Chen, L.; Liu, J.; Liu, H.; Liu, Z.; Song, Q. Synergistic effect of graphene oxide and glass fiber on mchanical and thermal properties of composites: Experimental and simulation investigations. *React. Funct. Polym.* **2025**, *210*, 106201. [CrossRef]
23. Feng, R.; Yang, Q.; Dai, H.; Deng, M.; Wang, H.; Hou, C.; Cheng, Z.; Gao, H. Effect of glass fiber on the mechanical and thermal insulation performances of kaolinite-based thermal insulator. *Case Stud. Constr. Mater.* **2024**, *21*, e03879. [CrossRef]
24. Zhu, X.; Sun, Z.; Wang, Z.; Wang, H.; Lin, N.; Shan, C. Thermal runaway in commercial lithium-ion cells under overheating condition and the safety assessment method: Effects of SoCs, cathode materials and packaging forms. *J. Energy Storage* **2023**, *68*, 107768. [CrossRef]

25. Wang, Z.; Yuan, J.; Zhu, X.; Wang, H.; Huang, L.; Wang, Y.; Xu, S. Overcharge-to-thermal-runaway behavior and safety assessment of commercial lithium-ion cells with different cathode materials: A comparison study. *J. Energy Chem.* **2021**, *55*, 484–498. [CrossRef]
26. Mao, N.; Gadkari, S.; Wang, Z.; Zhang, T.; Bai, J.; Cai, Q. A comparative analysis of lithium-ion batteries with different cathodes under overheating and nail penetration conditions. *Energy* **2023**, *278*, 128027. [CrossRef]

MDPI AG
Grosspeteranlage 5
4052 Basel
Switzerland
Tel.: +41 61 683 77 34

Batteries Editorial Office
E-mail: batteries@mdpi.com
www.mdpi.com/journal/batteries

www.ingramcontent.com/pod-product-compliance
Lightning Source LLC
Chambersburg PA
CBHW040751250926
51850CB00007BB/1470

* 9 7 8 3 7 2 5 8 6 3 9 6 9 *